# Defining Singularity:
## AN OPEN  ACADEMICS LETTER

# # 1

Written By Peet (P.S.J.) Schutte

## ISBN-13: 978-1499688511

## ISBN-10: 1499688512
## a Theses Introduction

This book uses the flowing work as
reference :

© KOSMOLOGIESE EN ASTRONOMIESE TEGNIKA

**An open letter**

# TO SELECTED ACADEMICS

## ISBN 0-9584410-9-X

An open letter **TO SELECTED ACADEMICS** is THE ACADEMIC PROLOGUE AND AN ACADEMIC INTRODUCING LETTER TO ACADEMICS PRESENTING A NEW COSMIC THEORY AS IS STATED IN MATTER'S TIME IN SPACE THE THESES, ISBN 0-620-27041-1 which consists of the following seven books in title:

1) **A Cosmic Birth...Dismissing Nothing** I.S.B.N. 0-620-31609

2) AN OPEN LETTER ON: **XEPTED ASTRONOMICAL MISTAKES** ISBN 0-9584410-1-4

3) AN OPEN LETTER ON: **INTER GALACTICA SPACE TRAVEL** ISBN 0-9584410-2-2

4) AN OPEN LETTER ON: **CORRECTING COSMOLOGY** ISBN 0-9584410-5-7

5) MATTER'S TIME IN SPACE: **THE HYPOTHESIS** ISBN 0-9584410-6-5

6) AN OPEN LETTER ON: **" STARSSTUFFN'** ISBN 0-9584410-3-0

7) **" SEVEN DAYS OF CREATION"** ISBN 0-9584410-4-9

---

© KOSMOLOGIESE EN ASTRONOMIESE TEGNIKA

**An open letter**

# TO SELECTED ACADEMICS

## ISBN 0-9584410-9-X

All rights are reserved. © KOSMOLOGIESE EN ASTRONOMIESE TEGNIKA
**WRITTEN BY PEET SCHUTTE**
No part, parts or the entirety of this book may be reproduced by publishing, electronically copied, duplicated by whatever means that forms reproduction or duplication of any description, without the prior written consent of the copyright owner.

**THIS LETTER IS THE ANNOUNCING OF THE BOOK IN SEVEN VOLUMES** CALLED **MATTER'S TIME IN SPACE:** THE THESIS ISBN 0-9584410-8-1 VOLUMES 1-7

This letter was the letter that was sent to near enough eighty Universities through out the world in regard of announcing a new cosmic theory. It now is turned into a separate and individual commercial book.

TO WHOM IT MAY CONCERN,

An open letter **TO SELECTED ACADEMICS** ISBN 0-9584410-9-X Is THE ACADEMIC NOTIFYING OF

**MATTER'S TIME IN SPACE: THE THESIS** ISBN 0-9584410-8-1 Written by PEET SCHUTTE

**Professor, Sir, Madam, Dear readers, why would I send you this book and inform you that you are about to read science you have never encountered before? You know the accepted version of science that holds the doctrine of mass or material pulling other material with mass. With that idea I disagree full heartedly.**

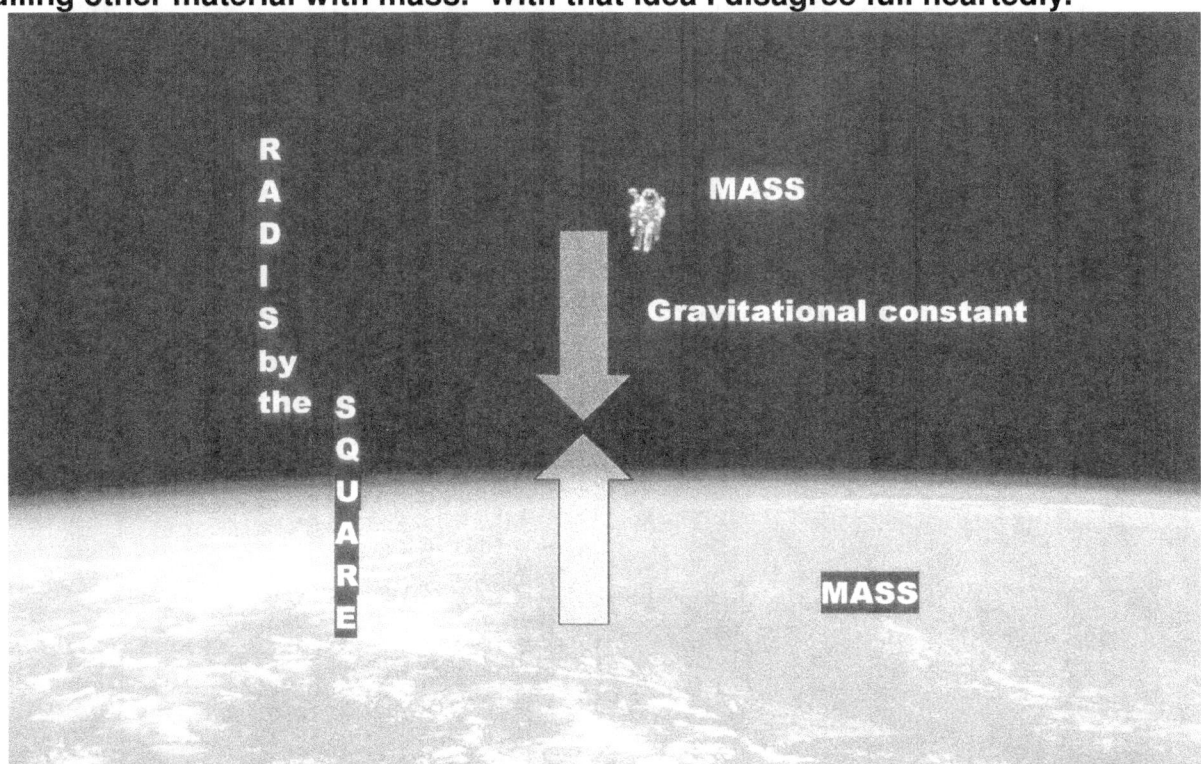

This is part 1 of a 2-part book. In this part of the book I show that it is not material pulling onto other materials but gravity is material collapsing space by turning and thus the turning compresses the space into ever-smaller and less volumetric occupying parts.

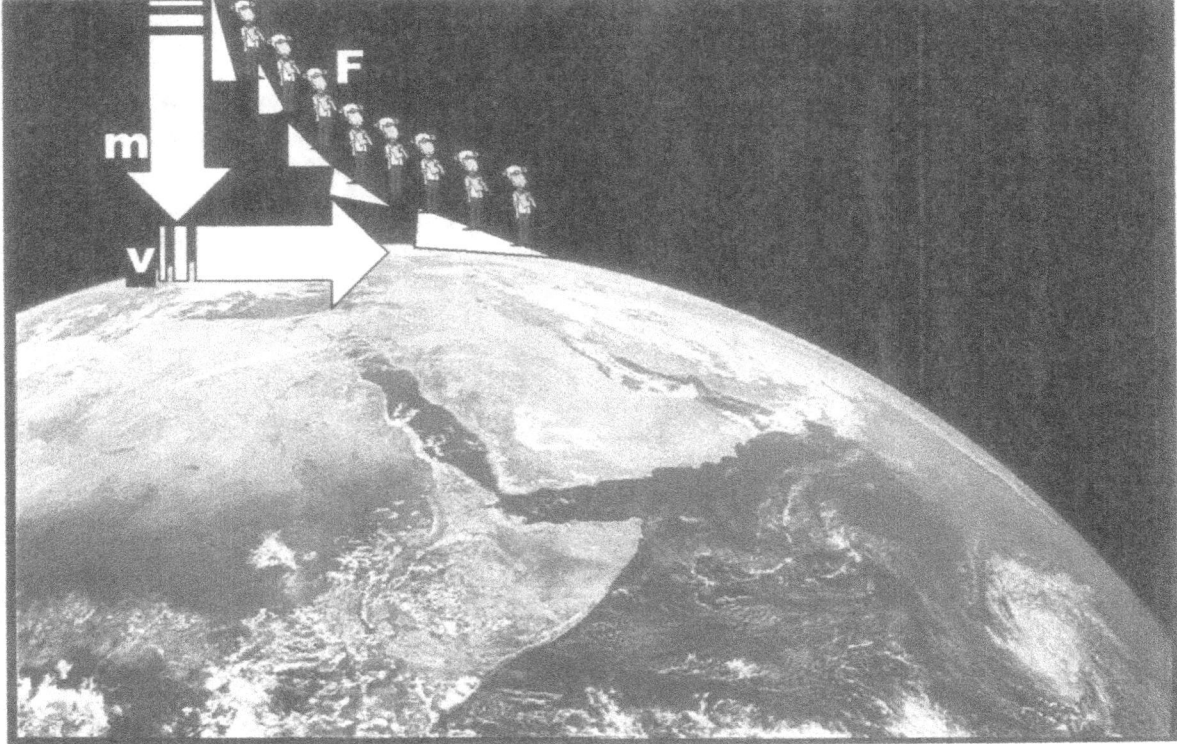

I concluded that gravity is result of movement that compress non-material holding space. Please let me explain in as simple terms as I am able. The earth turns applying gravity...

Space being compressed by the rotation of the Earth and being pushed onto Earth forming the factor mass where the object is blocked onto the Earth by the space coming from above

is not a force but normal movement moving the object from $V_1$ to $V_2$ as $V^2$ or $g = 9.81$ F is the repositioning or relocating of an object from where it was to where it will be the next instant in time

The rotational direction of the Earth movement is drawing space downwards while it is moving space sideways according to the rotating direction of the gravity of the Earth and thus moving the compressing space onto the turning Earth where the Earth's turning is compress the space which we know as the atmosphere. The turning motion reduces the space and the reducing motion is gravity $g = 9.81$

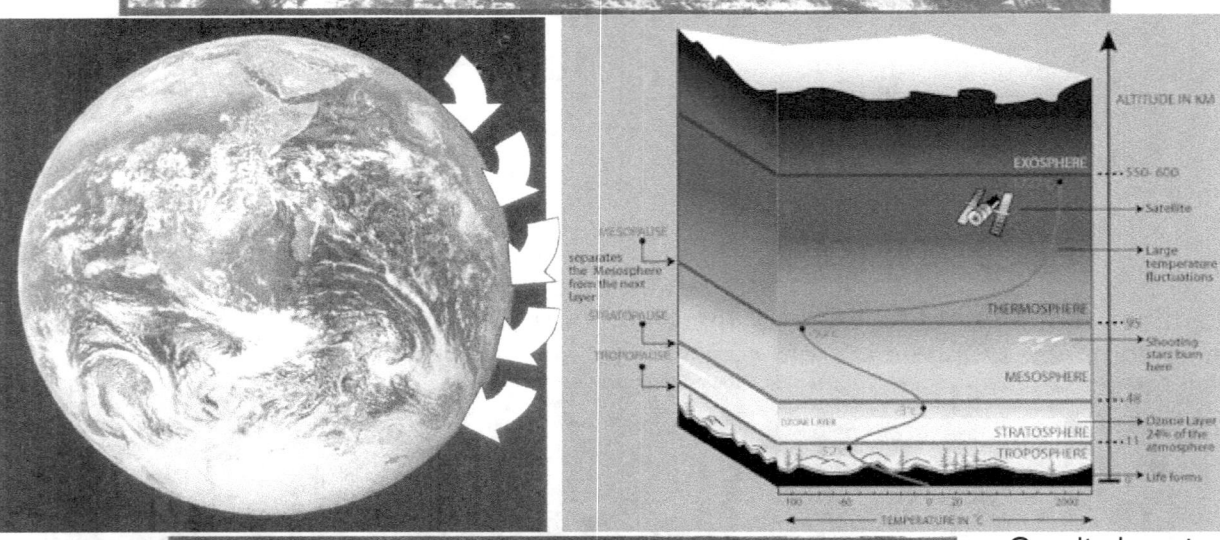

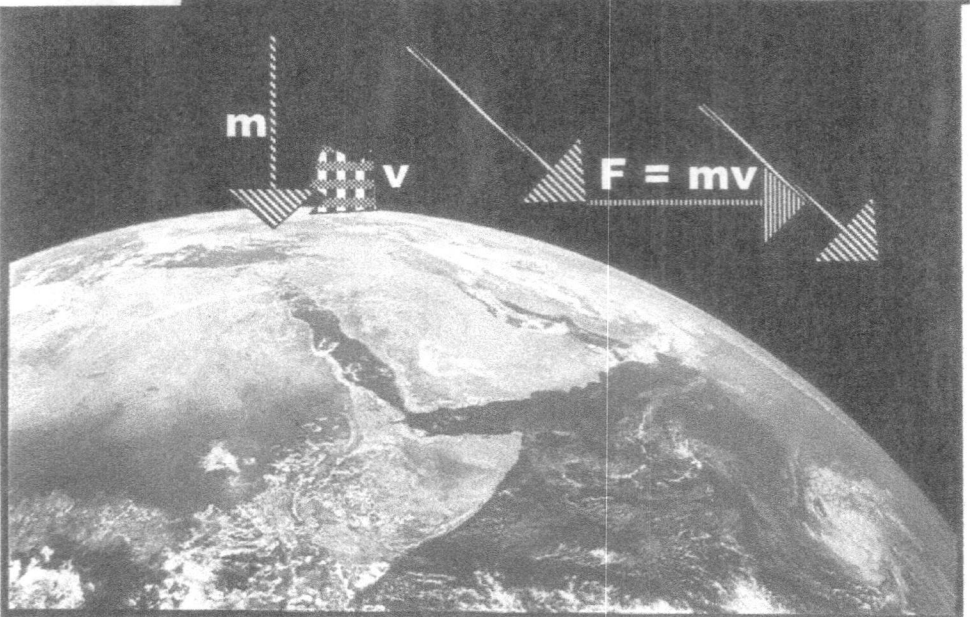

Gravity is not only about movement going down but also movement going sideways in a circle. That places the atmosphere in layers, evermore going denser as it gets closer to the earth. This is not where this increase in density stops. Going towards the centre the heat increases and that increase in heat by contracting space is all forming the increase in density the of non-material composite that forms the Universe.

Take a truck loaded with cargo totalling of 15 tons into an airplane.

Put next to the truck a petite little dancer weighing 45 kilograms.

Then to keep the dancer on her toes, put a frog of 150 g next to her.

Now we will have the mass of each object "pull" by "gravity" as this lot falls down

We all know from seeing it on TV that all things fall equal when under the same conditions and this had me thinking: if mass pulls mass this would not be possible. Take a truck a dancer and a frog and throw this lot out of an aircraft at the same instant. Think of this scenario again because what I am about to say disproves Newtonian science altogether. Take that the truck ways 15 tons, the dancer ways 45 kilograms and the frog ways 150 grams. Now apply the Newton formula of $F = G \dfrac{M_1 M_2}{r^2}$. Remember on the formula $F = G \dfrac{M_1 M_2}{r^2}$ rests the entire formulation defining physics. The earth has a mass and the truck is 150 000 000 grams, the dancer is 45 000 grams and the frog is 150 grams. If Newton was correct the truck must land minutes before the dancer and hours before the frog. The mass pulling on the truck is millions times more than what the frog is and thousand times more than the dancer. Consider this idea again please…

Whatever those in science try to say, there are two opposing opinions in science that contradict each other. Newton holds the opinion that mass pulls mass and therefore object of different mass cannot fall equal because of $F = G \dfrac{M_1 M_2}{r^2}$. Then Galileo Galilee said all things fall equal notwithstanding mass or size. This then formulates to the expression $F \neq G \dfrac{M_1 M_2}{r^2}$ because nature proves Galileo correct and that is that all objects fall equal when under the same conditions. This then made me think that it is not mass that pulls but it is space forming a buoyancy in which all space, either holding material or being empty, falls and that makes that whatever represents space condenses.

I first started my studies in the field of Cosmology as a spontaneous development of my natural curiosity spawned from childhood interests in the field of cosmology, which I developed even before I went to school. The studies were a reaction (I would imagine) that was part of my personal childhood development in how I was forming a personal concept of a lifelong interest that followed me into my future. At first I conducted all my earlier studying mostly on the basis that inspired me to find out more about what made the Universe tick, with no intention ever on my part to reach a point where I would be writing books on the subject. At first I was investigating cosmology on a part time basis. This went on, on and off, or the best part of twenty odd years (*as* time and *when* time would permit). Then in later life with my health deteriorating I committed myself to more intense investigation and my effort developed onto involving a study using time that is only permitted by a person when that person is involved in such a quest on a full time basis. That quest has now been going on for the last seven years in full devotion and if one includes all the years invested on my part including the twenty odd years before, part time, then the time I have spent in completing my theory when adding all in comes down to almost twenty eight years. This is to say that I did not come to realise what I am about to introduce on a light-hearted conclusion. I mention this because I wish to ensure the reader that he should have no doubt about my most sincere commitment in producing a cosmic theory on matters concerning the start and the working of the Universe during and before the Planck era. At first I began by arguing that there is a something that is blocking our progress. There is some barrier preventing humans passing a threshold whereby our understanding will pass such an obstacle. If there were any way that any one may break through that barrier there is that is preventing normal research to go pre-Big Band, it would be accomplished by finding the barrier whereby then our vision we use to focus would pass such a limit. If we wished on progress in our pursuit of the very first cosmic moment then we have to find and cross the barrier that blocks our view. We have to look deeper and in another direction should the desire driving us be strong enough to commit us to reach into the very birth of the cosmos. We have to rethink the strategy that we use. Max Planck was one of the most brilliant men of all times and even he, notwithstanding all his personal brilliance, accomplished little. There are parts missing in what we have and that which we have at our disposal to use because if there was no such an obvious barrier then the Wise-Men involved in science would by now have found the way to break through the seal that is locking us out of the critical past which will uncover the origin of the Universe's infancy stage. I went about trying to find what everyone since Adam, (meaning all of the rest of mankind and myself) were missing throughout the ages of speculating and interpreting while philosophising about whatever we find inspirational. The obvious we saw; that was clear. Therefore I had to find a route that would lead into the not so obvious that all of us were missing, notwithstanding the best efforts of the best qualified to accomplish such a breakthrough. My effort involved trying to accommodate that what was in the cosmos available to use by the cosmos in all phases of developing. If I had any hope of finding the answer, such an answer had to be simple because I am not very inclined to unravel what is deemed as complicated. The simplicity had to be locked in what was not yet understood about that which was in the cosmos as it formed part of the process used in forming the cosmos. My realising this brought me to focus not on that which we understand. There is not a lot we actually understand because even gravity is very poorly understood. In fact gravity is so poorly understood that there is not one person alive that can claim the prestige of understanding gravity and among the dead there is even less that can make such a claim. There are several phenomena that are presented in nature and acknowledged by science but also discounted by science and therefore not presented as accepted science. By admitting that that what we have available to us to use concerning our research of cosmology in an attempt to better our understanding of cosmology, is useless to use, then one realises that not having what there might be makes what we already have useless. It then is useless to use what there is as part of the big picture we are trying to paint because what we use is not really part of the picture. This leads one to believe that the picture of the cosmos Mainstream science is painting, is being painted without painting a full picture.

In my first attempt to understand the full picture of what science was painting I found so many colours missing there was no picture painted that anyone could appreciate. This is what made me decide to go on researching the 'unknown' in the hope it might clarify the 'known' and as the book unfolds you as the reader may agree that I was correct in pursuing the misunderstood and rejected phenomena. Finding the missing phenomena helped me to place the phenomena mentioned above in a theory where the principles also mentioned above form a part of the overall gravity used in binding the Universe. I believe what is in the Universe is not able to be coincidental because of too many influences contributing to what there is - notwithstanding the fact that that is the manner which science uses when they refer to the Bode law. What is in the Universe has a role as it had a role, which is the same role that phenomena has had and in future will have. This is establishing a very new idea about the working relationship between particles and in explaining it by using Kepler's studies. Redefining the work of Kepler's views brings a new Universe to light involving new concepts that are based on old principles but principles in updating man's view about cosmology are very new in that capacity. Through that new vision I was able to come to realise what the reasons might be why Kepler never saw it fitting to include the measure of $\Pi$ in his formula. I do not suggest his neglect thereof was intentional, nevertheless the formula he devised without using $\Pi$ proved that there was no need for the inclusion of $\Pi$ since his figures brought about a correct answer in the final end result leaving a well concluded fitting answer. The numbers he produced brought about a specific space $a^3$ contained in a circle $T^2$ at the distance of **k** from a defining centre thus the calculations did not require the use of $\Pi$ to find a meaning. In that Kepler did not see a need to include $\Pi$. I would not go as far as declaring with absolute certainty on his behalf that he did it deliberately, however there never arrived such a necessity. It is prudent to agree on whether or not such a need is necessary, because if one is agreeing about such changing not being required a new Universe emerges. The circle that Kepler discovered came about without ever forcing $\Pi$ into the frame because it is clear that the circle formation came about as a natural consequence and came spontaneously delivering an equation while he was working. In this book I prove that the reason for adding $\Pi$ to the rest of Kepler's formula is unnecessary. This unnecessary addition is because when going one step further in the investigation one will find that **k** and **a** and **T** are symbolising the same value with the only difference being that each one represents a different dimension to our six dimensional or six sided Universe we enjoy. In fact I shall show that $\Pi$ replaces "**a**" and "**k**" and "**T**" and that $\Pi$ is the true value that should be replacing each factor as to indicate the correct value to the sides nominating $\Pi$. We humans work on a numerical base using ten as a basis where we count to nine and re-establish a new decimal numbering line by adding a nought behind the number in value. This is using the numerical basis of ten, which I suspect we took from ancient knowledge about cosmology and not from using our fingers and toes as the earliest calculating processors. In this book there is unfortunately no room to explain my suspicion but another fact I do prove is that the cosmos uses $\Pi$ in the cosmic numerical basis as a means to measure and quantify. Therefore in fact the Kepler formula should read instead of $a^3 = T^2 k$ as it does it must be $\Pi^3 = \Pi^2 \Pi$ where I shall show that $\Pi$ represents singularity wherefrom the entire Universe sprang from $\Pi$ and by forming as $\Pi^3 = \Pi^2 \Pi$ it is confirming that space is equal to the motion thereof. Kepler's greatest achievement was showing that the cosmos is space –time $a^3 = T^2 k$ while time is the motion of space in space. The value of Π is the primeval and most basic of measures applying as an accepted cosmic legal value that the cosmos used exclusively in the very beginning and as it does today. The measure of Π in the Universe, values particle development that brought about all development ever conducted in the Universe. Only after this stage did the rest come including mathematics and went on to freeze spilled singularity into frozen material. Reading this statement may sound suspiciously senseless but as the book unfolds the sensibility will become apparent. The full implication of such a statement will become clear when one dissects different facts coming from studying Kepler. My discovery of this fundamental basis of legal valuing ensured me again that there was no need for someone the likes of Newton to add Π in any form to the work of Kepler because Kepler discovered the ultimate Π in the Universe, the Π giving the Universe form and gravity. The concept of Π that is the only single form of all other forms available that can by duplication of Πs assemble the value of gravity. When replacing the symbols with Π the facts of the Universe become self-explanatory because the most basic form that forms the cosmos has a definitive and uncompromising value.

But getting this far took me down roads overgrown by ignorance and which I had to uncover myself as if hacking away miles of overgrowth with a machete chopper. All of the disbelief science showed to my work in the past and their refusal to see past Newton made any and all attempts on my part as bad as they could be, strangling and smothering my attempts to announce my uncovering of the newly found insight on my part.

For decades I tried to come to terms with the inability there is in science to explain the cosmos in real terms, when using the science of official reputation. That which there is makes a mockery of science because the undisputable clues left in the cosmos makes what little correct explaining there is available, seem like a comedy of errors, when it is mixed in with all the other near Dark Age errors we still use after so many centuries that provided countless opportunities to revise the old muck. By applying current accepted Astronomy as such the phenomenon found all over the cosmos is still beyond the explaining ability of Mainstream science. This is true and it is a shame because it also is an undeniable fact in spite of the vast knowledge and progress in other forms of science taken in the manner science uses when it approaches cosmology. Cosmology truly lagged behind while the understanding and advancing of physics, mathematics and chemistry as subjects were flourishing. By comparison I saw how little there was available in explaining cosmic phenomenon and how much improvement in understanding the other departments such as chemistry, electronics, medicine etc. could offer as results were coming about from research. Even where there is a little explaining available in cosmology it turns out that such explaining is confusing to say the least and at best it highlights the manner in which science is applying double standards. For decades photographs were the only progress forthcoming as an addition to improve the meagre field in cosmology and that improvement was artificially stimulating cosmology. By providing a false impression of advancement, everyone missed what and how much was missing…To the connoisseur desperately looking for more than the obvious stirred in with some out-dated misinformation dating back to the Middle Ages, it all seemed as if it was a picture portraying the ridiculous to make the sublime look good. The pictures only proved the opposite of what progress in cosmology will represent. In truth and as such in cosmology the cover up that was hiding the lack of progress about the science of true cosmology was only forthcoming in the improving of electronic optical telescopic advances and spectroscopic progress. There were only photographs carrying beautiful pictures which pleased the less informed except the photographs did not bring progress to cosmology at any intellectual level by promoting insight. The explaining that the photos demanded about the subject had the opposite effect of installing hope because what it did do was underline what lack in any notable progress there truly is in our understanding of cosmology and laws in the cosmos.

While such Hubble telescopic images might seem to be as clear as daylight it was more than clear there was little academic value to them. To the person in need of more stimulation than being impressed with pictures of God's marvellous Creation and the sightseeing that always accompanies such pictures, such persons always felt very disappointed. The pictures did give satisfaction to those more easily impressed, but the rest of us seeking knowledge accompanied by understanding the images left us despondent. Although they leave the vast majority in total amazement there are those less impressed about not knowing the 'why' and the 'how' in such amazing pictures. I know the group I fall into may be the greater minority and the majority may only demand the portraying of the images, which is what that easily satisfied group demand. The rest of us rouse with anguish at the lack of information about what is known and what lies behind what those pretty pictures are conveying. Nevertheless there can be no real progress in scientific understanding about the images portrayed by the Hubble telescope, and others, if no one is able to show the slightest clue of a deeper understanding of what is going on in the Universe. Everyone is almost breathless waiting on the commentating by the most informed, which accompanies the magnificent cosmic portraying of God's Creation. When we are portraying the new images, we should also be investigating that what we see that the cosmos is at the moment portraying. The lack of actual believable explanation coming from investigating by means of telescopic imaging should impress one and all, but the impressing must not be based on the colours in the images but the sensible information attached to the image investigated. It is *that* that we wish to see. What we wish to see must at least be accompanied by scientifically backed information, which provides the proven understanding coming from science. When science is employing new explanations with such photos it should also be discarding senseless baggage carried over from the past. Most images contradicted Newton and for saying that, every Academic I ever came across in the past ostracized me. That bothers me little! I know I cannot possibly be the only person absolutely discontented with what Mainstream science accepts as science. Here I refer to the out of date theorising Mainstream science still accepts amongst many others as how they suggest stars and planets are forming. One cannot promote cosmology in honesty and advocate scientific fact whilst dishing up such fairy-tale nonsense to students. Moreover I hold the opinion that amongst Academics in particular there must be many if not most that share my personal serious doubts or have an inclination to share some of them. This I say when considering the overall doubtful picture painted about what there is and what one believes there should be. I just cannot believe those forming the most intellectual group of mankind are unaware of the mismatching facts seen over the broader picture because the contradiction and lack of a plan, makes what there is so very doubt provoking. Newton dismissed the formula Kepler presented as all factors forming motion. That is where the apple cart derailed.

In honesty we have to realise that we cannot dismiss the whole formula that Kepler produced as being motion. It is so much more than just motion. It is $a^3 = k / T^2$: That is what Kepler brought into civilization for all time to come. He saw space $a^3$ being in isolation due to the time it uses to move $T^2$ claiming such space forming independence according to the lines $k$ indicate. Let us look at the factors in more detail before we proceed with the rest of the book.

$a^3$ symbolises a mathematical interpretation of implicating the three-dimensional space.

$T^2$ is representing the period or time that Kepler suggested we should use to calculate time that holds the orbiting planet in direct contact with the space in relation to a very specific centre.

**k** is the space taken from the centre to the end of the line from which the planets must have grown if one accepts the Big Bang growth of particles and the affect of the Hubble constant on all cosmos material. The specific value about the centre is most important because from the specific centre gravity always applies the strongest influence.

One cannot justify Newton's dismissing of Kepler's formula as that all factors only contribute to the motion indicated because that is misleading. We all accept that the true cosmic form *would be* and most probably *is* a sphere. Everyone accepts the Universe as a whole as a sphere...but why would the sphere form? What would be the reason why the original form that we devote to the Universe would take on a sphere as a natural form? Apparently our imagination grabs the sphere as form. In all natural events the gravity in that space which stands apart and independent from all other space takes on by cosmic pre-casting the sphere as form of shape ... **it is because gravity chooses the smallest space to hold the strongest force**.

I am of the opinion that gravity is about dismissing space to the advance of heat increasing in such a specific and concentrated space using the concentration as measure for the heat as well as the space holding the heat in space. According to Kepler that is what he found to be true. Space $a^3$ will always be circling space around as $T^2$ in any position from the centre **k**. That is what Kepler said when he said $a^3 = T^2 k$. Kepler indicated space $a^3$ will forever fight for independence and show separate individuality in remaining apart as identifiable cosmic components by means of motion. Every space will cling to independence indicated by **k** through fighting off the integrating of another coverall unifying unit by applying the motion of $T^2$! The problem we have to solve in this book is what will the cosmos use to secure such independence between all particles? What sets space apart from the rest of space? First we have to admit that Kepler was the one that introduced the following.

Kepler gave us the answer to the following but no one ever took notice!
Kepler was the one that discovered **space / time** as $k=a^3/T^2$
Kepler was the one that discovered **singularity** as $k^0 =a^3/T^2k$
Kepler was the one that discovered **gravity** is holding **space-time** relative by the measure of distancing **k** as $k = a^3/T^2$ and $k^{-1} = T^2/a^3$
Everyone able to read mathematics has to realise that Newton suggested collisions between cosmic structures must eventually come about as gravity erodes the distance separating the cosmic structures multiplied by the product of the mass of both structures from both ends. Newton said the multiplying mass of both structures destroys the distance between the structures by using the eroding force of gravity in the square. The cosmos then must end in a Big Crunch with all material joining together but that joining is not forthcoming at all...and that only indicates how much insufficient understanding there is on offer in cosmology by the educated–to-be-wise-about-these-matters. There is precious little available to explain about their field of cosmology amongst the ranks of Astronomers. So...let's us return to the beginning of cosmology before every one became oh so wise and see what there is to see.

The cosmos informed Kepler of another gravity, which the cosmos applies much more widely and is used by nature all over the Universe.

The picture we see coming from the Hubble telescope shows why, in the perfect Universe...but can the Universe be perfect when... we see a radius between the Sun and individual planets is not using a regular distance as one would expect of gravity in being a force driven by the mass and in that sense the mass is producing the gravity that always remains even because the mass doesn't alternate. As the mass is never changing on either side, that steady mass has to keep the gravity steady. But in our imperfect understanding of the Universe we find that the radius that should be constant varies considerably proving either that mass somehow adds by measure unnoticed while the structure is in orbit and later allows the same amount of mass to escape undetected; or it's the seasons adding and removing mass at will. This is an absolute contradiction to reality if mass was the factor determining the radius we find between the Sun and the planets. This suggests strongly that we'd better be getting very suspicious about the idea of mass contributing to gravity.  But in contrast to this, science is unshaken about their confidence in the perfection about facts they use in terms of correctness. It is well known amongst all persons that science only uses dependable and ultra reliable facts coming from sources beyond doubt. Referring to any work done by any scientist will find a remark about science only accepting facts they use to work with. It is accepted overall by all communities that in science those in science use one hundred percent accurate facts or they use no facts at all. If our view was as perfect as science would lead us to believe it then must be the Universe that is imperfect as it otherwise would not behave so mystifyingly.

**While we are in gravity the manner in which gravity applies in our use of gravity makes us part of the Earth by mass forcing us onto the Earth as a semi unit with all other Earth belongings. Is that which we have truly gravity?**

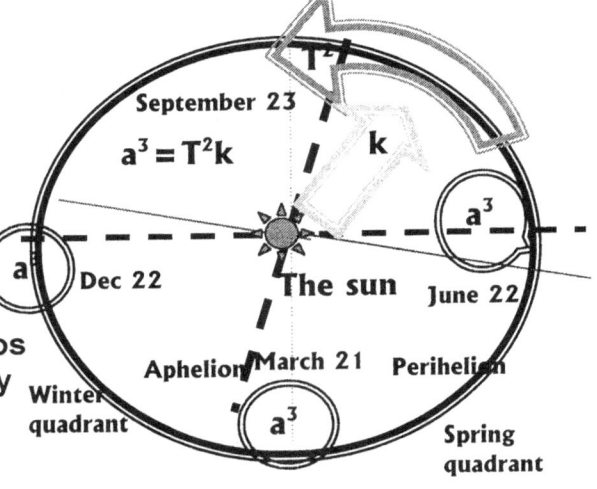

**By using mathematics, the cosmos spoke to Kepler personally and by the use of mathematics as the medium, it provided Kepler with information about the cosmos coming directly from the cosmos.**

The unshaken confidence science uses has us believing at first consideration that the drawing of gravity should produce an even diameter positioned between the Sun and the planets because of ever dependable evenly distributed gravity… but I believe there is a perfect Universe and our understanding carries the doubtful suspicions. Delving deeper uncovers even more contradictions and the level of accuracy contained by our scientific understanding then arouses more suspicion about the correctness of science. Remember Newton changed wha the cosmos told Kepler leaving much suspicion as to how far the misdirection takes science. We have to correct the facts we doubt because when correcting the facts they use in science concerning our view about science such correcting brings along a better understanding and then the Universe has to become ever more perfect as one learns to understand the perfect Universe even better.

But it does require an open and clear mind and it needs no culture driven preconception that should confirm interpretations about facts surmised even before they are carefully studied. It becomes obvious that Newton never gave careful attention to Kepler's findings because if he did he would have seen what gravity is. Kepler described gravity without using the name that later was given as 'gravity'. Kepler did not give the name gravity, but

Kepler's studies gave Kepler the insight to coin the concept of gravity. Nevertheless it was a name and not the concept that was later named by Newton. The naming was the contribution of the Englishman. The concept that Newton later introduced is totally incompatible to the concept that Kepler introduced. What he (Newton) introduced as the force of gravity, he connected to mass, which diverts totally from Kepler's findings. With giving a name, the Englishman also changed the concept that Kepler introduced.

Kepler made no mention of size or mass as part of the phenomenon that later was named as gravity, yet it must be gravity that holds the Universe together. The concept the Englishman changed when he introduced what he introduced with the name he introduced. That what he introduced, he corrupted beyond recognition. The concept that accompanied his new name strayed completely from what Kepler introduced. Newton brought in something that was mismatching what Kepler saw in Kepler's view of the phenomenon that holds the Universe true to form. The name was dominant but even more dominant and totally inaccurate was the other concept Newton introduced. In truth Newton only gave the world a name of an idea, which he then corrupted as far as cosmic physics are concerned. It is important to admit that as far as cosmology is concerned Newton gave the concept the name but *only* the name and not the concept of gravity. Newton's persuasion on matters of gravity as gravity functions between cosmic structures orbiting one another as we find in outer space is inaccurate. What Kepler saw, Newton saw differently and used the opportunity that Kepler left by not giving any name to the process he (Kepler) and Tycho Brahe worked on for two life spans. Newton did seize the opportunity to name what he, Newton, saw but that what Newton saw did not include that which Kepler uncovered. In Kepler's era the name or title was lacking but Kepler established the concept of gravity and the formulation thereof.

The concept came from Kepler even before the name gravity was used by Newton to describe in the concept of whatever we today (after Newton) became accustomed to believing what the concept of gravity is about. With the help of Newton everyone since Newton confused Kepler and Newton on the issue of gravity and this confusion even begins with Newton. Gravity might not have been named but became a proven concept and factor after Kepler formulised it, which is before Newton named it. The concept of gravity that Kepler saw is about the manner in which the structures orbit because there is a space that circles around a centre and this process has kept planets secured, connected and rotating around the Sun which is the same concept that is keeping the Universe secure and comes about with a process Newton later named as 'gravity'. This what Kepler saw is not the same as what Newton saw when he saw two objects drawing closer by pulling on each others mass. Then later on Newton named what he thought he saw as the force that Kepler saw but introduced another completely different concept. Kepler saw cyclic formations keeping the Universe together and never approaching each other. Newton ignored what he wished not to see but he changed as he saw fit and what he thought that should be. His experience as a young man drove him to establish a process he formulated as the process that is keeping the Universe together.

In that act he corrupted as much as ignored the work of Kepler, which he also named as the same gravity that he saw as a young man. Why he chose to ignore Kepler's findings on gravity we shall never know but why the world still chooses to ignore Kepler's findings about gravity almost four hundred years after the fact I shall never know. My saying this has literally made Academics ignore me, as they would avoid the plague. I am not pretending nor do I exaggerate when I say there were those in Academic institutions that questioned my mental development. Some went as far as seeing me as a joker of sorts and I have correspondence to show evidence to that fact. I know by now while

Newtonians are reading this book I have aroused the tempers of every Academic reading this far, therefore let's see what is being ignored by the Academics which I blame to do just that.

Kepler said gravity in space is about

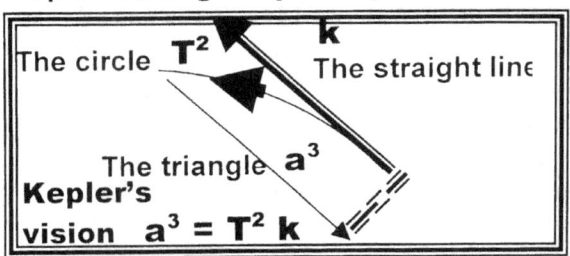

the area $a^3$ that would always keep equilibrium with the time $T^2$ it takes to travel the distance of the full circle position placed by the indicator **k**, therefore adjusting **k** as the need arrives. With k shifting in length $a^3$ will have to readjust and therefore $T^2$ will find a new relating value each time. This was the finding of Kepler and came after his intense study of orbiting planets. The line formed in rotation is never straight but always part of a larger circle. The line is eternally connected to a specific domineering centre that always insists on a circle by rotation.

We live through seasons which comes from being that at one point, $(a^3_1)$ the distance between the sun and the earth **is less than** at another point we call $a^3_3$

Let us put a value of $a^3_1$ = one and $a^3_3$ = three. This means that each year, for the past 4 500 000 000 years the effect of the common gravity between the earth and the sun has a greater effect than at another point six months later. That means at one point the earth should be drawn or pulled closer to the sun  and after another six months interval the earth should stand less effected by the sun's gravity, therefore it   should move away from the sun.

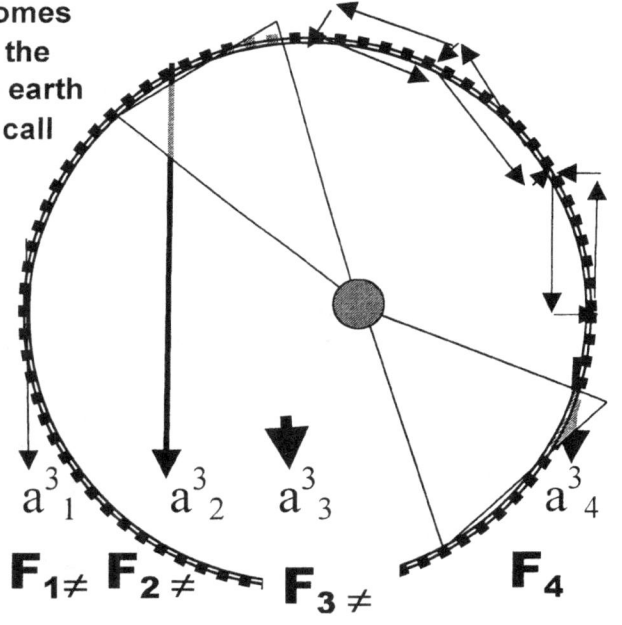

May I remind you THAT NEWTON'S OWN LAWS ARE IMPLIED, and again the planets disobey these laws completely!! **In the modern age all evidence point away from Newton's vision**

Before I attempt any investigation into this matter there must be coherence in our agreeing about what gravity is. If you the reader insist that the falling of objects is the only gravity found, your further reading will convince you little. Anything we do decide upon must support the fact that it is gravity that prevents planets from dislodging from the grip the Sun has on them. Gravity is not about the Sun trying to catch the Earth by attracting the Earth...no, there is so much more to gravity. We must be under no illusions about what gravity is and that being the focus of our discussion and where that gravity is because we have to identify and not confuse the gravity we are looking at. We are now discussing the gravity, which is keeping planets circling around the Sun, and stars around specific galactica centre. In that we do not find one example to use as proof in connection to stars coming tumbling down on galactica centres and crushing into galactica centres. If that is gravity keeping structures in orbit around specific centres we must look at the behaviour of the structures in gravity. We have to find a reason why the planets do not reduce the radius between them as Newton suggested but we must trace the reason why it is gravity, which is keeping them apart because if anything, they are departing as they extend the radius connecting them to the Sun. That is gravity because it applies throughout the Universe. The gravity Kepler found is the general gravity that is keeping structures from colliding and in that the principles are avoiding collision or on the other hand avoiding abandoning each other. It is about confirming respect for one another's independence and clearly staying at a predetermined distance while at the same time both are sharing a common space unit. That then must be the defining of gravity we have to study to find the Universal enticing gravity holding the Universe together. By close investigation one will find three factors in urgent need of investigation. There is firstly a centre that draws the object closer. This gravity is clearly a synonym to what Newton saw as gravity. If it were not drawing the object closer the object would not be orbiting around the centre and applying motion. It will draw and absorb all rotating things in its field of gravity.

The fact it does not draw the object into its ranks is because there is another gravity standing alongside this first mentioned gravity. Our recognising the first gravity forces us to accept the presence of another part of gravity. This forces us to recognise the second gravity. When saying this we are not using Newton's cosmic formula concept $F = G (M.m)/r^2$ because that can barely be what is out there happening. What Newton saw was falling. If that what Newton saw is the only gravity then whatever Kepler saw including all other parts of everything out there that are spinning around some centre must come closer to one another and connect in collisions. While that is not happening we must start to look past Newton to new grounds we can investigate. We have to go beyond Newton and admit there is more than that what Newton had us believe because it is clear that what Newton had us believe…is not happening. That confirms the presence of the second gravity. The fact proves that everything is departing and not arriving. Even the moon is drifting away from the Earth and this information comes about from the most advanced investigation up to date, including a moon visit and the placing of measuring devices there.

Looking at the gravity intensely we find the roving structure travels in a straight line, which repeats another circle around another centre but because of the influence of a centre keeping the roving structure attached to such a centre the motion allows a circle to form by reforming motion from the original straight line to that of a partial circle. There is a centre; a connecting line travelling between what the two points establishes the specifics of a centre within a circle and the end of the circle. According to Newtonians the centre supposedly draws the rotating object closer. That is half the story.

I suggest we do some deliberation and in deliberating may I remind you THAT NEWTON'S OWN LAWS ARE IMPLIED, and again the planets disobey these laws completely!! In the modern age all evidence points away from contracting and favours eternal expanding.

**The latest news confirms that the lot is apparently not coming any closer!**

In our manner of considering gravity as a phenomenon we find there are three factors interacting and together the three factors form a balance, which produce and are responsible for a balance between all particles in the Universe. This must be gravity since it seems to be the glue that is holding the Universe intact. We can visually see that as the object moves in a straight line because it counteracts the pulling from the centre by a line that indicates the repositioning each time. In parallel with this it also moves in a circle. One can only interpret this action as being caused by another line just as strong but counter-directed through motion. The circle comes into action as a counteraction that is trying to accommodate two opposing directions being evenly strong and from that counterbalancing eventually forms a rotating motion trying to satisfy the direction coming from the straight line in one direction and another straight line counteracting the first straight line. In the motion the straight lines coming from opposing values also forms an immediate circle (though only partly) but the overall complement forms a triangle. This shows a very different picture to that which Newton saw.

| | |
|---|---|
| **The lot is more evidently moving further away from the sun.** | $F = \dfrac{M_1 M}{r^2} G$ |

This is the suggested formula confirming the behaviour of planets used by Newtonian scholars underlining the argument that contraction is coming about between all cosmic objects. What Newton witnessed, if my memory serves me correctly was an apple falling from a tree where both the apple and the tree were part of the Earth and this did not constitute - or lead to - or come as a result of - a catastrophic cosmic event happening. In the mathematical sense it does not make sense when Newton's argument is taken out and used in outer space. What Newton saw with his falling apple was a mass influencing another mass to reduce the distance as the influencing involved motion that came about. In outer space there is another gravity where in the case of those cosmic structures in outer space there is no mass pulling each other about or pulling one another onto each other. In the case where there is particles falling from space onto the Earth, that falling also results from gravity, as much as it varies from the cosmic gravity. There is another type or form of gravity different to the concept Newton introduced. That, which the concept Newton introduced, is not the cosmic gravity Kepler formulised. What Kepler introduced is a duel where both objects are clearly in an eternal compromise therefore neither party relents its position. Newton saw just the opposite...Newton saw both compromising their individual as well as each other's position.

Since the mass in both cases is unchanged and the mass is the factor that is establishing the force that is used by the circle to hold the radius steady and in place, these facts point to a balance that formed bringing about the above-mentioned steadiness. In the view of science however it is the mass that either draws the orbiting objects closer or is keeping them apart. The mass does not change and since that mass of both produces the radius between both, the logic is that there has to be an even and steady radius that develops. The radius has to be equal all the time since the mass never changed throughout the rotation. The radius must be the same from any and all given points that form the rotating circle which must keep the radius equal from every angle...yet we know that Kepler proved this not to be the case even before Newton's naming and changing of Kepler's work came about.

$$\frac{M_s \times M_c}{100} = 1 \times F \quad (r^2 = 100)$$

$$\frac{M_s \times M_c}{50} = 2 \times F \quad (r^2 = 50)$$

$$\frac{M_s \times M_c}{25} = 4 \times F \quad (r^2 = 25)$$

$$\frac{M_s \times M_c}{5} = 20 \times F \quad (r^2 = 5)$$

There is this comet coming towards the Sun and everything Newton said was coming just as Newton predicted. Gravity was pulling, the comet was doing all the coming because the Sun was to massive to come, so the comet had to do all the coming and came on behalf of both and the issue seemed scientific as was expected. We all know how gravity works. Gravity pulls what ever it pulls towards the centre of the gravitational pulling object. So the comet was coming towards the centre of the Sun and the immanent collision was unavoidable. Also by the reducing of the radius the mass of both increases many times over, therefore this comet's chances of escaping and not surviving became lesser and lesser as it drew close to the Sun. Newton says two pieces of rock will draw each other closer by reducing the distance keeping them apart. That we all can see by merely jumping in the air. No sooner have you lift off than you are back on the ground. That is what Newton said about three hundred and fifty years ago. Even trying to tell the Official Brainy Bunch, those with the Hi Que that Galileo said mass of an object has nothing to do with the falling, seemed to pass the Official Brainy Bunch comprehension by miles. I was told on so many occasions that I did not understand Newton, but there it stopped. No one could explain to me what it was I did not understand about the comet missing the Sun by miles, whereas instead, it was supposedly to hit the Sun with a dazzling impact. That never happens! On this point, I cannot get through to them as much as they cannot get through to me. It seems to me as if our understanding of the same issue is so far apart, we do not share the same planet, and yet after all my arguments and investigation no one, and I repeat: not one could once clearly tell me what it is that I do not understand. They all claim though that I do not understand. In addition they can't explain to me what it is I don't understand...

My doubting this idea about gravity came when I was still in school. It was my inspecting the route the comet travelled in the milieu of applying $F = \dfrac{M_1 M}{r^2} G$

As a child I knocked on many doors for an answer but everyone spoke through me, above me, around me and not to me but never did one address the issue by explaining to me. In short this was my problem no one explained to me. At first you as the reader may think I am trying to create a mountain from an ant heap, but in scientific terms the human race is preparing for the start of the cosmic journey.

By completion of this book you will realize how "Xepted Science" believe they built science on a solid foundation, and, boy are everybody in for a rude awakening. Compared to the leaning tower of Pizza, science is about to start with the next section of a much bigger building adding many levels and already the view at the bottom where I am looks far worse than the leaning tower does.

You have the Sun and you have a tiny piece of rock covered by water also better known as the comet. There are thousands of them flying around, but never aimless. At first

Newton's formula makes pretty much sense. The Sun draws the comet towards the Sun, as Newton said it does. The comet responds by speeding towards the Sun, also as Newton predicted. Anyone can see a collision coming ten miles away. The Sun applied gravity, the comet applied gravity, the Sun is far too massive to fly to the comet, so the comet with much less mass does the flying on behalf of both objects. This is still what I understand. Every person with even the least of knowledge about science knows how the gravity application works.

However in all the excitement and the pulling something odd happened. This comet was starting to break some Newtonian law because at the time when the gravity started to become overbearing, the centre point of gravity, which the Sun suggested to be focused in its centre shifted and the more the comet came closer the more it shifted from the Sun centre. As the Sun / comet gravity radius reduces, the radius between the Sun and comet effectively increases the relativity of the mass influence on each other in the form of gravity. The mass of the Sun and the comet increases by the factor of reduction of the radius separating the two objects. That will produce a growing gravity force as the comet / Sun radius becomes smaller. By the time, the radius to infinitive proportions, making the radius infinitely small, the relevance to the mass of both structures will raise a force with eternal power. At a point, where the comet / Sun apply a force of immeasurable strength, the comet find the strength to brake this immeasurable force. Remember the direction of gravity always point to the centre of the object, and that is where the collision was heading. As the objects draw closer, the distance reduces, but in accordance to the relevance of the radius reducing the objects also become that much bigger in drawing power. It depends how one consider the relevancy to grow by the approaching nearness diminishing the distance between the objects.

Then out of the blue, when it really matters and where the storey should end, the comet finds the ability to eliminate the eternal powerful force of gravity, and keep at a safe distance while gliding safely around the Sun. At this point, Newton goes sour. Nothing Newton predicted is happening. Newton loses all credibility except in the eyes of the science community. The comet and Sun not only stabilized the force, the force begins to decrease as the radius between the comet and the Sun is on the increase. AT THE POINT WHERE THE FORCE IS THE STRONGEST, THE COMET BRAKES FREE AND SLIP AROUND THE Sun, UNSCATHED.

Then, in complete defiance of the Newton Law on gravity, quite the opposite applies. At the point where the radius that is separating the two cosmic objects is at its strongest, it will also bring about that the gravity force is at its weakest. At the point of almost no ability the gravity force suddenly releases enough strength to break resulting in the parting of the two structures. The force now curbs the rebel comet on its way heading into the deep dark yonder while escaping the Sun gravity seemingly for the very last time. At the point where the force was the greatest, the comet overcame the force, but where the force was the weakest, the force overcame the comet's rebellion. I fought hard as I fought bitter but Academics never could see this point. The correctness of my argument is no longer the issue. It was twenty-five years ago, when I still held the impression that I was missing some point here. I do not try to correct this phenomenon any longer in the hope of bringing across some flaw in my understanding. The flaw in my argument is not in my inability because the flaw is science as a whole.

I could never understand the reason why "the ordinary", like others and me with my development level, can see what I can see, yet academics that has more brainpower in their heads, than I have life in my body, were unable to see such an obvious conclusion.

You; those Official Policy Protectors are my superior in every sense a human can have, with the brainpower to bust a dam wall by brutal thought, and yet you cannot see how far the tower of Pizza is leaning over. I make the point to help you the reader to judge yourself. If you are able to see the validity in my argument, you are not brain dead. Education has not yet bashed your thinking ability out of your scull. However, if a cloak of not understanding role over your brain, and a numbness sets in on your ability to reason about this phenomenon, beware, you are a Newtonian…and Newtonians are unable to see.

Newtonians should read this book very slowly because the effort you are about to launch, may be the most painful you shall ever experience throughout your academic career. You are going to suffer from reconditioning and Newtonian withdrawal, not that dissimilar to that of an addict in rehabilitation. You are going to reject me, hate me, despise me, loath me as you never felt about anybody else. If you think I am sarcastic, I am not.

You will reach a point where you will abandon the reading of the book. You have my sincere sympathy and with all the soothing it may bring, know that you are not the first I saw suffering such painful Newtonian rejecting. Let's get back to out comet in collision. Once more, this phenomenon should not occur with Newton's presumptions about gravity. These bodies will and must collide and destruct, without a doubt.

When the formula $F = \dfrac{M_1 M_2}{r^2} G$ apply, there should not be any force which is able to keep them apart especially when r reduces to almost infinity compared to what it is at maximum. However, the in discrepancy does exist because both maintain a certain distance apart.

With the "force" of "gravity" "pulling" the stars closer using the accumulative mass of the stars and multiplying that value with both objects by the mass component, this will reduce the radius $r^2$ progressively until $r^2$ reduces to the infinite.

Seen from this view, it is little wonder that the significance of this was lost in the notion that this is yet another "mystery" of the Universe. The scientists of the day (and the past) lost the importance, which this holds for us as Earth dwellers. A most surprising aspect of this is that it is not that an unfamiliar or rear phenomenon. However, any answer to this would clash with Newton's presumptions, and before the scientists allow that to happen, they would much rather ignore what is obvious. However, what is the obvious? Obviously the lot is not coming closer…

# The lot is more likely moving away from the sun. The lot is not coming closer!

$F = \dfrac{M_1 M}{r^2} G$ This is the suggested formula confirming the behaviour of planets used by Newtonian scholars underlining the argument that contraction is coming about between all cosmic objects.

Later Newton introduced gravity to the vision he had about gravity in his youth when he saw the apple fall from a tree and landed close to him. This event impressed him at first and later on the entire world. What he witnessed was an apple falling from a tree where both the apple and the tree was part of the Earth and not a cosmic event. In the mathematical sense it does not make sense when Newton's argument is taken out to apply in outer space.

## The lot is more evidently moving further apart

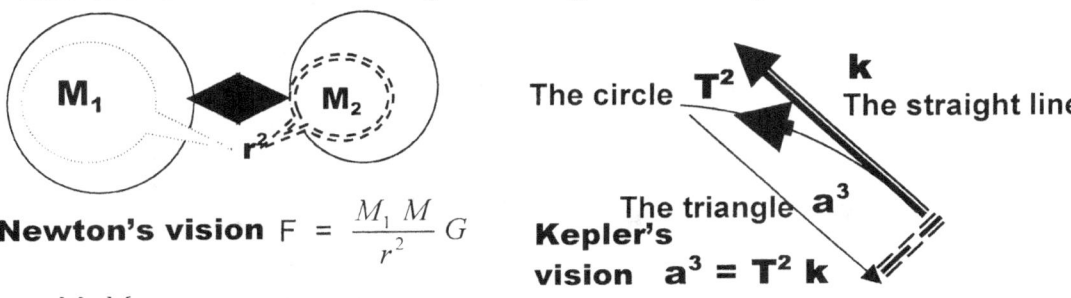

**Newton's vision** $F = \dfrac{M_1 M}{r^2} G$

The circle $\mathbf{T^2}$ $\mathbf{k}$ The straight line

The triangle $\mathbf{a^3}$

**Kepler's vision** $\mathbf{a^3 = T^2 k}$

$F = \dfrac{M_1 M}{r^2} G$ this is referring to one "force" but actually there are three:

▲ The pulling away of the smaller space. $\mathbf{a^3}$

▶ The double counter-acting referee. $\mathbf{T^2}$

▼ The pulling towards within the larger space $\mathbf{k}$

Newton saw just the opposite...Newton saw both compromising their individual as well as each other's position. However, since the mass in both cases is unchanged and the mass is the factor that is establishing the force that is used by the circle to hold the radius steady and in place, these facts point to a balance that formed bringing about the above-mentioned steadiness. In the view of science, however it is the mass that draws the orbiting either objects closer or is keeping them apart. The mass does not change and since that mass of both produces the radius between both, the logic is that there has to be an even and steady radius that develops. The radius has to be equal all the time since the mass never changed throughout the rotation. The radius must be the same from any and all given points that form the rotating circle which must keep the radius equal from every angle...yet we know that Kepler proved this not to be the case even before Newton's naming and changing of Kepler's work came about.

What we see is that there is one factor that is trying to run away being a lesser space within the pulling powers of a larger space (the second factor) trying to capture and control and a referee (the third factor) is seeing to it that the even-handedness is at all times applying in the fight. That gravity which I am familiar with and know is there. In some part but not in all out representing all the gravity there might be because I cannot see the jerking, as much as I do not feel it. That is then most probably another gravity I can see and which is Kepler's gravity which $a^3=T^2k$ represents. We have a motion of pulling...yes and that is what Newton saw...but then there is another motion of establishing a motion trying to depart, leaving the centre by tearing away from the centre and thirdly there is a motion that sees to it that the balance evolves as rotation. That is what Kepler said when he saw all three factors whereas Newton saw but one of the three. The one space is filling the next space as the space duplicates the position it had in the next moving moment that brings about the next position through motion. This eventually will have confined the next point by using a circle motion, which at first was intended to be a straight line, which is stopped by another straight line. The quest in this book is to find out why the other two factors apply in outer space as only one of the factors comes about on Earth under normal applying conditions.

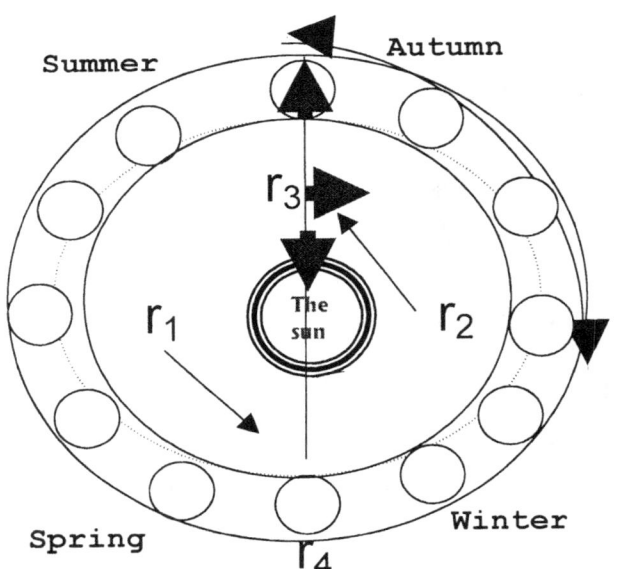

**Kepler's investigation indicates to the fact that the orbiting structure is in a motion that is going on where one strength is in a fight with a second strength and the two are pretty much matching in strength because not one of the two is very much winning the dual so no one is winning or losing the fight.**

**As the two factors are in a motion directional dispute there is obviously one of the two factors or strengths fighting to cut loose from the other one's grip and run off. If there were not such a force trying to escape, the first force would have a quick and decisive victory by reeling in the loser just as Newton predicted. The fleeing object and its matching fighting partner has a third party referee that allows the fight to go in a specific direction as long as there is no decisive victor.**

This book is on a quest to find the missing two factors and I can declare with some delight and with even more certainty that I found the missing factors. By Newton's introducing gravity as a force with the formula $F = G (M_1.M_2)/r^2$ a precedent was set of gravity being a contracting force forcing distances supposedly to grow smaller. Apply Newton's view to comet behaviour. Newton insists that the Sun has gravity reducing distance between the objects and while lecturers are teaching this during the day, at night they all witness how the comet follows this principle in detail showing Newton as a prophet. No sooner does the final conclusion draw near by orchestrating the final demise of the distance separating the two cosmic components when the opposite changes all concepts taught by institutions of science, the next minute out of the blue with no pre warning of the comet changing its mind, the comet defies all logic in scientific circles that apparently even included defying Newton and his logic. Because at the very point you'd think there is no chance of any return where gravity supposedly should peak because the comet is so close to the Sun and due to that fact makes the collision unavoidable...then the comet chooses that very point to dart away into the blackness of outer space, missing the definite collision by miles. By the time the collision is truly unavoidable with the radius between the Sun and the comet being as small as it realistically can be the comet starts gaining on the radius distance in spite of Newtonian denial of any possibility that such an event can in fact take place. The radius that should be shrinking further is instead enlarging. The radius that now begins to stretch proves Newton incorrect and it even depicts Newton as possibly being a fraud.

The gravity applied that focussed on the comet reducing the radius between it and the Sun was not acting predictably by maintaining the reducing of the distance until collisions come about as Newton insisted on. In our reading the Newton formula in English it says that $F = G (M_1.M_2)/r^2$ which when one translates that which is said in mathematics to a verbally spoken linguistic dialect, the translation then suggests that a force is committing the material that forms the factors involved, and forcing the material into a path that is leading to a collision. It says that the two will eventually collide because of the non-retractable mass inside each one that enforces the pulling which by the mass in each case is creating the force. The unchangeable ability of the mass and the unavoidable pulling each mass creates would bring about such a collision. The mass contributes a force making a collision imminently unavoidable. The collision is beyond any attempts of diverting any oncoming objects away from the inevitable possibility of contact. The force that mass contributes is ruling out all possible evading each other or avoiding the destruction. By enforcing a mass created force removes all chances from diverting away from the collision that is about to occur. Such a force then removes all possibilities of avoiding the oncoming collision.

The force will not allow any attempt to try and bring into the equation other possibilities in as much as rerouting the approaching object and changing the course in the imminent collision that is due and in due course will come about between the comet and the Sun. That which I explained is what Newton mathematically suggested with the formula. That is not what Kepler said notwithstanding so many arguments with Academics that I had in the past who tried to prove to me that the two visionaries views were equal and the same. Well…it's not the same because when we go onto translate Kepler to the verbal English the letters that come out do not even spell the same words.

Translating Kepler's mathematical expression $a^3 = T^2k$ correctly to the verbal statement in English, Kepler said that there is a **space $a^3$** which is **equal =** to the motion in the **time duration $T^2$** thereof between two specific points which is a straight line **k** that holds a relation from a centre to an end where the two ends run from the beginning of **k** to connect at the end of **k.** I might not be the smartest boy on the block but I'm not that stupid either. I know how to translate… and I translate as follows:
$a^3$ must have a volumetric interpretation because the third dimension is sure evidence of multiple conjunctions of dimensions put together in three sides opposing three sides having the third dimension in place. The fact that any symbol uses a value to the **third power $a^3$** indicates **space** or a volumetric established and separate unit. Using a cube by three dimensions symbolises a cube, a room, a space to be filled, a unit able to hold other ingredients on the inside when empty or partly filled. It is space because it is volume using the third dimension.

$T^2$ is an indication of something having a cubic nature other than the square forming motion that is provided by the motion the square indicates, which is where the moving object is representing a third dimensional object that is moving from point to point and it is this point to point that multiplies into the square. The space is moving as a unit from one point to another point and the moving between the points are represented by a flat square or following a flat distance between two points. The cubic space was in one instant in one place and then the second instant in the other and because time can never stand still or become single dimensional (this I am about to prove as the book unfolds) insisting that time must always support the motion it consists of or time cannot be. It is motion that is taking time, which is motion in the second dimension moving the space in the cube.

$k^1$ is the symbol used to indicate a straight line between two points with a definite beginning and a specific end position. It is the location where the cube is holding space and where the space was and where the cube in space is going to be in very the next split instant that follows to which will then in multiplying form the square that indicates the time the journey took to move the cube of space from one point where $k$ is indicating the location of the space to where the next indicating of $k$ will shift the space being the cube pointing at the end of $k$, but since time represents the square and with $k$ being the distance that proves that the $k$ represents the distance the space representing the cube went to take the time represented by the square through the motion. It is the distance moving space in the cube to complete time in duration in the square of motion; therefore $k$ is permitted to be in the single dimension.

There are infinitely more implications in the statement Kepler delivered than what is merely a contribution to motion and only motion as Newton was of the opinion. What is there mathematically not correct in my interpretation of Kepler's manner of translating mathematics to English and why is any changing thereof by Newton or any other person necessary in any way?

We can test any of the following symbolic values in the mathematical expression and also test the principals behind the expression in which Kepler stated them. By such testing we will find that time after time there were never any corrections in the translations required since the translation thereof was never incorrectly presented and in that a case asked for no alterations to secure the correct reporting of the cosmic information being translated. By taking the formula on face value it can change as follows: $a^3 = T^2 k$ can become $k = a^3 / T^2$. Let's take the Kepler dynamics back to the route the comet follows while orbiting the Sun

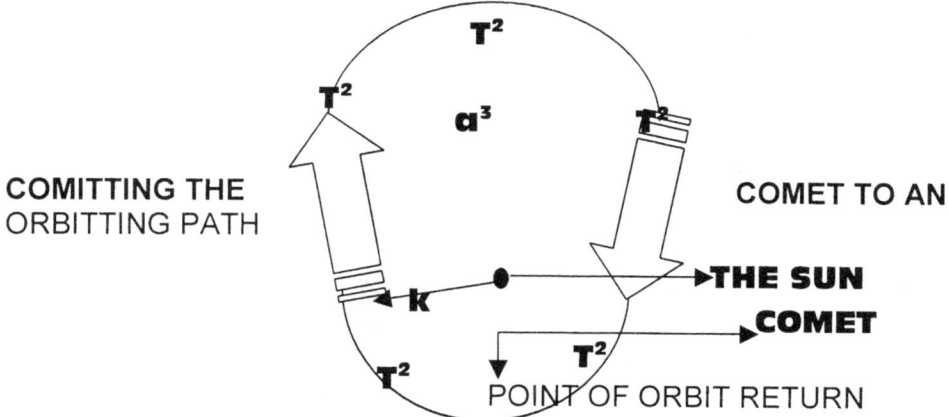

**COMITTING THE** ORBITTING PATH

$T^2$

$a^3$

$T^2$

$T^2$

$T^2$

**COMET TO AN**

**THE SUN**

**COMET**

$k$

$T^2$

$T^2$

POINT OF ORBIT RETURN

With this mathematical reality what then later formed the grounds for any individual to develop any need to change Kepler's translations from the cosmic given to mathematics and then from mathematics to English while the guilty party is renowned for his superior skills in mathematics?

Kepler translated what he found to be the cosmic given to mathematics which we humans are able to interpret from the mathematical expressed to the verbally pronounced and written but Newton still saw a need to change what the cosmos said about how the cosmos is presented and by no one less than by its own interpretation of its self structured composition.

When viewing my interpreting of what Kepler said I might have asked myself countless times what did I not translate correctly from the mathematical expressed to English after encountering a battery of Academic onslaught and resentment on my Newtonian views because after all it is directly diverting strongly from the teachings presented by Mainstream science and the diverting is not coming in a small way.

In truth from my diverting I came across very new ideas I am able to prove. By my translating Kepler's work correctly I came upon answers not yet uncovered by Mainstream Science

Kepler gave the World mathematically translated cosmic answers he received from the cosmos that Kepler uncovered long before Newton, Einstein and others got wise about cosmology...and later the wise came up with old news (old views as far as Kepler expressed their views before they, the wise were born with the purpose of coming to the conclusion that those wise men eventually did) and where the conclusions that the wise concluded brought much surprise to the world with the originality of the later Masters' initiative while Kepler said the same thing ages before...! )

Such is the advantage of recollecting Kepler facts that it does answer many questions, which went unnoticed and therefore not spoken about up to now and some were previously never even thought about.

**Newton said a sphere is $a^3 = 4/3\, \Pi\, r^3$, which is mathematically correct, however**

**Kepler said the cosmos told him a cosmic sphere is $a^3 = k\, T^2$** There is the two distinct possibilities which Newton saw and which Kepler saw and both are most valid. Between the two concepts there is literally one Universal difference and the two can never be mistaken as promoting the same principles. 'Ever try to answer facts about the Universe in as much as...what brings about the expanding? Kepler said the Universe plus it entire content is expanding centuries before Edwin Hubble realised what he was seeing through his telescope.

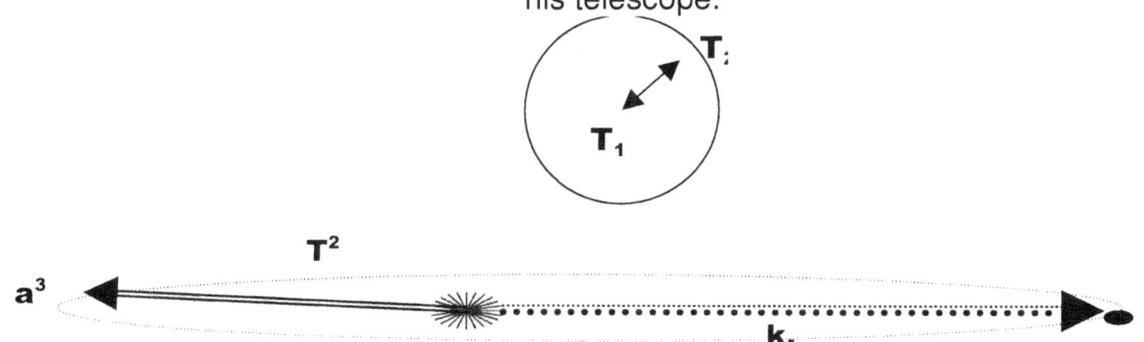

Kepler was the very first person to mathematically introduce **space $a^3$ centre k** and **time $T^2$**. Not only did he introduce **space-time $a^3 / T^2$** but he also placed **space $a^3$** and **time $T^2$** in a relevancy long before Einstein did and placed **gravity in space-time $a^3 / T^2$** even before Newton named gravity. He showed that space **k** is growing in the measure of what means the Universe attend to by promoting space-time as **$a^3 / T^2 = k^1$**. Kepler was the person who placed gravity as the ingredient in the Universe that determines **space $a^3$** and **time $T^2$** and much more. Kepler was the first one that said that gravity comprises of two factors being **k** or linear gravity and **circular gravity or $T^2$** as gravity keeps space in form while all is staying together.

Although not one Academic has ever openly admitted to me that they as members and part of Mainstream science are more aware than I am of all the facts and doubts I point out to them, such evidence then becomes clear whenever I mention the matter to them I get more than the impression it does not come as a surprise to them and hit them like a brick between the eyes. The lack of surprise and initial doubt they should show at first when they discover the incorrectness of evidence in their theory is a telltale sign confirming my suspicions about their evidently knowing all this information all along. They clearly seem very agitated about every detail I show when I bring the mistakes and double talk to their attention in the hope that they may confirm my doubts. Never is there a whisper of a surprise or a hint of a suggestion that would initiate an argument carried on by the bewilderment or the astonishing surprise they should feel confirming my arguments because there is a mild complacency in their voices. My jumping them total unexpectedly about matters they never contemplated in the least leaves them unturned. The rush in blood pressure that should be a factor on their part and part of the instant where total surprise will bring about some confusing thoughts that will inspire the unleashing of an argument in defending their holy grail should at least carry a surprise in an attempt to save what they believe as being the Gospel in science and with that defending their honour. They lack embarrassment, which they should have in their disputing of my claim as they fight off my allegations with a countering of denial claiming foul on my part as they are in shock when finding out about any doubts. A lack of true emotion on their part is a telling sign that they also may have some serious thoughts on the quiet about any inclination presenting a flawed view about what they always thought they knew to be true. There is only that eerie dismissing of the seriousness and the lack they show in excitement that would deny or support my credibility as I present my findings. If they know about the inconsequential facts in science why is it not generally acknowledged and pronounced as a matter of fact? Why is there the covering up and hiding facts that we associate with some professional criminals such as politicians. The fact that Academics are aware of this evidence in general terms about the misinformation and doubting evidence about Newton's cosmic vision but moreover underlying this is their total denial of knowing about it and that is what is so seriously unforgivable. The fact that all Academics are aware of my evidence even before my presenting them with such evidence is beyond doubt. If that is the case then why are they forever trying to kill my viewpoint and forever try to silence me where I am only the messenger because I bring the solution and the answer?  Please note that the answer and the solution are unbelievably simple and unsophisticated. It lacks all the splendour and grandeur expected by all Academics concerned. It is because it is so simple that it went amiss for four hundred years. It is because it is so simple that it misses the grandeur that will entice them. Instead every academic accuses me of not understanding Newton while they can't show me what part it is that I can't understand and I on the other hand can't see what there is not to understand..

Newton said that it is the reducing of the distance between the objects that would bring about the un-reversible reducing that will end in a total demolishing of the radius that is between the cosmos structures, but instead we find the gravity applying in outer space is one of the instances where gravity provides an orbit circle that gravity seems never to completed as the orbiting objects follow from closing any circle that is leading into a following circle up to where the circle is completed in cyclic precision. That is not the gravity that Newton identified        although Newton admitted that there is a presence of a centre forming a point in the middle between the two objects. He was unable to know what caused or even the presence of the Coanda principle, which forms so critical a part of my theory. The formula concerning cosmic balanced gravity however leaves no room for the admitting of such a point and by not leaving a possible inclusion of such a point in his formula Newton did by such gesture in principle repeal his admission of such a centre. This had me cast doubt on what is taught at institutions of learning. It motivated me to venture back to an era before Newton came to influence science. **I came to acknowledge Kepler as I came to understand Kepler. The accepting of that what I understand in Kepler involves much more reading into what Kepler said by finding what Kepler did not say in the way that he did say what he said than the reading about what Kepler said as it is written in the precise detail and to the letter used in his statements. He never directly stated what he said. Again I must stress this point: when I refer to what Kepler said it most likely means reading into the part that is being a part of the part that he did not say when he was saying what he said but I accept that he meant to say what I am reading and translating from Kepler as part of what he did not say but meant to say.** I have to read more with my mind than with my eyes. This comes as a result of interpreting Mathematics to the verbally expressed. I had to learn to read with my mind and not my eyes and I found that that is the manner in which one has to approach cosmology. From the first time I discovered what manner one should use if one wished to read into Kepler's findings I saw Kepler was all about uncovering the unknown. Realising that, the conclusions I drew by reading in such a way cemented my better understanding of Kepler's work, which then helped me improve my insight into Kepler's work as it increased my understanding about cosmology several fold. This helped me to realise what implications were to be found underneath Kepler's discoveries. From my realising what approach I should use, it helped me to improve my cosmic realising by using the method of reading Kepler and from that I could come to appreciate what Kepler introduced.

Only then did it bring insight and proof to me as a student of Kepler and this proof I found by dissecting what Kepler **did not** say instead of what he **did** say, which I now present to you with this book, you being a superior intellectual person. Kepler said $a^3 = T^2 k$ and that correctly translates to a mathematical expression $k^0 = a^3 / T^2 k$ which in the verbal statement in English translates that Kepler said that there is a **space $a^3$** which is **equal =** to the motion in **the time duration $T^2$** thereof between two specific points which holds a relation onto a centre $k^0$ where from there forms **a straight line k** that is centred on the spot where space begins from $k^0$ **that produces k** as well as producing the circle therefore that spot $k^0 = a^3 / T^2 k$ has hold $k^0$ at a value of having the least space. The line **k** is centred onto a spot where space begins specifically at $k^0$. This point not only produces the line $k^0$ but represents also the space that forms the eventual circle $T^2$. Therefore from the centre holding $k^0$, $k^0$ leads to **k** that forms the roving space $a^3$, which is rotating at a distance **k** where $T^2$ forms the outer limit of $k^0$. Mathematically $a^3 = T^2 k$ will be $k^0 = a^3 / (T^2 k)$ because $k^0 = 1$. But $k^0 = 1$ also present the single dimension where all factors are a product of one. If one can locate $k^0$ one will find singularity. That is where gravity is because gravity is strongest where space is least. Then that suggests that gravity is strongest at $k^0$ because space is least. That is gravity because that is what keeps the orbiting object s in orbit but also that is what Newton completely missed when he changed Kepler's work. Newton failed to recognise gravity as the only ingredient in Kepler's formula. He admitted he missed this because he admitted he did not know what gravity is while Kepler explicitly showed what gravity is. Gravity is what keeps the orbiting object s orbiting. $k = a^3 / T^2$ is **distance$^1$ = space $^3$/ time$^2$** forming from a pivoting centre $k^0$. That is a cycle and moreover it is a cycle formed **by space/time**. What Kepler said is that space is $a^3$ **in motion $T^2 k$.**

That says **space$^3$ ($a^3$/)** relates directly to **time$^2$** that uses the symbol $T^2$. This is also what I refer to when I say one has to read what Kepler did **not** say when one wishes to see what he **meant** to say. Kepler introduced space$^3$ –time$^2$ long before Einstein's date of birth appeared on any calendar although Einstein is credited with the formulating of the concept of space-time and giving it a name. Going even further Kepler stated that the space $a^3$ is on the move $T^2$ around in a circle at a distance **k.** That is what that comet we are discussing is doing. The space$^3$ (Comet) is circling the Sun using a radius **k** to establish the cyclic time$^2$ as a period of continuous motion and continuous motion is gravity. That reads much more correctly and closer to the truth than what Newton predicted what according to him (Newton) was happening in space. Remember in this statement I am separating cosmic principles applying from the way that gravitational principles apply on Earth. I distinguish that which is the rule in the cosmos from what we find ourselves trapped in on Earth. The two just don't mix. I am removing cosmic physics from normally accepted physics because the gravity concerned is not the same.

The proof I bring is real however simple it may seem. It has none of the mind-blowing complexities normally associated in the presenting of investigative analyses of Astronomy. I realise the information in this book carries the arguments in a childlike manner which are very simple to follow, and for that in the past I have been blamed over and over again as being unprofessional. In my answer to that I can only reply by using another question: Are only professionals adequately equipped with minds that make them (the professionals) the only ones able to think? We being part of the human race are all thinkers. Everyone as a human being can think. Every person on Earth is a thinking thinker that uses his brainpower by exploring thoughts mainly and normally to his or her personal benefit. It is what we think about that produces the results of our efforts by which we accomplish what ever we are thinking about. I have met professional Academics that I found foolish as much as there are other cases where the so-called amateurs can credit themselves with much wisdom and insight. Albert Einstein as a patent clerk was that much but to name one. Please understand that I do not compare my achievements or myself in any way, shape or form with the likes of a Master such as Einstein although I speak my mind when not being totally in agreement with some of his or other views. My unsophisticated retracing of Mainstream physics concerning the Big Bang in detail helps to reinvestigate established principles and moreover investigate proof in the light of modern evidence. In principle I distinguish between Kepler and Newton in that Newton is one hundred percent correct concerning gravity on Earth but as far as outer space forms gravity the conclusions of Kepler and Newton do not match and they had totally different ideas about what they saw in gravity. I am in disagreement with some basic principles that science acknowledges and I divert strongly from all accepted roads Mainstream physics follow. By my doing that those who are considered and accepted as self-proclaimed members of Mainstream Physics have categorised my views in the past as incoherent. That I do not accept. I admit that my line of thought is extraordinary and controversial but only to Mainstream science and not to the standards lay down by nature. Since the concepts I follow start at the beginning, and I take Kepler at the point where modern cosmology began and in that mindset I re-evaluate Kepler's work. I start by tracing a new approach as to what I see Kepler found. The main condition of my investigation is to establish a divorce between what Kepler said and what Newton thought to add to what Kepler said. It is this divorce I create that Mainstream science finds repugnant or even in some persons' opinion repulsive. I believe the repugnancy does not come from or is not manifested in any part of my work to the book as such, but rather what my work suggests and who is doing the suggesting. To my view in cosmology such adding to Kepler by Newton was unnecessary and it diverts Kepler's work away from cosmology. But as the generations moved on Newton became religiosity in the mind of science wherever science was taught. To students there is little or no choice in the matter since the only choice left to them is one of understanding by forcefully accepting or die an academic death since Newton is academically accepted without asking questions or raising an opinion. For the second choice, the less accepting students are greeted with a

Dear John good-bye book sending them off into the unknown Sunset that such a future outside physics will bring them. That is brain washing.

From studying Kepler I saw that we have to gauge what we find in the Universe. What we find is not that what we realise with our eyes but that what we observe by using our minds to translate from visions coming from our eyes to our minds. We have to test the part that we are seeing much more than merely accept what there is to see on face value. We have to not only see what other life beings blessed with much less insight most probably also should see. We must stop using our eyes in the same manner as animals do and start seeing with our mind, as humans should do. Being the superior evolved species that we are gives us the ability to read into that which only we can see and that we only can see by using our intellectual mindset. By seeing with an intellectual understanding what there is to see when we see what we can observe, we should therefore have the ability to be in understanding by looking at what we can see but moreover understand that which we cannot see. It is the same as playing chess. See what there should be moved instead of noticing object not having an ability to move on own account. This I first found to be true about Kepler's work and when I started projecting this method of observing what the Universe is, as it scattered most previous perceptions I found that using the new method brought along answers so fast I could sometimes hardly keep up with the interpreting thereof. But as is the case with Kepler so is the case with the entire study of cosmology: One should see what there is about the cosmos which is unseen to us and then we may find so much more in the cosmos unseen to us representing that which we cannot see and that which we cannot read because we have to learn to read what is not written in light. Armed with this realising I then proceed from that point by further arguing and debating the full implication of Kepler's contribution. Kepler placed cosmic structures in relevance to one another and so does the Big Bang Theory. The backbone of the Big Bang is that relevancies apply in dynamics and such dynamics are placing all structures without any reservations independent from each other. As the Big Bang progresses all inside the Universe is in the same Universe that will always be the same, however the relations that the elements comply to bring across new relevancies with new positions to fill. The father of the Big Bang concept is a person by the name of Father LE MAÎTRE, GEORGE ÉDOUARD (1894-1966) who was a Belgian priest and cosmologist. He was the first person to embrace the fact that the Universe expanded from an infant stage. His model of an expanding Universe (1927) was superior to that of W. de Sitter in that it took into account mass, gravitation and the curvature of space. Similar models were proposed in the early 1920s by the Russian mathematician Alexander Alexandrovich Friedmann (1888-1925) but Friedman compiled various such possibilities. Lemaître argued further (1931) that the quantum theory supported an origin in the explosion of a 'primeval atom' or 'cosmic egg' into which was originally concentrated all mass and energy. As modified by A.S. Eddington, Lemaître's model provided the springboard for G. Gamow's Big Bang theory. In the wider picture of science in general a lot changed to just allow such turnabout in thought since the day of Isaac Newton. From Newton's attraction and contraction many things came into place that allowed change in the most hardened minds. Accepting facts about the Big Bang concept is quite radical. By promoting expansion the Big Bang theory contradicts gravity and our accepting of the Big Bang has to change all other concepts. By accepting the Big Bang other changes are also involved.

KEPLER, JOHANNES (1571-1630)
The German mathematician and astronomer KEPLER, JOHANNES (1571-1630) became Tycho Brahe's assistant in Prague in 1600 A. D. where he undertook to complete the tables of planetary motion Tycho had begun. Kepler first calculated the orbit of Mars. He spent much time trying to reconcile Tycho's accurate observations of the planet with a circular orbit, but concluded (in Astronomia nova, published in 1609) that Mars moved

instead in an elliptical orbit. Thus, he established the first of his laws of planetary motion. A theory that the Sun controlled the planets by a magnetic force led him to the second and third of his laws, which were published as part of his treatise on theoretical astronomy, Epitome astronomiae Coernicanae (1618-21). The Rudolphine Tables (named after Tycho's patron, the Holy Roman Emperor Rudolph II) of planetary motion appeared in 1627 and were still in use in the 18th century. Kepler also wrote De Stella nova, on the supernova of 1604 and Diptirce on optics and the theory of the telescope. The overall view followed in this book **A COSMIC BIRTH>>DISMISSING NOTHING** places the true significance of his work in true contents. In KEPLER'S EQUATION is the equation that relates the eccentric anomaly of a body in an elliptical orbit to its mean anomaly. The equation is $E - e \sin E = M.$, where $E$ is the eccentric anomaly, $M$ the mean anomaly, and $e$ the eccentricity of the orbit. It is important as one of the mathematical relations enabling the position of a planet about the Sun, or a satellite about is planet, to be calculated from the orbital elements for any time. However this only relates to the solar system, and KEPLER'S LAWS only apply in the contents of the solar system. The three laws governing the orbital motions of the planets, discovered by J. Kepler is as follows: The first law states that the orbit of a planet is an ellipse with the Sun at one focus of the ellipse. The second law states that the radius vector joining planets to the Sun sweeps out equal areas in equal times. The third law states that the square of the orbital period of each planet in years is proportional to the cube of the semi major axis of the planet's orbit. The first law gives the shape of the planet's orbit; the second describes how the planet must continuously vary its speed as it follows its orbit, moving fastest at perihelion and slowest at aphelion. The third law gives the relationship between the planets' average distances from the Sun and their periods of revolution.

Instead of studying the true value and contribution of to Kepler's laws an Englishman going by the name of I. Newton placed his own interpretation to Kepler's laws, and in doing this, he wilfully destroyed the principle working of the Creation. Saying this I hear the alarming hooters announce Newtonian dismay. In the past my experience was that all the revered Academics lost their appetite for any further investigation of my work. That is sad as much as it is regrettable. Through Newton's tunnel vision, he applied his own misinterpretations to the correct presumptions of Kepler and through the Newtonian tunnel vision Academics did not move an inch away from repeating the same procedure. In the past it was this that had Academics shying away from me because at the point where I raise criticism of the Newtonian viewpoint I am rejected. The point where I declare my suspicions concerning they're accuracy and the correctness about they're theorising, which is where I should then be raising their doubts about their way of thinking is the point where in stead I raise their suspicions about my way of thinking. That is what caused the rejection of my criticism about Academic Newtonian science and evoked their criticism in the past about my views instead of they're following the logic by investigating what I said. Their rejection of self-investigating got me and my work rejected to a point where the applecart lost its wheels on every occasion. It is where Academics read my remarks and what brings (seemingly in an instant) wrath to Academics. I say this because I realise that reading my remarks or hearing me remarking about this notion brought much resentment on their part and if the reader at the present moment is a Newtonian, boiling his/her blood. It is blood boiling because I believe they see my remarks as belittling that which they feel they have accomplished. This is not the case but still my remarks have the same effect on the Academic as pouring icy cold water down the back of his shirt. I mention this because I know it has happened many times before and if possible I wish to avoid this response. Therefore I ask you kindly to please be warned about the negativity you must feel towards me where you are the Newtonian and I am not. Before you lose interest in reading this book any further please allow me to finish. In the past Academics thought me to be presumptuous and that normally became the point where all the Academics find their interest vanishes. That should not be because if Newton's work is as utterly accurate as those with faith in his work believe it is, then every aspect about Newton should stand above any and all reprimanding or any form of doubt causing a notion to reprimand. The testing of Newton's work should withstand all testing notwithstanding the person or the prominence of such a person's social or academic standing in the Academic society or even the prominence that such testing will deliver. From what I see about Kepler's work it is a flow of circumstances that lead to Academics neglecting Kepler's work and the realising of the theory I suggest is not forthcoming due to my personal brilliance. I do not consider myself to be the brilliant in any way as to be the one that can remove the verbal splinter from the eye of the Academic. Yet...if there is a splinter what else should I then do...Newton reduced the implication that Kepler findings hold by introducing to the law of gravitation. He then went about and changed it to three laws of motion. It is clear that while he formulated the laws on motion he missed the way Kepler introduced gravity as space $a^3$ coming about through motion $T^2$ and that gravity is space $a^3$ within space $k$ within motion $T^2$. Newton also missed the fact that gravity is at its strongest where motion and space cease to be. This is most important to recognise about gravity in one of the two forms it has. I. Newton generalized Kepler's first law, verified the second law, and showed that the third law should be amended to the form; $4 \pi^2 a^3 / T^2 = G (m + m_p)$. In this, the value of "T" and "a" are the period of revolution and semi major axis of the orbit of a planet of mass $m_p$ about the Sun of mass m, and G is the gravitational constant.

It should be clear to any person investigating Johannes Kepler and his work that Isaac Newton hijacked Kepler's work and any time there is the slightest referring to Kepler about the research Tycho Brahe and Johannes Kepler did such referring to Kepler always lead to and always include the mentioning of Isaac Newton changing the work of Johannes Kepler. It is as if the World never could acknowledge Johannes Kepler because the work of Johannes Kepler would be completely wrong and misleading if it were not for the intervention of Isaac Newton saving the skin of the less admirable Johannes Kepler. This comes in the midst of every one realising that Kepler used the information he received directly from the cosmos. I do stress this on many occasions throughout the book because the embarrassing part is that Newton changed the work of The Universe and not of the man called Kepler. Should you reading the book entertain the opinion of Newton and feel any urge to defend Newton you should ask the question as to who is standing corrected, is it Kepler or is it the cosmos that gave Kepler the information he concluded? The cosmos supplied all the information by using mathematics, which Kepler then had to translate. But Newton destroyed the accuracy by altering what the cosmos said and directly by adding to that what he (Newton by name) thought that the cosmos left out. This set a precedent by Newton in cosmology and also set a trend, which was retained in all future cosmological development and it lasted in cosmology for three hundred and fifty years. In this book you are reading I am about to show that such practise should no longer be accepted in cosmology. In the process the world of Mathematics developed and the world of cosmology stood still for almost four hundred years. Faculties contributing to cosmology and feeding off cosmology improved as much as they developed, but when cosmologists see the Roche limit in action in the lens of the Hubble telescope and refer to the event as "stars blowing bubbles" being the ultimate response coming from those persons who are supposedly the Masters of cosmology affairs, then the truth of what I just said comes down on you like a ton of bricks. Everyone having any remote interest in cosmology will find they are being very disillusioned by such "official" testimony about the evidence the Ultra Wise report on. This book is about showing how great Johannes Kepler was and how enormous his work was. It will show he preceded all ideas of everyone that came later and officially introduced the novelty of such ideas. Back during the time Kepler was introducing his work the stature and the magnitude of his work was beyond any person's understanding (including Isaac Newton) and this prevailed for most of half a millennium. I do not say I am the brilliant one to uncover Kepler in the face of everyone failing that came before me, but as I am not a Newtonian such bias was not part of my repertoire and denying me the fortune of being a Newtonian added to my fortune of realising Kepler. Yet as you will notice, the work I contribute is much below the sophisticated norm of modern investigative research and the levels that modern research accomplishment demands to better the effort of the understanding ability in the splendour that investigative research work should deliver in view of our modern times. It is only pure neglect in science circles that moved science past Kepler. Not seeing and therefore not investigating through almost half a millennium has paved a road past the inferior levels that the researching of Kepler's work holds because it was rocket science four centuries ago but the brilliance of it has faded since then. My contribution holds no astonishing flair that may add to science in general. Only failure to notice what I see on the part of those truly brilliant can explain my being able to present my contribution about my work in investigating Kepler. Only by their passing such degrading levels of the Academic establishment in the past and the present can bring the blame for such an obvious discrepancy because any involvement in the work at such an inferior level as that which I bring cannot interest and excite a salted Academic and when thinking about it, the idea is totally unthinkable. This book, although it is on this inferior level is about correcting this tendency and has in mind the effort to put in writing what would place Kepler in the greatness and glory he deserves. As I already said, if Kepler was wrong then the cosmos was wrong about facts and applying relevancies and tendencies in the cosmos. I yet again wish to reiterate we should never for one moment forget that Kepler received his information directly from studying the cosmos so how could the cosmos stand corrected? In spite of all the brilliance attributed to Newton nonetheless if Newton had the mind to change Kepler's work and my saying this includes all persons agreeing with such changing by Newton of the work of Kepler those persons admit that he or she or Newton never took any time to really and truly investigate what the cosmos told Kepler. From my reading into the work of Kepler I prove gravity, the Titius Bode law, singularity, space-time, space-time relevancy, the Lagrangian system, the Coanda effect and the Roche principle, the sound barrier, the principle behind the Black hole. The precondition for my ability in doing so is that I have to remove Newton's opinion about Kepler's work from Kepler's work. Whenever cosmology comes into question and all the phenomena, which I mentioned just now remains unexplained and by that token alone it shows to what degree did cosmology remain undeveloped. Whenever there is any mention of Newton, Kepler is never mentioned. But the reverse is always applying. Mainstream physics holds the opinion that Kepler may only have an opinion if Newton can change the opinion. Kepler gave space-time, gave gravity, gave singularity, gave the Plank theory, gave the theory on relativity but no one ever found Kepler's work deserving enough to launch any investigation such as I did. I belabour this because of what revulsion my rejection of Newton unleashed. That is one barrier much unnecessary but it has been an insurmountable barrier this far.

NEWTON, ISAAC (1642-1727) and NEWTON'S LAWS OF MOTION
An English physicist and mathematician who developed his principal theories about gravitation, optics and mathematics between 1665 and 1666. In 1668, he made the first working reflecting telescope. Most of his work remained unpublished for long periods, partly because of criticisms by c. Huygens and the English scientist Robert Hooke (1635-1703) of his early work on the corpuscular theory of light. However, in 1684 E. Halley persuaded him to organize his work on the celestial mechanics of the Solar System, which was published as the Principia. Newton's other major work, Opticks, was not published until 1704. It contains his corpuscular theory of light, and the theory of the telescope. His greatest mathematical achievement was his invention of calculus,

independently of the German mathematician Gottfried Wilhelm Leibniz (1646-1716). His profound influence on physics and astronomy is reflected in the phrase 'Newtonian revolution'. Three laws published in 1687 by I. Newton concerning the motion of bodies.

1. A body continues in a state of uniform rest of motion unless acted upon by an external force.

2. The acceleration produced when a force acts is directly proportional to the force and takes place in the direction in which the force acts.

3. To every action there is an equal and opposite reaction.

4. However there is one more law on motion that went undetected by Newton...This book is not about trying to disprove Newton...it is about adding too science more than there now is available without removing any that science already accumulated.

In this book I use Kepler's formula to either prove or to disprove the following accepted principals in cosmology and if any person in the past gave only the slightest attention to Kepler's work, many statements would have come much sooner delivered by someone else or may never have come at all. By applying Kepler's formula correctly in this book I can either agree with or in other cases deny the following principles.

It began with NICOLAUS COPERNICUS who changed the status quo. COPERNICUS, NICOLAUS (1473-1543) was, according to the Anglo Americans, a Polish churchman and astronomer although this is just more politically inspired propaganda because his parents were both German (in Polish, Mikolaj Kopernigk). While he was completing his studies, he had realized that the Earth revolves around the Sun and not vice versa. Such a view was in that time, held to be heretical. As I pointed out in the first few articles, the Church regarded the geocentric world-view of Ptolemy as consistent with its doctrines. Copernicus set down his basic ideas around 1510 in the Commentariolus, which he circulated anonymously, because of the Islam link. In 1512-- 29 he conducted his study and concluded the observations that he needed to support his theory, while carrying out ecclesiastic and local administrative duties. In this time, he had to defend his mother in court on charges of witchcraft. In 1539, the Austrian astronomer and mathematician Georg Joachim von Lauchen (1514-74), known as Rheticus, became a pupil of Copernicus and began to spread his ideas. The published work was openly spread as the Copernican system, in spite of the life-threatening dangers connected with such a "crime", in 1543 in the book De revolutionibus orbium coelestium. However, the reality of a heliocentric Solar System was only commonly accepted, after the work of Galileo and J. Kepler. The ideas introduced developed along and proved to be correct until such a time it met a solid wall with the investigation of Max Planck.
PLANCK CONSTANT
(Symbol h) A constant that relates the energy of a photon to its frequency. It has the value $6.62076 \times 10^{-34}$ Js. It is named after the German physicist Max Karl Ernst Ludwig Planck (1858 – 1947). PLANCK ERA. In the Big Bang theory, the fleeting period between the Big Bang itself and the so-called Planck time when the Universe was $10^{-43}$ s old and the temperature were $10^{34}$K. In this period, quantum gravitational effects are thought to have dominated. Theoretical understanding of this phase is virtually non-existent. It is named after Max Planck (1858-1947). PLANCK'S LAW

A mathematical description of the energy radiated at different wavelengths by a black body: $E = hf$, where E is the energy of a photon and f its frequency. It was formulated in 1900 by Max Planck (1858-1947), who realized that energy is radiated in discrete packets, which he called quanta, and it formed the basis of quantum theory. The quantum of light is a photon, the energy of which depends on its wavelength.

There is one rule which is well established and which Mainstream science all agrees about. It is one aspect, which forms the very principle that holds the theory about the cosmic start together under the covering of a verbal blanket. All in science agree that it all started with singularity but I manage to go one step further where I prove that it is also where it ends, as singularity reunites space-time, which is from where Creation split in the very beginning.

Singularity is as follows: Singularity: a mathematical point at which certain physical quantities reach infinite values, for example, according to the general relativity, the curvature of space-time becomes infinite in a black hole. In the big bang theory the Universe was born from singularity in which the density and temperature of matter were infinite. From singularity flows space-time.

Space-time is as follows: Space-time is a four dimensional position of the Universe where the position of an object is specified by three coordinates in space and one position in time. According to the theory of special relativity there is no absolute time, which can be measured independently of the observer, so events that are simultaneous as seen from one observer occur at different times when seen from a different place. Time must therefore be measured in a relative manner as are positions in three-dimensional Euclidean space, and this is achieved through the concept of space-time. The trajectory of an object in space-time is called world line. General relativity relates to curvature of space-time to the positions and motions of particles of matter.

## SPECIAL THEORY ON RELATIVITY

A theory proposed by A. Einstein in 1905, based on the proposition that the speed of light in a vacuum is constant throughout the Universe, and is independent of the motion of the observer and the emitting body. A consequence of this proposition is that three things happen as an object's velocity approaches the speed of light: its mass goes up, its length shortens in the direction of motion, and time slows down. Hence, according to special relativity, no object can ever reach the speed of light because its mass would then become infinite, its length would become zero, and time would stand still. In addition, Einstein concluded that the mass of a body is a measure of its energy content, according to the famous equation $E = MC^2$, where c is the speed of light. This equation describes the conversion of mass into energy in nuclear reactions within stars.

## GRAVITATIONAL COLLAPSE

The collapse of a body that is unable to support itself against its own gravity. Gaseous bodies undergo such collapse if they are not hot enough for their gas pressure to balance gravity. This can happen in the early stages of star formation, or when nuclear burning ceases in a star's core. The time taken for such collapse decreases rapidly with increasing density, varying from about 100 000 years for the birth of a new star to less than a second for the formation of a neutron star. Star clusters may undergo a similar collapse if the random motions of their constituent stars are insufficient to offset gravitational effects, either during their formation or at an advanced stage of their evolution.

## GRAVITON

A hypothetical particle or quantum of gravitational energy, predicted by the general theory of relativity. Gravitons have not been observed but are predicted to travel at the speed of light and to have zero rest mass and charge. A graviton is the gravitational equivalent of a photon. It is this anti-photon-being-a-graviton by just merely swapping direction and all is proved that I find not very indigestible in modern science. One of the main issues that I wish to protest by my writing of this is my argument that if the Universe can be compressed back to the size it had at the point of $10^{-38}$ seconds after the Big Bang the daily outdoor temperatures of $10^{27}$ K will also come about once more. The expansion was the result of compressed space, which then formed into heat and in turn resulted in finding a Universe with all the insufficiency of space less ness prevailing throughout and wherever space was needed. By that it forced space-time to come into being. Space-time came about at the time of endless time duration without space availability, which brought about the period of the Big Bang wherein space growth was the converting of such heat to space. If the Universe was in a vacuum as big as being available now then what was the temperature of the vacuum while it was empty before material filled it later. Then I presume the vacuum was there present as it is now in this present day. If the Universe then employed the space of say one atom, the impression comes through that from edge to edge and from Universal border to border the space occupied was the same as one atom will claim in our present day and age. Normal gravity started at $10^{-43}$ seconds. The Universe was the size of a neutron or somewhere in that vicinity. The Big Bang began and GUT, or the grand unified theory, produced the attempt to describe the strong and weak nuclear forces and electromagnetism in one single mathematical theory. Somewhere before $10^{-12}$ seconds of counting the Universe cooled to about $10^{15}$ K the electromagnetic and the weak interactions acted as one single physical force. Science reckons that unification may come about at temperatures of $10^{27}$ K, which was the temperature of the day at $10^{-38}$ seconds after the Big Bang. This statement echoes my viewpoint but one has to look carefully for that to surface.

In the suggestion the presumption claims that all the space that the Universe made available at that time was the total space one atom might take up today. If that might be the case then where was the rest of the space that now fills the Universe? Or was the rest of the space we now find in the Universe and what is now explained away as the vacuum, also available back then. Did the Universe only have that one tiny hot spot it filled with huge volumes of heat? Was the rest of the space vacant being out there all along during all the time running to the present date but filled with emptiness standing around as a big vacuum with no better to do than sucking on the Universe while the Universe was exploding at the speed of light. Then that statement suggests that in this hot Universe there were light-years upon light-years of vacuum waiting to be filled by the intense heat soaring in the smallest spot. If that is the case then why did the vacuum not fill in the blink of an eye by all the exploding expanding material growing at the speed of light? Was the Universe overall bitterly cold where the vacant space was locked in with one spot of the vacuum filled with temperatures so hot we can only produce it in numbers suggesting a value but never claim to be able to digest the reality thereof in the human mind? If so what happened to the natural consequence that heat flows in the direction of cold and equalise between hot and cold. Was the space being available at present available then or was the hot space the only space available at the time. If so what prevented the heat from instantaneously filling the eternally cold vacuum because with the rules controlling vacuum in affect, it should have filled in such a manner in less than a heartbeat?

I believe that singularity formed space-time and space-time developed from the overflowing of space-time at the time is extending by marching onwards and outwards to this day. Space-time developed another product that everything in the cosmos has to have. It must be in such large quantities everything imaginable in the Universe has to have it and that is space using time to move about. I suggest that it is space that is holding heat in a quantity providing density and ratio to space available and in relevance to the space being available to quantify the presence of the heat and which then proves to form the time factor. The container and contained all together mixed by motion. From that very first separating of heat and space, which is what formed from singularity to produce space-time. The Universe was full... It was overflowing by the speed of light in the beginning...so where and when did vacuum or nothing enter the Universe as a factor if and when the Universe was so full.

The answer to that is absolutely crucial because how did the Universe decide to fill some parts with a variety of something and decide to fill some parts within the in-between with nothing? If that is true why did gravity not prevent the vacuum filling because no gravity that came about since can beat the force that gravity had back then? This leads to another question following the previous one in asking why did gravity at the time when it was so strong with $r^2$ so much compromised not fill the nothing immediately as it entered with something that could absorb the nothing. At the very beginning the mass that was pulling on the mass by force was immeasurable and none quantifiable. Even more to the point is the question to be asked in how big was the radius between the materials with the immeasurable mass placed in such a little space. This is all the more important in the light that the smaller the radius is the bigger the force will become from the immeasurable mass pulling...

With the immeasurable mass that was producing the first gravity between the particles divided by an almost non-existing radius the gravity produced had to be in gigantic proportional quantities and with the separation of the radii being in the infinite measure that it was at that point then how did the Universe establish the chance to expand. It did expand, as we all are witnesses too in spite of this contraction of gravity that had to have been compromising the expanding factors. Still the expanding filled the unknown part of the unoccupied Universe, which at the time was there or was not there and it was there it was then filled with nothing. If the nothing was not "nothing" then the nothing that was not being nothing was also filling the rest of the vacant Universe that was or that was not because if it was it was filled with nothing and if it was not then it was nothing.

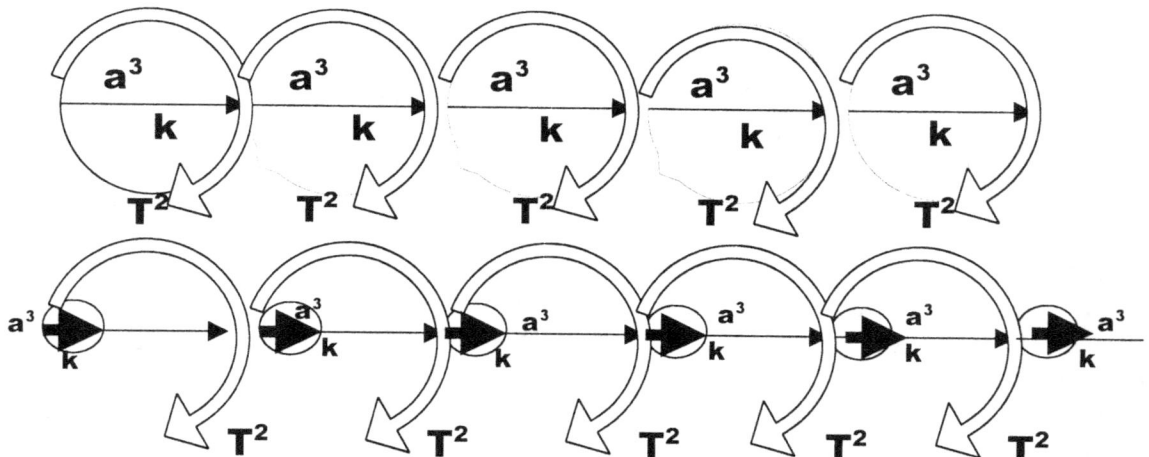

This is then taking into account that then all the reducing that is resulting from Newtonian contraction and that was going about in the space available at that time was something filled with nothing and surrounded by more nothing? With everything in the Universe being that much crowded and crammed where and how did nothing enter the Universe and fill the rest that was unfilled? What factors introduced nothing into the picture since the entire Newtonian concept finds its base on the principle that matter reduces using gravity by force which then bring about reducing or the removing of the many nothing between particles, which will then lead to nothing that has to vanish even before nothing can enter the space. This question may seem small-minded belonging to the mentality of a child or to that of the mentally impaired with not much factual appreciation developed yet. Please do not see it that way. If you think in those lines it will be because you do not have an answer to challenge these silly questions. Beware, silly as they are they represent official backing by the Wise-and-Informed. If the space is nothing and if the space was as large as it is at present then there was no need for such a small area to fill with something leaving only the rest filled with nothing at first since all the space we know about was there present and by being present it was there then for the taking. What ever filled the Universe had to start at the centre of the Universe and fill the entire Universe all over from a centre as it moved outwards filling from the inside outwards. This is a natural human instinct realisation but is beyond proving by using Accepted Scientific policy. But that leaves Newtonian science with a massive unsolved problem: where is such a centre at the present time and where does the centre produce the limits or border it apparently has to form as it expands.

By expanding there is an additional contribution too that that was when that that was, was receiving more than there was before the addition increased that which was and by then becoming more than there previously was it had to be improving the border from where it must have been before the adding took place to where it was after that that was added was added. When that was less than it became when it was added too it was at the limit that was there before it was added too and that limit there was, was a limit that is the limit that I am referring to as a border being there. The cosmos is filled with unrecognised borders. The expanding has to be an ongoing filling that is at the same time expanding from the inside towards the outer limits of the Universe. Since nothing can enter from the outside where nothing is, the filling of nothing as a substance that would take up vast quantities of room had to fill from the very centre spot where all other filling came from. This filling of nothing with material has to be well mixed. The truth about cosmology is that space forms no borders but by using any Newtonian centre from where mass is attracting we must find a point where there has to be the ultimate Universal centre which is the cardinal point in the entire Universe and it is the first, the prime position to locate coming before any other concept one wish to put forward because all concepts has to start with locating that cardinal centre. There has to be the ultimate $r^2$ radii located precisely between the ultimate mass drawing the other ultimate mass closer. If there was a Big Bang then there has to be the spot where from the Big Bang developed therefore there has to be such a centre connecting the past to that ultimate centre with the line of development flowing onwards to this day.

The fact that science is Newtonian proves that in the meantime Mainstream science is still of the opinion that there was the specific centre in the Universe that is nowhere to be found as it was filling the unknown with nothing coming from nowhere, but which somehow is still somewhere in the centre of all of that which is something. On the opposite side of nowhere there is an outer border in space producing a limit to nothing and serves nothing with a specific point to stop being nothing because that point is precisely where nothing ends and forms a beginning of a Universal border or a Universal end. How one will stop vacuum being no longer nothing was a question everyone comfortably missed to ask therefore no one ever seemed to deliver any form of answer.

One night some years ago very close friend of mine had a meal at his restaurant and as the conversation progresses he asked me about space and where it must end. I tried to explain to home what I believed in comparing to what Mainstream physics believed but soon saw I was not gaining in his understanding. Then I decided to jot it down on paper and he could read it at his leisure as he saw fit. That led to the first book written by me (in Afrikaans my native language). What I tried to explain to Johan Boonzaier that night is that if the Universe was the size of say even a tennis ball with only the size of a tennis ball being the very all of space there is available, then yes, it must take time too expand from that having the excessive heat there was back then in all the space we have at present. It then is converting heat into space bringing about the expansion. But one will find most expanding within the atoms, as the atom must grow since the Universe in all was the size of what one atom is today. The space in the atom pushed the space outside the atom but there must be plenty more too the growth. Something outside the atom contributed in it own rite because there is more expanding than there can be blamed on coming from the atom. But the space then also developed as the Universe developed and if space developed then it cannot be total vacuum filled with nothing because "nothing" cannot develop. You the reader must judge whom is correct between my view that space developed with the Universe as part of the Universe and reject the official view about space being nothing or otherwise you the reader must then decide that I am wrong, but should you do that, then find a reason why the Big Bang started out small and filled all the available vacuum or what is contemplated as vacuum that we have with the motion of time. When Mainstream science accepted the Big Bang as the principle that will take science into the future the view about such a Big Bang concept unlocks a different door to another view on the cosmos from birth to end. It calls for revising all aspects of the entire history on cosmology and change what dead wood needs chucking out. Most of all it was my following the lead I got from Kepler that unlocked the doors I now present to you. I claim there is no graviton as there is no gravity forming weight or forming mass. I hope the sketch helps with my explaining effort:

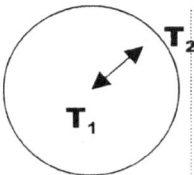

$T_2$ There is a point where the two points forming the relevancy unite in shared singularity. It comes because of shared motion

In all space-time, one finds at least two relevancies where one is at the centre.

That part Newton saw and formulised

Newton missed another part. Crossing a limit of inclusion is the limit of division and such limits are in distinction by motion producing the gravity, which is parting the two objects. Motion brings about a relevancy where two positions no longer share a common point in singularity. That is what Newton missed.

That is the gravity aspect Newton and all other Newtonians miss.

Two objects of substantial size differences are travelling at the same time but one has a space, which it has to move when it travels that is considerably different from the larger space. The larger space will produce an extending line equal to the space it moves while the smaller space will also produce a line in ratio to fit the space holds relevant and that it has to move.

Mass has precious little to do with the whole affair except to be an obstacle intended to restrain the motion of the hosting space. The difference in size between the one in circular motion and the space in contracting motion must bring about that the smaller object has to move about a circle much closer to the centre because the larger space form the centre hosts it. However there is no large or small in the cosmos but only those better developed or those poorer developed. By duplicating there is more to duplicate in the better developed than in the lesser developed. When the lesser-developed space is duplicating the less developed space would hold a lesser extending from point to point forming a shortfall by distance in comparison. The motion being extended needs less extending and should therefore be closer to the centre in relation to what the better developed space would need in extending by a duplicating effort.

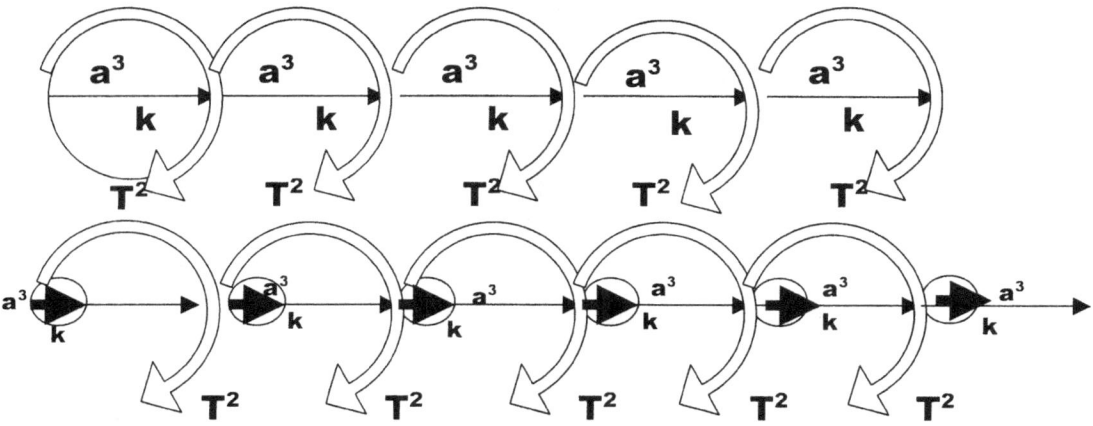

Two objects of substantial size differences are travelling at the same time but one has a space, which it has to move when it travels that is considerably different from the larger space. The larger space will produce an extending line equal to the space it moves while the smaller space will also produce a line in ratio to fit the space holds relevant and that it has to move.

> ⟶ Is the line that $a^3 k$ representing the larger space has to use to duplicate while using $T^2$

> ▶ Is the line that $a^3$ $k$ representing the larger space has to use to duplicate while using the same time constraint $T^2$

> ⟶ ▶ The difference there is in length brought about by moving the larger space in the same time $T^2$ as the smaller space is what brings about mass. There are other factors too which I shall touch on as the book develop.

The difference is between me being in mass and me being in the correct position in the space-line the Earth has will place me in the correct position but the heat that then will surround me will fry me into non-existing. Fortunately for life the soil forms a barrier through which I cannot fall any further as to correct my location. But being where my position would have no mass would allow me to float there in that location in the same manner as I would float in water. I would be buoyant. It is because I do not harmonise the displacing frequency as I should that I have mass.

This is the principle we find that is behind the sound barrier. The motion the aircraft produce forms an increase in the duplication of the aircraft which extends the duplication splitting the Earth producing a extension and the aircraft producing an extending that goes beyond the attempt of the Earth's extended by such motion. I know this may sound barely believable but please hear me out. While we use gravity the use of gravity as such makes us part of the Earth. We see gravity as some influence or force producing mass and that mass is forcing us down on the solid ness and onto the Earth. By having the mass we become a semi unit with the Earth. That is how we on Earth see gravity but when investigating gravity in outer space we must come to a basic question: Is that what we experience as gravity on Earth truly gravity? Much of the proof about gravity is part of our perception about gravity because we experience certain conditions with gravity while we find ourselves bogged down on Mother Earth. But are our perceptions about gravity truly correct? We experience mass but are the mass the result of gravity or are the mass the product of gravity. We only experience gravity, as a factor from the position we have on Earth and the conclusions we form is a product of a perception we formed while we are being forced to be part of Earth. It's as if we are upside down and has to decide on which route we should follow. I want to make a suggestion, which I aim to prove in the following pages. My personal being on the ground and having mass that is keeping me on the ground comes about because of the speed that I travel through space being the very same as that which the Earth has.

By me not applying a speed difference I then inherit the speed the Earth places on me. But my space which I use $a^3 = T^2k$ to travel and the space which I use tot ravel through is much smaller than that which the Earth burdened with to move and to move through. By me having a smaller space to move $a^3 = T^2k$ the space $a^3$ being moved $k$ in the time it would take to move $T^2$ will produce less space $a^3$ to shift $k$ and therefore a smaller distance $k$ to replace all the space $a^3$ that is moved in the time $T^2$ the space $a^3$ needs to enable it to move $k$. To duplicate by motion the smaller space requires a smaller distance to shift the space but the motion will take up, as much time to complete than would the larger space take to complete though the space the larger space has to duplicate will require a longer distance to complete the total duplication of the larger space. A large space $a^3$ will produce a large extending $k$ when using $a^3$ the same time duration$T^2$ when using the same time factor as that which the smaller space is required to use when under obligation to use same time constrain. Behind this is the most basic principle hiding which allow us the fortune to be able to fly using a flying machine. It is all about motion supplying relevance and forcing on time constraints.

Because my body that I have is travelling so much slower than the Earth is travelling due to my size in relation to the size the Earth has and although I am using the same time as the Earth does to move, such a speed difference is not in the time differences it takes to complete but in the space differences that has to be completed in the same time but is unable to fill and the space is trying to crush me into the Earth where I am forced toward the centre. If I were able to penetrate the soil solidness I would reach a point where my speed as zero would equal my space I occupy.

The space I duplicate by moving from one position and placing the space I hold in the next position while keeping my space I move as it is identical in the next spot but located in the next position. Such moving by duplicating takes a certain time to move from one spot to the following spot and it will use a certain frequency that will have the same ratio in bridging the gap from one point to the next point as that which the Earth has. My speed of duplicating by motion has to be even in frequency because I am within the duplicating space, which the Earth is duplicating and as part of the space that the Earth is duplicating but the duplicating of my space I do myself. But in size there is a massive difference between the space I hold and the space the Earth holds but to duplicate will take me as long as it takes the Earth. Notwithstanding this common factor the Earth has to use equal time in duplicating its massive space, as I have to duplicate my small space when we both have to share a frequency that will keep us duplicating evenly. Therefore the frequency of duplicating using the same time period will be a lot different to my much shorter frequency of duplicating space.

When I travel through the air my body has the same driving force propelling me as any other body falling through the air. When I am on the surface of the Earth the driving of my body takes on a factor because a larger body requires more driving than a smaller body. Not so when falling because then a centurion battle tank and me will fall at the same rate and hit the ground at the same time. The impact we make on the ground would be very different and the damage we make to the ground would be very different but that mass I only acquire once I hit the ground, but while the tank and me are falling through the air using free gravity the tank and my mass is the same.

My having weight is what Mainstream Physics use to give me my gravity. Science purposely switch my having mass and confusing my mass with my having weight to explain what is beyond explaining. It is said that while I float in outer space in state of suspending hanging above the Earth in the weightlessness I still have all the mass that I had on Earth. Butt in order to prove that those in science will give me a mass even in outer space whether I deserve it or not. By that token science first have to cheat all logic by reasoning in some bazaar way that I take my mass up there to where there is only micro gravity. They firstly claim that all of a sudden I take my mass to outer space and in their next argument they say I have micro gravity in outer space since my body is floating as if it is in the sea. But if I stop floating and start falling to the Earth my body and I did not gain any mass. My falling then comes as the result of my motion being much smaller in relation to the space I claim and my motion then is being less than what is required to keep me in the position I have which in I maintain my orbit up there. By moving to slow I fall. I do not fall because my mass grew. But science has been proven wrong by their work without any of them aver admitting to such a defeat. All the satellites fall if the satellite motions are not reset. The satellites do not gain or lose mass. They gain or lose motion

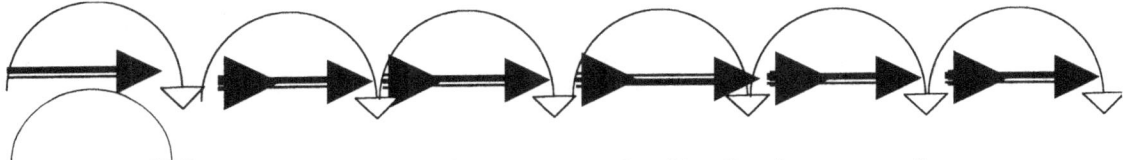

The duplicating frequency the Earth shows as $k_1$

The frequency of motion duplicating my body maintains as $k_2$.

$k_1$ minus $k_2$      The frequency of motion difference my body has minus to what the Earth has where that difference in motion becomes my mass. It is the sum total of the reducing of the motion that my body has ir comparison with the motion capability of the Earth that is the mass value In $k_1$ as well as $k_1$ the symbol represents motion, however in the case where $k_1$ minus $k_2$ that shows an incapability of motion, which is motion frustrated or a more commonly used name would be <u>mass</u> is created

By amplifying either my space (using a Hot Air balloon) or by accelerating my motion that I have in relation to that which the Earth forces me to have, I will break free from my weight or mass. I shall become airborne and float as if I am in outer space. By pretending my mass can be multiplied many times over in using a process, which then is called not gravity but momentum. But motion and gravity is all the same because motion is gravity that is redirected, which then forms another part of gravity where gravity again is also only motion applying. Science maintain the argument that when I am in outer space and am no longer part of the Earth I then will only have mass. But since there is only micro gravity I will be in a state of weightlessness. My mass is what gives me gravity and while being up there I take my mass long with me. But with my mass up there I will only have micro gravity. I am floating with my mass and it is my mass that is responsible for my gravity and I am floating above the mass of the Earth, which is rite down below me, but still I have micro gravity. That is true if I wish to incorporate the dubious use of double standards by separating mass from weight. The mass my body will have in a Black hole will be a billion times (at least) more than what it is on Earth. With that the Black hole destroys the fact propagated in science that my mass will be the same everywhere. That is more than permitting double standard. Because our motion is much slower than the Earth is spinning we place a breaking effort on the velocity the Earth has and that breaking effort we accept as the mass we have. The truth is that my mass comes about from the lack of motion I have in relation to the space I occupy and has nothing to do with any gravitons pulling me down. If I increase the motion I have there shall come a point where my motion will be sufficient to pull me into the air, as I then will have the required velocity to lift from the ground. That motion being in excess of what I have and is complimenting the motion that I receive from the Earth counteracts the motion of gravity that is containing me. The motion I adopt then release me from the motion containing me and if motion can release me by only becoming more then gravity is my motion not being enough in the first place to keep me onto the Earth. Nowhere and at no time does my mass ever gain by having more protons that will get me back to the ground as if I am bigger or carrying more material or does my mass reduce to get me into the air as if I am smaller or carrying less material. Please note that this is my way of explaining to you about the fact of bodies having weight or mass. It is not mass or the lack thereof or any means to measure occupied space within the atmosphere of a larger body that pins me onto the ground. My body is claiming space by motion in space. Gravity is the result of motion because it is in the motion that bodies have that gravity affects them. This is proved because by adding motion the mass does get more but the body never gets bigger or hold more material, and in defiance of that statement by increasing the motion my body lifts and flies. The reality is that my body in motion has more mass being momentum but still my body lifts when motion allows my body to lift. This statement confirms Kepler that $a^3$ becomes more (massive) when motion $T^2k$ becomes more (moving).

Mainstream physics admits all along that nobody, human or otherwise knows what gravity is. While investigating Kepler's work with employing much motivation and detail in order to give his work the much duly credit it deserves it will also serve a valiant purpose when by the same token we try to establish what gravity is, because I believe Kepler possibly answered that mystery. We have to start with the person that introduced gravity or so does everybody acknowledge. Newton saw an apple fall from a tree and he subsequently realised there is some force pulling the apple to the Earth.

Although he still was a student he announced his findings and became a genius on the spot. The concept he introduced as gravity gave him instant admiration from which he became the legend he is today and that reputation he gained there at that moment would last him from that day he instantaneously unveiled his mastermind, and that same genius still serves him in his honour to this day long after his death. He found that this force has to have some thing to do with the weight and the mass of that particular object and the mass of the Earth. There is some force pulling that apple as much as the force is pushing the apple and the same goes for the Earth because the mass the Earth has is doing the same to the apple. Between the two objects facing gravity there is a force that develops where such a force is pulling the apple on a constant basis towards the Earth even after the apple is already in a steady state on the Earth. That forms the mass and the mass forms gravity. He concluded that the mass is responsible for the pulling. Remember this observation came three point five centuries ago when knowledge and brilliance carried a much different defining than what such defining of brilliance is worth today. He realised the pulling on that apple brings about weight that brings about mass because the apple departs from its location and arrives at its end location when the falling is completed. Then he went out convinced all that was in line of finding the needed convincing because no body before Newton thought of what Newton thought quite in the way that Newton thought about gravity.

Newton succeeded because he found a way in presenting science with the fact that objects move closer because of some force. He went one step further and named the force he fathered as gravity. But there it stopped! Any and all other further defining the matter or going into any possible observations of whatever magnitude concerning the topic never realised any motive to go further. Inspiration to further commitment just flew out the window as the essence to do so immediately expired as far as the rest of science is concerned. What might he have missed if he missed anything? We all fall down when we are unstable and out of balance. He never realised that balance is more crucial than brutal gravity because that part is the defining part about gravity. No one ever gave a thought about the balance part even centuries later even as we grew into all the sophistication we now enjoy. What brought about the balance that secured objects in an upright stance and supplied some form of control over the managing of a position? Any other position than being flat on the floor would have a better defining than being just at the mercy of the force gravity. Standing tall is a stance that defies gravity so there is another force other than the pulling of gravity. Admit tingly the force would first and foremost have to aspire to the rules of gravity and then comply with other demands. True enough is the fact that that position would ultimately and firstly by all accounts have to satisfy gravity before any further motion could commence. Yes but then by balance motion defies gravity by changing gravity's force of pulling everything straight down towards a visionary centre between the objects. In affect this means somehow there is control over gravity and gravity does not leave objects beyond outside control. Gravity is manageable and can be controlled; we just have to find a way…

Years later some one came up with the novelty of hot air ballooning. Ballooning proved that there is antigravity but that part was missed by all even to this day. Some people speak of antigravity as if that is some mystifying mysterious concept that is so well hidden in the secret annals of the hidden Universe that only Ali Baba and his magic words can reach it. Please consider the following statement. If gravity was bringing the object down, because of the affect of gravity which is that what we experience as the gravitational sensation and that is what we interpreted as gravity by our sensation and observation, then that is only coming about by our bodies that is in a state of being dragged down. The dragging down of the body is in the direction of the Earth centre. That sensation of being firmly locked onto the ground constitutes to what we believe we experience as gravity.

When some influence bring about the very opposite affect, which then results in establishing the opposite result it deserves to be anti. In example we feel dragged down but anti will be the lifting of the body into the air. Anti will be going in an opposing direction of the motion that gravity inflicts. It will counter the influence that gravity apples. Such motion has to indicate antigravity. The counter acting of the mass dragging us down must be anti gravity pulling us up into the air above the ground. Antigravity must come from such an opposing influence that will bring about the lifting of my body. If hot air ballooning gave the object an opportunity lift, then ballooning must be antigravity. The balloonist and the entire balloon found a manner to counteract the pulling of gravity enforcing weight. The balloon can lift what gravity depress and if Newton said gravity is the falling then later Newtonians must agree that the opposite of falling is flying or lifting. A balloon is lifting-and- flying. If gravity is pulling down objects in the direction of the centre of the Earth then flying is antigravity. Moving away from the Earth by means of motion and in particular flying is using whatever means to defy gravity where the lifting can also be the hoisting of a body by a crane. Lifting by ballooning in a hot air such balloons escape from gravity where the balloon constitutes to bring about the effect of establishing antigravity. Climbing up mountains must fall into the antigravity department because parachuting down the mountain definitely falls in the gravity department. Nevertheless it still does not answer the question of what gravity is.

Let us look at antigravity because the antigravity is releasing the object from the gravity that controls the object by an Earth fed force. The balloon starts flying when the confined space of the balloon is veraciously and violently heated in access. The balloonist shows us that in order to overcome gravity we have to introduce heat. That is the only manner in which we can defeat gravity. Even by an engine driving an aeroplane such flying can only result if an engine combust solid fuel by creating motion as the fuel mixture is turned into heat. It is heat that makes the difference. That is the very thing that Kepler said. Expand the space $a^3$ and the motion $T^2$ will move further increasing $k$. Blowing hot air into the balloon is increasing space within the balloon $a^3$ which then results in providing the balloon with a larger distance $k$ from the Earth centre $k^0$ that still holds time with in the Earth atmosphere with the Earth $T^2$ within the space of the Earth $k$. Using Kepler provides us with insight and the ability to see what gravity is by showing us what antigravity is ($a^3$ gets bigger and that will bring in a larger $k$). But moreover the larger space in enough compensation to bring about extra motion that will defeat gravity by the extending of $k$. If that is not antigravity then we can forget about Ali Baba and his magic rhymes too.

The balloon assists us to escape the Earth's hold on our body, because there has to be the force producing motion countering the motion of the Earth gravity. The balloon shows that releasing enormous quantities of heat into an inclusive area excluding space such as that which the balloon canvas provides, which is establishing the release from the gravitated containing force on the body giving the body a means to escape by floating about above the ground. The motion is at that point breaking free from the containing gravity by moving in a specific direction, other than the direction the Earth gravity inclines the body to travel. By concentrating the releasing of heat into the balloon, the direction of motion starts to contradict the enlisting of the Earth gravity and the heat breaks the balloons confining properties while the balloon is released from the Earth as the balloon and us lift up into the air and away from our confining to the Earth.

At the point of explaining we arrive at the point where we can say what we think the difference is between the balloon floating in the air above the Earth and a body suspended in outer space floating above the Earth's atmosphere. The difference is the heat that is in the confined air per volumetric ratio favouring the heat being more in the space than what the heat is outside the confined space. If we had any method to put the required heat we need to escape from the limits of the Earth to outer space into the canvass of the balloon there was no canvass left to contain the heat. The heat is available to do the job but the means to do the job with the tools in hand is unavailable as far awe can use the balloon. By having more heat in the one area than there is in the other area beats of the pulling of gravity. Obviously it is antigravity that keeps the balloon in the air and what keeps the balloon in the air is having a larger volume of heat per space unit than what is in the atmosphere. The balloonist shows us that by applying more heat we can defeat gravity more. Someone took the advice, because the next minute the Germans had rockets. The launching of rockets brought about the ultimate defeat of gravity but it involves almost the ultimate releasing of heat.

In antigravity we find heat more concentrated in one definitive area than the heat concentration is elsewhere. The more the heat is that we release into space the more the antigravity is that we achieve and the more release such antigravity can produce. But what connection can gravity have with heat and if there were any connection between heat concentrated and gravity, what would such connection be? The history behind Carl Benz should bring the answer but more so would be the story behind James Watt and steam although the James Watt story may not be that thought provoking because it is much less filled with the ever popular cheap thrill only sensational gossip can provide...Still both stories cover the same principles. In the Carl Benz story a housewife leaves a pot of benzene fuel on a coal stove. The pot with benzene heats up where the pot with benzene becomes hot and under pressure. This performing of heat increase releases the heat as newly creates space, which then removes the housewife with her house from the neighbourhood she used to regularly frequent as her residential address. Afterwards almost the entire neighbourhood is not there to tell the tale or ask why...

It was a stupid tragedy that brought about the end of steam and the rise of the internal combustion engine and on Earth billions on billions of human souls are in torment not to please or suffer for the advantage of coal Barons any longer but now they are dying and suffering in agony to please the wishes and desires of oil Barons. How much did the world not change...While it is no longer the coal Barons shackling us in chains and telling us democracy broke our burden of slavery, we have now the pleasure of the oil Barons enslaving us with democracy and telling to be happy because we are the fortunate slaves, there are others circumstances in which they can enslave us that will leave us worst off. All this came just because the pot of fuel created a houseful of space that was enough to remove the house from the address the house previously enjoyed. But Mainstream science neglects to appreciate this. They see the heat, they see the antigravity but they fail to add the heat, the anti gravity and the space that no longer housed the house of the naive and rather impractical thoughtless housewife. They call the tragedy an explosion but then again everything that expands while using a noise during the expanding is an explosion. Adding of new space to the space holding the house at first altered everything that was previously proportional positioned in the space where the house was. Such exchanging of heat to accumulate and introduce more space in the process referred too as an explosion was bringing in more space that came directly as a consequence from the explosion which was producing more space where the increase in space brought disorder because the well organised material distribution and placing was before the event filling just enough of the required space arrangement that was holding every object in a prearranged order of tidiness.

Then suddenly out of the blue the space which held the house in a tidy arrangement had to accommodate more space therefore the ratio of material per space volume increased dramatically many times over in the favour of the space in the balance. That part no one ever acknowledges. However the losing of the house was not much surprising to Mainstream science back then and even today because who cares about old news. All of Mainstream science was at the time as they are today very familiar with all explosions because of wars and bombing that leads to maiming and killing and all the unspeakable monstrosities we associate with war so that the dirt poor can suffer and die to leave the disgustingly rich even richer. The poor has not the means to pay science to be clever and devise methods to save their lives so the rich does the poor the favour of paying science to find methods whereby more poor could be killed as long as the rich saw it as a good investment with great capital gain on the part of the rich. Therefore science is well established in the method of creating more elaborate and destructive explosions that the rich pay them to invent. In the explosion caused by our housewife no one put up money to investigate what happed during the explosion but money went to why the explosion happened.

That inspired an investigation in connection with the fact of the finding more about what takes place during the carnage as more money goes to finding means to create more carnage per money unit spent. At least that is why the poor were invented and that is why wars are invented. It is invented so that no money goes wasted on saving the poor people except if the poor has the money ready and available to pay the rich for medicine to enable the poor to stay alive. So science goes out and develop more fuel for carnage but fails to find out why the housewife and her house is no longer part of the neighbourhood she use to frequent. With the loss of the presence of the ignorant housewife with her house her neighbourhood and all was a normal way of leaving us with a new way of tapping and harvesting energy and untold riches which was born with the death of the absent minded housewife. But according to the mindset of science they saw not what the incident presented in space producing for to their view nothing new came about since it was just another exploding of fuel...so no body bothered as to finding out how. What they missed was the part that the coal stove played in the whole tragedy. Without the intervention of the coal stove producing the heat that turned the liquid fuel to liquid heat liquefying the space that turned the liquid space into a gaseous space where the liquid space revealed its true incentive in nature by turning out as space and the newly created space that was in fact liquid space that went onto become more space, well that space was providing the one main factor in space-time relevancy. The stoves heat producing space by transferring heat leading to the expanding of the fuel as such expanding was creating new space that is transforming all other surrounding space and is rearranging every aspect that contains space or that space contains. It will bring a much different looking end. Everything about this concept is missing from Newtonian science because Newtonian science failed to investigate Kepler. Kepler said space $a^3$ is equal to the motion $T^2k$ thereof and then that says without Kepler directly saying it, it says that if space $a^3$ goes bigger as a result of the explosion then such increasing in space will constitute to more space $a^3$ which has to produce an increase in motion $T^2k$ where more motion $T^2k$ will bring about faster displacing space. This is one small fact that Newton robbed the world of realising with his ignoring of Kepler's work.

We are now serving time in the twenty first century. One Professor once told me I must realise that Newtonian science took man to the moon and back several times and in such a view I am rather annoying presumptuous to criticize Newton. The Professor missed the point. I criticize Newton on what he did not give us, which he gave us as incorrect by his own admitting that it is mostly guesswork on his (Newton's) part and his guessing about the facts where later that guesswork became institutionalised facts believed by all concerned to be correct and to be proven to a degree of correctness that is far beyond doubt. Newton gave us gravity but Newton never gave us the explanation about gravity. At the time Newton met strict opposition from his colleges and piers because others felt his introducing of an unexplained force was taking Science back in time, which of course it did. Many scientists at the time accused Newton by name of dragging science back in the wrong direction of progress by introducing unexplained forces acting in a superstitious and mediaeval manner.

I went one step further by asking myself the question: If space becomes more when heat becomes uncontrolled why can space not become heat when space is under control? If space becomes more as we see with every explosion of every kind and such heat forming space releases energy, then why would space being managed not form heat being under control and produce energy. We only have to see what Kepler said gravity is. Motion gives us energy.

Where space is the least, which is in the centre of the circle gravity is the strongest. The gravity located in the circles space less centre holds not only the sphere together but all that is in the surrounding of the sphere outside the sphere as well. It is from there in a giro action that gravity bonds all atoms forming the structure of the sphere as one unit together in a unit as well as distributes a specific alliance in shape and form. How the atoms manage that we will get to in a while, but there is a law allowing for that to take place. Gravity is the strongest in all cosmic structures holding the form of the sphere and gravity controls all around from that very centre where space is the least therefore the more any star produces gravity. The smaller the star is as far as volumetric occupation goes, the stronger the gravity is that is coming from such a centre. The less the space there is the less the motion is and therefore the stronger and more deliberate the motion is evoking gravity. From the centre in the middle where space is absolutely at a premium the gravity grows stronger as it draws all material.

From every point there may form on the outer circle line of every part of the circle structure and all structural positions of the circle in all circles, all circles refer to the centre in perfect aligning. Every point wherever located on the sphere has a matching and equal but an opposing point on the other side of the circle but in equal position on the other side of the circle. Between the two controlling points runs a precise straight line connecting the two opposing points in counter balancing. When drawing the connecting line between the two controlling points and connecting such points on further edges of the circle by lines formed, the lines will all cross the centre. From wherever a line may cross and from every point forming a line to the other side of the circle rim holding the connecting points there has to be a counter point located on the very opposing side, that when connected by a line, such a line crosses in the centre. In the middle the centre spot bonds all sides coming from and every direction there can possibly be. The line will run to an equal point on the other side across the same distance from such a centre and that then has to be where the strongest gravity can be located.

**All connecting centre of individual connecting lines between opposing points**

The motion is one of confining the space to a centre by the moving or trying to move the flow of space and whatever is in the space into the centre where the space is least. Take the Neutron star and the Black Hole as an example and compare that with the Sun and the answer is self-proving. I claim that gravity is all about reducing space and not attracting matter but that I explain a little later on. Therefore the matrix of gravity must be permanently located in the location where space is the least. Looking at a sphere we find that what holds the sphere true to form is placed in the centre of the sphere, which then has to be the most intense point of gravity.

Gravity is confirming the round shape without favouring any specific point. Such evenness of gravity come from what is applying at such a centre and is in control of the surroundings. The centre that secures all of the space and material in the space holding the specific form has to be round if it is anything. That shows that in the sphere one can see that the sphere as a form is dominated or controlled from one specific location in the centre. The explanation about the reason there is control coming from the centre has a very childlike simple answer.

The Big Bang was where gravity held the Universe in the least space there ever was. To find the original gravity we therefore have to reduce the sphere to the circle and reduce the circle from there narrowing the circle down to as far as one can go. The Universe is a magnitude of spheres constructed by a complexity of circles. This is because everything sprouted from on matrix singularity. To narrow any circle down will be the same as narrowing down the Universe. In our reducing of the Universe we must first acknowledge that the Universe constitutes many spheres, which is giving the Universe gravity as a combining unifying part which is the part of the sphere giving the sphere form (or gravity) and that confirms that the sphere is a circle in many times over multiplying the positions from where gravity secures form. If we wish to go back in time by taking the Universe back down the same route and at the same time maintain some coherency we must concentrate on a single circle because a sphere is a circle by millions of possibilities linked together by just a name that changes the concept.

When one takes this accepted route in thinking that by reducing the connecting line to the connecting circle point in the centre of the lot, it must take us back in time at the same time as the circle reduces to the time during the Big Bang. During the Big Bang where all circles was as small as they can get we run into an unknown substance we came to know as antimatter. This theory is propagated according to Mainstream science but what is most surprisingly I do agree with this part of the statement. All material produces gravity. I go one step further and say all material apply motion where some motion may be to contain by using gravity attributing to the contracting that leads to the reducing of their space. Then as everything in Creation has an opposing the restore and maintain balance, there had to form another or other material that did not by our lamentable standards produce gravity because those material produce antigravity, a concept beyond human discernment.

Antigravity must be the expanding in counteracting contracting. A counter action to contracting is where expanding provides pappy to that which has no gravity. Forming pappy provides more space by losing density to the advancing of their space. Material either have gravity by solidifying or concentrating the space they hold in ratio to the material within the space they hold whereas others lose their solidness by entertaining more space within the ratio of material to space where such material becomes liquid and in more extreme cases they become gas. Being a gas they float which gives that material a high degree of antigravity being airborne. It is however not clear if antimatter produced gravity as it did when it went to lunch on and ate up all material in the immediate surrounding. It was cannibalistic but the unanswered question is this: was it a gravity producing predator or a non gravity-producing carnivore. Did material find a comrade in their gravity forming of form or did the gravity it produced bring on the demise that subsequently followed the event as is reported by the highly informed.

The Accepted statement on antimatter reads that matter composed of anti particles where such subatomic particles that have identical rest mass to corresponding particles of ordinary matter but opposing charge and are opposing in other fundamental properties. One example given is that an electron would have a positron, which then functions as the anti particle and has a positive charge compared to the electrons negative charge. That is put bluntly in its utmost simplistic form. Unanswered and tough questions arise from such a statement. What kept the electron bonded to the atom since the protons must by implication produce expanding or by definition be repelling the atom and surroundings instead of the normal contracting or confirming of form.

What is a positive compared to a negative charge, because it is human concepts that put the directional qualities of material into a positive or a negative contexts as we did with hot and cold. It is human standards that humans brought about to make all human inadequacy by lamented human understanding better but it is not applied cosmos principle. If there is extracting electrons performing in the capacity as antimatter, then there better be protons by other name in service to the anti electrons, which then of course serves the anti electron in the capacity of an anti proton with an equal but negative charge to that of the proton. When matter and anti matter meet the two opposing particles annihilate each other until one vanishes from the Universe. I have to add that at the time this theory was devised the first computer games became a crazy fashion played by young and old, those wise and those foolish all alike. This game was called the packman and the packman ate up all the skulls and after eating left nothing as evidence.

The theory about antimatter has some very striking similarities to that packman game. It still does not answer the most ardent questions: What makes a positive electron different from a electron in the working place each has and can any person show such an object found in nature. Can people take a positive electron to an investigative bureau and are awarded for such evidence? It is unwise to substitute nature with human concepts just to further mathematical equations. This was apparently presented as normal as nature was when nature developed with the Big Bang and nature then did behave this oddly just after the Big Bang came about. But one huge misgiving in this argument is declaring that everything the antimatter had as a meal vanished and even moreover then antimatter went and vanished too. Where could the combination that was produced when the matter and antimatter collided go after it disappeared and did it form the by-product of antimatter science is talking about, which since then apparently vanished too. What a bloody none-intellectual fairytale that is on the bargain too one of those made–up-as-they-go-along stories, which is told by persons that supposedly should know of better. Since there is no place other to find a location to be within than being in a place inside the Universe it is hardly possible to vanish from the Universe except in fairy tales because for one simple fact: there is no other home to have but the home we have which we call by the name the Universe and we have no where to escape too but within the walls that the Universe provide for such a purpose.

There is one Universe containing all and preserving the lot. Mainstream physics is accepting this fact. But then by the same margin they accept a principle that allows property that once was part of the Universe to leave the Universe and go somewhere outside the only Universe. They create a loophole whenever it suits them to misplace what they cannot explain readily and logically. In Creation to their and my thinking there can be no hiding of anything but in the Created Universe. This they admit and confirm although with the same breath those very same intellectuals also admit that there is another place outside of what we are able to find in the Universe. When someone comes up with the marvel where such a person can declare in all honesty that the product of antimatter or singularity escaped from the Universe to God knows where that person should leave the field of science and go for fantasy writing such as fairy tales or reporting about politicians inner deepest chastity and integrity. That is what we can find outside the spectrum of what the Universe can deliver.

With such a statement of any Universal product disappearing from the Universe alarm bells should go off in the mind of the trained and professional Scientist working with such matters. Yet those in charge do not once belabour a question of the validity of a statement that involves a stating of factors declaring the possibility that there was are now an outside of what once was part of the only place there ever can be. They can read mathematical calculations and agree on an outside the Universe without stating it in an explanation what happened to the lost and found or they're ability of introducing the concept as a reality, which they claim it is. That such factor as anti-matter can go outside the Universe and leave the Universe by causing a Houdini vanishing act of never–to-be-repeated-again status. Science would have us to believe this antimatter went into hiding in a manner that where it is now, is out of the Universe we have. They applaud this thoughtless presumption while fully knowing that at the time they do this acknowledging that there is no other place for anything wishing for a place to be within then having to be in another place other than inside the Universe. If it was ever anywhere it still is within the Universe merely because there is no other place to go than to be inside and part of the known Universe! There cannot be some factor and then misplace it as if a valid factor calculate the value can prove the disappearance and by disappearing it no longer is. If it was in the Universe it must still be in the Universe somewhere. Then we better start looking for it.

Another big issue is that what ever the Big Bang produced must be in equal terms everywhere. The Big Bang was a process that had the Universe act as a high-speed cocktail mixer of no repeating ever again. Whatever the Big Bang was of all that it was, the most it was in the beginning was that it was one massive mixer mixing everything in it at the speed of light. With all the mixing time and time to mix that there was going on with nothing better to do than mix and match the mixing was done thoroughly. That we can count on. The relevancies might change slightly and balances may change favouring opposing ends...yes and known appearances did change...yes. But in the end all the factors must always be present everywhere through out the Universe. By this lacking of a fundamental explanation about what antimatter will look like when found Mainstream is incredibly poorly judged by scientific standards. Those mathematicians calculating physics suggest that science should take antimatter as a cosmic fact and then in disregard of other realities they dispose the truth by discarding its properties onto the unknown. That hardly suggests plausible science by any one's admitting.

By that Educated Scientists of High Standings are discarding even more of the old fashioned basic elementary science taught as science principles to children in schools in science classes through out the world. One thing surer than any other fact is that matter in whatever form consists of the purist energy there ever can be. In the cosmos is, was and will be all the material there can ever possibly be. Our concepts we put forwards can be faulty but nature cannot ever be at fault. Our arrangement of our ideas can be at fault, but we cannot pull a vanishing act on certain cosmic products and in doing that then dismiss the existing of such a factor or factors, which we then claim, have vanished in the further developed Universe. Our concepts of what they became may be at fault and by changing some basic principals such changing may produce a better understanding about what we think we read into mathematics. Mathematics is purely a language and mathematicians are purely translators.

Mathematicians translate from the language they read to the verbal equivalent they speak and as in all translations made certain concept may become misinterpreted. The terminology used to explain this is "lost in translation" Mathematicians must see what there is in the translation and try to incorporate what there is available in the cosmos to what the Mathematician sees in his mathematical calculations. The Universe was full of heat and it was full of material but it was not full of free space. If that is the case then where did the heat come from and where did the heat go? Hiroshima and Nagasaki taught us many things about the horror of human nature but most of all it taught us that material is heat secured in atoms and atoms are heat tightly wrapped in a cocoon, which we named the atom. Heat in any form cannot have anti in another form. The package holding heat wrapped can unwrap as it does with nuclear atomic demise. But the anti to heat is cold and cold is space.

The undeniable fact about the Big Bang theory is the accepting of a growing state in which the entire cosmos seems to be in. With all the expansion that went on we came to the point where we now are at and in such growth all aspects in the Universe must grow in relation to quantifiable progress in all different aspects, which takes us to that which is seen and that is unseen and which came along as products in the Universe where everything took everything on a growing spruce by unveiling space. That is where we now are. Such expansion include all there is including everything and not just with outer space growing.

The dynamics of outer space alone cannot grow by leaving the growth of material behind. Should we wish to see where we came from we have to reduce that which we now see in our surroundings to apply to the measures that once applied in all aspects of the cosmos.

Mainstream physics is over pronouncing the growth of space and with that suppresses the part matter must play in such growth by simply ignoring the issue. That is the reason why they prefer to ignore the evidence that material is growing notwithstanding that material is growing or that their disbelief about the matter of material growing do not change that material is growing in any case. Because they cannot find any reason why material should grow they refuse to admit that material does grow. This is hiding from the truth by hiding the truth. If space grows and the Universe is getting bigger then all space grows to allow the Universe to get bigger. That includes matter and space not holding or filled with matter.

Space can only grow if materials that also hold space also grow within the space that is growing with the growing space. It means that stars get bigger by the cosmos growing from the Big Bang onwards and outwards to the moment in which we are at the present. But if stars grow then the atoms forming the stars are doing all the growing as they secure more space within the space they claim. If Hubble saw space grow, the growth of space must include the growth of space holding material as well. In studying the Hubble's expanding theory we come across evidence that makes it clear that all material expand in a manner as if the expanding comes from the centre of each and all particles within the expanding space and the expanding grows outwards from every particle centre. It is using every star centre to grow from in all directions proportionally in all directions evenly. This leads one to believe that gravity is this securing of space in the material just as Kepler showed it to the world. It proves a connection with deliberate implications coming from every as well as in every specific centre. It proves that the centre $k^0 = a^3 / T^2 k$. It becomes apparent if and when separating Kepler from what Newton thought about the work of Kepler which Newton accepted as being inferior and all incorrect.

To find our birth we have to take back all growth that brought development in the mean while but the only way that that can be done is by man drawing the cosmos down to what man may perceive which forms mans ability in understanding. That is making the Universe small and as man grows man allow the Universe also to grow in relation and corresponding the man's ability to comprehend. We see the cosmos as a circle and we accept the circle because the circle is what gravity implement when the choice of form is coming from material that has all options to freely choose from. By taking the circle back one will follow or said even better we will trace the rout of the cosmos to where it then started.

All stars are many circles in many dimensions, which form when all circles join into what we call a sphere, but that leaves us only with the circles in the plural. Taking the cosmos back can only lead to one point and that Kepler told us we will find singularity $a^3 = T^2 k$ which is $k^0 = a^3 / T^2 k$. We can only reach $k^0 = a^3 / T^2 k$ if we repeat $1/k = T^2 / a^3$ in a continuing manner indefinitely. When one does the effort of reading this correctly, it says that when distance $k$ brakes from singularity $1 = k^0$ that is then $(k^0 = 1) / k = T^2 / a^3$ where the space $a^3$ produced a time $T^2$ equal to singularity $k^0$ and singularity $k^0$ is equal to eternity which was where all was equal to a never changing cosmos that was holding the single form into one dimensional space that included all the filled and vacant material filling in from all sides.

This is one way of looking at the issue and by doing that I am about to prove that singularity is $\Pi$. I am about to prove that not only is the planets adhering to the Titius Bode rule of seven over ten and ten over seven in relation to the Roche limit but that the Roche limit explains the very, very first instant the Universe experienced outside eternity. The atoms relates to space in the very same manner of seven singularity positions to ten points and from this motion of material interacting with space is securing material on the inside as well as on the outside.

By that motion gravity comes about finding the value of $\Pi^2$. Gravity uses the relation of the Titius Bode seven on ten and ten on seven as well as the Roche factor to form gravity and gravity is always $\Pi^2$. This I see by reading Kepler's work as Kepler produced the work and introduced the work as $a^3 = T^2 k$. With this formula $k^0 = a^3/ (T^2 k)$ must also be true because $a^3 = T^2 k$ is a relevancy that has to be in relation to singularity and therefore singularity must be $k^0 = 1$. Where will we find $k^0 = 1$?

All motion brings about results as the motion eventually ends in spin. Even our linear motion travelling along the surface of the Earth by sea or land seems to us as going straight but it is eventually following a circle around an axis. There are as many axes as there is always an axis. The axis provides a partition between the rotating directions that the spin of the material is securing at the location of where the axis will follow. The spin will have motion and the spin will have direction although the axis will forever instantly change the direction of the spin continuously to fit the linear part of the spin. By going straight the directional change singularity used because singularity is what it is, is continuing to eventually become the circle motion. As the direction will forever change the linear then will forever remain steady due to the eternal changing of the direction.

Our gravitational falling to the Earth is a result of a circle going straight and forcing us straight down to an everlasting directional alternating circle we have as we spin with the Earth as we spin around the Sun. As we fall straight down we change direction while we are falling straight down because that point we are heading t what we are falling to is changing too. But from at the centre of the axis everything seems neutral. The axis does not spin at all because the axis brings about spinning motion changing eternally. That is in nature and not with man made motion.

**The linear remains linear because the linear redirects its intentional direction because of the rotational change that the linear motion always ends up doing. The line forms an eventual circle because the linear line must constantly entertain the centre.**

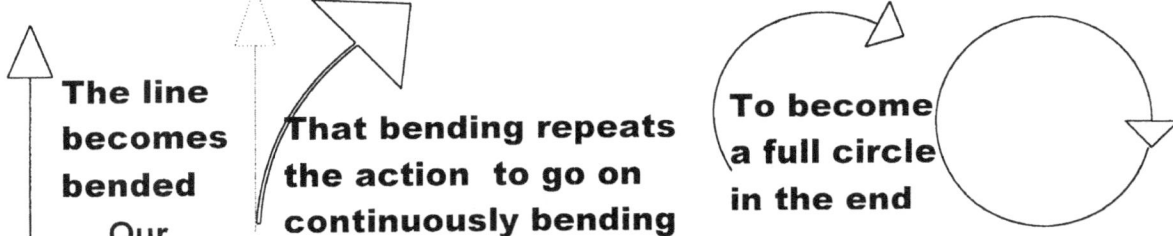

**The line becomes bended**

**That bending repeats the action to go on continuously bending**

**To become a full circle in the end**

**Our** gravitational **falling to the Earth is a result of a circle going straight and forcing us straight down to an everlasting directional alternating circle, we have as we spin with the Earth as we spin around the sun. As we fall straight down we, change direction while we are falling straight down because that point we are**

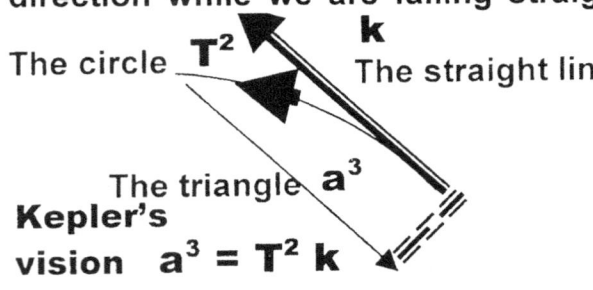

**The circle** $T^2$    **k**

**The straight line**

**The triangle** $a^3$

**Kepler's vision** $a^3 = T^2 k$

**heading to what we are falling to is changing too. From the centre of the axis, everything seems neutral. The axis does not spin at all, because the axis brings about spinning motion changing eternally. That is in nature and not man-made motion.**

Because the axis provides and demand direction changes to secure everything in motion around such a centre to such a centre, such a point forming the axis is beyond dimension. It has no side as it has no space and it has no motion. It cannot be detected because it does not contribute to any space the Universe has but it can be located because it does contribute to all the forming of space the Universe has. Without ever moving and because it never moves, the centre forces rotation by being in the centre as that centre is also commanding from the centre. That point allows motion to apply where such motion act as the partitioning between objects. When the spin comes about say in a child toy such as a top, the top gains independence producing (as long as the energy will last) an independent motion in spin but when the motion dies the independence is lost. Previously I mentioned that all circles in the plural forms a sphere by duplication but never repeating opposing controlling points connecting to a joint circle that confirms all possibilities and re-ensures all possibilities. In the final analyses there is one centre on will reach in reducing every radii.

We accept that the time it takes a planet to move between two points is time $T^2$. Having that space $a^3$ in relation to the time $T^2$ is space-time $a^3 / T^2$ and that is precisely how Kepler expressed his findings $k = a^3 / T^2$. This indicates space-time that is growing through the extending of $k$. While it does prove the Hubble shift it underlines that that is not what the gravity which we experience, because $k^{-1} = T^2 / a^3$ (Newtonian gravity) dominates by contraction where the gravity permitting expanding $k = a^3 / T^2$ is not inclined to absolutely favour contraction. Newton's gravity is totally about a decline in $k$. But what Kepler shows that in outer space through motion of space performing space-time But our gravity does not exclude the implications of growth because $k^{-1} = T^2 / a^3$ (Newtonian gravity) allows material growth by extending.

The gravity we feel that is dominating us which is also that which Newton saw $k^{-1} = T^2 / a^3$ (Newtonian gravity) cannot realistically accommodate growth in the Universe. This should therefore remind us living a life of splendour on Earth that we must remember that we are part of the Earth and not part of the cosmos. We may find some ability to reach outer space and remain there for a very short period but then we have to return to the Earth. The returning part is compulsory and that we must accept as we accept breathing. There are many suggestions of how we can achieve the ability to distinguish mans superiority by extensive and elaborate travelling through out the entire Universe vindicating our millions of years of being confined by the dooming gravity that the Earth grips us with and committing us to our revenge by knocking off our shackles as we cross once more yet another barrier similar to that of Columbus but infinitely more, wider and holding unlimited vastness just to secure our seemingly unstoppable ability to travel through outer space at the speed of light. Here comes the shocking part: Those that cherish the hope such inspirational thoughts may bring, those thoughts as inspiring as they may sound are no more than blatant useless daydreaming that is at best and at the worst only promoting wishful thinking.

We have as much a chance of achieving that dream of visiting the next star let alone the next Galactica as we have of never aging, never going sick or never dying. Those thoughts belong to the mindless thinking pattern one will find in the muttering atheist, which is bent on proving the improvable by reasoning idiotically. The atheist practises a religion tempting them to think that if there is no God, they can take the role of God and be God. To do that they have to remove all barriers that divide the sane from the mentally incapacitated. Travelling through outer space on the breeze of a light bulb is just is not possible to do. We are born on Earth and we are part of the Earth. Only through our attachment with Earth do we become part of as well as involved in the cosmos but that is strictly because of our surety we find with the Earth and we are secured because that is the Earth which is comforting our needs like a good caring mother should. We are not naturally part of any location other than the Earth and any visiting of other cosmic locations is artificial

There is no doubt that such visits will be very short lived and even such a possibility is yet to be proven because from what information I can gather there will be dire consequences to follow which are to be avoided if man adheres to sanity instead of manic madness by promoting such attempts of visiting other locations. The Earth represents us in the cosmos and represents us in the cosmos on our behalf. The cosmos does not know life and the relation the cosmos has is not the relation we have with either the cosmos or the Earth. That relation the Earth represents us in the cosmos is that gravity, which Kepler introduced while Newton saw only how the Earth jealously hold us captured by applying the gravity that Newton saw. This far man could afford speculating with his dreams because the part of science that Kepler covered was up to now just a blind spot to science. I uncovered that blind spot with the aid of Kepler.
 Let's now proceed by using this information as we chase down gravity and find what more there possibly can be which it could hide. Gravity is space moving in a circle holding that space that is moving towards a centre in relation to motion. The space is identified by another space moving in the opposite direction. Between the two there is distinguishing differences and what is in space at a distance, which cannot sustain the required motion needed to maintain the gravity that is the separating second space factor that is giving independence through motion. The motion is completely different and totally harmonised holding equality by differentiating motion. The differentiation provides the equal sustained ratio in motion. If such a ratio in velocity comparison cannot be sustained the space removes as it shortens **k.**

Maintaining the distance of **k** from moment to moment is that requirement needed to keep velocity equilibrium sustained and velocity in ratio becomes the product and the result of gravity where that is prescribing the applying conditions forming equilibrium. Only when such conditions are broken by their inability to sustain the harmonised velocity ratio does space fall away and particles come crashing down to the Earth. Otherwise such conditions are maintained and an orbit comes about. But this falling comes from a lack of motion and not tucking each other's sleeves or pulling each other around. Performing a little science experiment such as the Coanda effect disproves the grabbing on theory. Gravity is about matter concentrating the heat in space through the spin of the proton spinning and reducing space. Such motion establishes an elected centre that houses gravity. The space holding the protons secure forms a demand on space flowing to replace space by filling from "outside sources" in order to replenish the point of space reducing.

The flow or motion comes about as a result of a need to supply space with more space as the proton diminishes space at the centre by killing of space as it nullifies the motion in the centre. There then is a vacancy forming as there in the centre is no space because there can be no motion to the space in the very centre. Because there is no space with motion that specific single dimensional spot has no part in the cosmos we know. Moving towards the centre there is a re-supplying of space equal to the number of protons, which brings on the reducing of space and accelerating movement of the space between the point of demise and the point of replenishing. This is one part of a group effort where all factors forming the group work together to provide the required gravity. This part Newton saw not. Being the master that formulised the existing laws on motion he had to detect the consequences of motion if he carefully studied Kepler's formula. He would not have brought in the idea of a force but would rather have recognised it as a natural flow of space bringing about the duration of time. Newton did admit he had no idea what gravity is and declared that gravity is a force. On that point I disagree and my disagreeing is not on the subject of gravity being or not being. I am emphasizing my disagreeing on the force aspect because every person on Earth associates the word gravity with the word force and confuse the two in concept. Life is a force but gravity is a natural and normal flow from the start of the Universe to finally bring conclusion to the Universe. Gravity is a natural motion of space $a^3$ in space $a^3$ and in that there is no force to be found.

One does require a force to resist the flow ...yes but that resisting of the flow then becomes the force, which counteracts the flow. His view about declaring the presence of an unexplainable force brought about much rejection from his fellow academics at the time because it enlisted a vision that Physics were moving back to the dark ages at the time. No one can blame the others about the direction they saw science move because admitting to a force without any ability to explain what brings such a force about constitutes much to the powers pagan gods, witches and other undesirable powers had at the time. However Newton did conclude that there is strength in the centre of a sphere that produces the strength within the sphere. I then decided to investigate his remark.

**What will be the reason why the original form that we devote to the Universe will take on a sphere as natural form? Yes... I have heard in Engendering that the sphere holds any point and every possible point pushed from the outside of the sphere secured by every other point the sphere has from the inside of the sphere which then is forming the sphere but that statement is as precise as it says a woman gave birth to me and not that my Mother specifically gave birth to me. It says I can be anyone's child instead of that I am specifically that persons child. It still does not reach the answer that will stop all other questions about the question. Apparently our imagination grabs the sphere as the only form of choice and that is as correct as it is true...but why...this is apparent coming from nature as natures choice to form when material is not pre-cast to have any specific form. In such an event the gravity in that space take on by cosmic pre-cast shape the sphere as form...it is because gravity chooses the smallest space to hold the strongest force. Such a point will also establish a line we call an axis. By reducing the radius there must come a point where the ring that is in decline from such reducing is infinitely small, where it can reduce no more, where it reached its ultra limit, but at that point cannot be zero, because the point is there for all to realise but nobody to see. From the point in the centre that is no point in he actual Universe there are in one space forming a unit two points separating the unit by holding relevance and without two**

**points there are no point.**

**When looking at what Kepler brought into science we find $a^3$ being equal to $T^2$ by the allocating of k. The mention of $a^3$ is referring to the space filling the space that is the space in at the very end of the point rotating where that point is indicating to the forming of a circle $T^2$. But $a^3$ also indicates a separate $a^3$ that pinpoints the allocated position of the space designated to have the smaller $a^3$ point out the precise $a^3$ that the smaller $a^3$ is claiming as a unit and that became the product of the motion identifying $a^3$ as a separate unit sharing one larger $a^3$ and one smaller $a^3$ of what all is brought about by a field invested to form the gravity. There is forever a larger space $a^3$ that holds a smaller space $a^3$ in relevance to the motion coming about in the form of $T^2$ and k. Then the relation between $a^3$ and the centre part of the larger $a^3$ there is a most relevant point being $k^0$.**

We all accept that the Universe uses the only true cosmic form there is, as an overall all-containing form we call a sphere. The sphere is that form, which the Universe has to be in to form the Universe and naturally the concept that immediately enters everyone's mind is thinking about it, would be and most probably is a sphere. Everyone accept the Universe as a whole will be the sphere...but why would the sphere form. If there was any one in the past that stopped for a minute to think about this question that philosopher then never stopped for that minute to write down as to convey his conclusion to the following generations. I have heard intellectuals explain it by telling students the form is used by the Universe because it is the strongest form there is, but that carries the same value in definition than to say the Universe uses the sphere because the sphere is round. The original question then still remains unanswered as being totally gone unanswered because the question still stands. Why is the sphere round and why is the sphere the strongest form one may find? So declaring the sphere as the strongest form leaves the question just as unanswered as before.

Considering the manner in which the expression of Kepler's formula read one may correctly be of the opinion that $a^3$ is in context with the broad space that covers all of the space indicated by the length of the radius which is symbolised by **k** from the centre $k^0$ to the point indicating the immediate border of the space **k**. Yes that presumption is very true but also true is the fact that if there was one point reserving the position for the smaller point $a^3$ that held a separate and independent space $a^3$ within the larger space $a^3$ which would without the smaller space $a^3$ not be identifiable as forming the unit $a^3$. If there was no such a smaller space $a^3$ within the larger space $a^3$ producing the outer limit to the larger space $a^3$ the larger space $a^3$ would have no independent relevancy in the overall totality that will distinguish such a space $a^3$ and to establish the containing as well as reserving position it holds. The larger space $a^3$ is there because of the motion of the smaller space $a^3$, which validates the larger space $a^3$ to be a factor worth of being calculated. Only by the motion of the motion of the smaller space $a^3$ can the larger space $a^3$ claim validation and on the other end also apply independence because as I shall show later on that the motion of the larger space $a^3$ validate the counter motion of the smaller space $a^3$. The smaller space $a^3$ cannot be in motion if the larger space $a^3$ do not contribute to a larger motion of space $a^3$ contradicting the smaller motion by direction where both accommodate each other by motion relevancies bringing individuality without bringing independence about. Kepler said $a^3 = T^2 k$ therefore if there is space $a^3$ such space $a^3$ has to be in motion $T^2k$ to allow space $a^3$ to be and have the other space within. Therefore by referring to $a^3$ one establish a relation of both in the context because not one of the two would be if not for the presence of the other $a^3$. When referring to $a^3$ one refer to the larger $a^3$ which is containing the smaller $a^3$ as much as one distinguishes the position of the smaller $a^3$ proclaiming the area of dominance of the larger $a^3$ in which the smaller $a^3$ takes up residence in space $a^3$.

Kepler's formula first drew my attention to singularity in the way he formulated his formula. The most important part of his formula is not visible from the outside or from the onset of investigating and one must look for that most dynamic part covered by the mysterious coming from way within. Kepler shows us that the truth is found in the darkness and not in the light. At a point where Einstein said gravity begins we will locate Kepler's gravity beginning because space (or as Einstein referred to it) the Universe goes flat. If $a^3 = T^2 k$ is a fact then there must be a starting point where **k** starts because there is a point where **k** ends. This then will change relevancies and will mathematically equate from $a^3 = T^2 k$ to $a^3 / T^2 k = 1$ and one can be any number or symbol to the power of zero. Mind you not to the value of zero but to the power of zero $a^3 / T^2 k = 1 = k^0$. That means one has to reduce **k** to a point where **k** becomes $k^0$, then in accordance with Kepler's advice I proceeded...Kepler said that from the smallest space within space $a^3$ there is the line **k,** which is connecting in a motion covering the spaces $k\,T^2$.

The space indicated by and that is a part of space $a^3$ in question wills run as the space-time unit $a^3 / k\,T^2$. That is where gravity will form being identifiable as a unit at a specific centre from $k^0$. Gravity lurking in the centre at the point **k** starts the line **k** where the line **k** holds space-time $a^3/T^2$ secure and in form. That has to form singularity and singularity can be whatever there is a wish for as long as the wish is to the power of zero. (Singularity) $k^0 = a^3 / k\,T^2$ which reads that in space-time has three sides on the one side and are opposing the first side by three other sides. If Kepler said mathematically the smallest distance between structures could at the least be $k^0 = a^3 / k\,T^2$ and we all know that $k^0 = 1$ then it should be some one's duty to find that point. One must then start by accepting that Kepler also stated there cannot be nothing or zero in the cosmos since the smallest distance between two structures is $k^0$ which is one and not zero. I wish to introduce an argument by disposing another Academics method in his disposing of my work. Some academic found a way through which that particular Academic was able to dismiss my arguments on the grounds that the solar system was not formed at the Big Bang period or that is the information that Mainstream science is promoting.

That was a loophole he suggested because he was unable or inferior or plainly just to lazy to interpret my work that I laid before him. I am not for one minute fooled by his passiveness because that is very typical of the New South Africa everyone outside South Africa helped to create. And in addition I really can't think that he thought me to be stupid enough to be discouraged by taking his arguments seriously. In order to circumvent such a loophole I shall begin my following argument by stating that the solar system, which I am referring to, is a hypothetic one. Notwithstanding that I know my argument is solid and serious as such about all aspects in the rest of the entire argument which include all other possible aspects, the following is the one part I use that as a part of my argument, this part I now identify as a part remains the only hypothetical possible fact in the entire argument that has a possible hypothetical truth as it stands.

There are those who avoid admitting to inconsistencies by arguing that my argument about growth of material in the cosmos throughout its entirety is invalid because the solar system was not in place at the time the Big Bang was in place. Please then keep in mind while reading the argument that I would like to point to the fact that my following referring to the solar system is actually referring to a similar solar system that is somewhere else and is now a part of a galactica we do not know about. That is where the dissimilarity ends. In all other aspects our solar system and the one I suggest is identical in every possible aspect I bring this in to disqualify any academic loophole that may come about from an argument about the solar system coming into place at a later stage of the cosmic developing and therefore I exclude any chance of using the counter argument that the Solar system became an eventuality long after the Big Bang was about to happen. The novelty in the forming of the solar system in the argument I am about to present is no longer an issue because I am referring to a solar system in another Galactica being precisely the same as the one we use with the exception that that solar system was around at the time of the Big Bang. I therefore hope that there is no more ambiguous loopholes that any one may use to unfairly dismiss my point of view when the validity of dismissing my point is as simple as raising such an counter argument as the one mentioned. That counter argument of the solar system not forming a part of the Big Bang does not apply.

To avoid such a loophole again we now use a hypothetical but real solar system in space, which formed when the Big Bang took place. To them we now present a solar system that is identical to the one we know and is a precise duplication thereof. Again I say to those the argument now represents a precise duplication of our solar system and was in place ever since the Big Bang. That means with the solar systems being apart in millions of kilometres at the present time there was a time when the planets and the Sun were apart by the same measure but only using kilometres instead of kilometres by the millions. The Big Bang shows a growth in space. If that is the case then there was a time when the Earth was 149 kilometres away from the Sun instead of the current 149 million kilometres, which it is at present. We can reduce the distance further to fit into a billionth of a meter but I hope my statement drives the point home. But that space we reduce also has to include the expansion in size of cosmic objects because reversing the expansion shows that any argument not expanding the orbiting structures is most silly. Then there must have been a time when all the planets were between fifty-nine and five hundred and ninety metres away from the Sun in comparison to the millions of kilometres we have today.

If there was no expansion of the orbiting structures in the diameter size they have then how did the planets being the size they are at present, which then was also the size they must have been back then, fit into such a space that small where the where that small space was keeping such giants apart. That is notwithstanding the fact that they still are with all the gravity they presently have but at the time was being apart only by the measure of 149 kilometres as is the case of the Earth? Material therefore too must be part of the growing in space. This line of arguing I suppose is much below the Academics pursuit of matters but since I am much lesser in mental standards of developing than they are, such reasoning prompted me to go on some investigating journey. Light journeys through out the cosmos and it will be sensible to follow lights travel in reverse to see where that takes us.

With objects being apart at some distances and light flowing in straight lines between them it must take light the size of a straight line to travel between cosmic objects. While I disprove this statement in future arguments I wish to stick with the officially accepted but as such a very simplistic concept for the moment. The distance the light has to cover depends on the radius there is between the objects and that forms the total distance forming the Universe. That put the Universe then in relation to lines forming differences about structures that is claiming independent space and space setting objects apart as such distances will be the radius standing between those objects. In reality that is what the Universe comes down too.

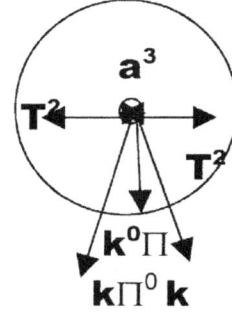

In dimensional terms, which I explain later on the value of **2k** relates to $T^2$. That relation extends to the next value where $T^2$ relates to **k**, which relates to $T^2$. The first space in the circle will then be $T^2$ **k**. From the centre being in infinity, one can realise by applying mental power the single dimension factor not seen but present all the same. Extending that into the 3D comes six **k** and any one of the six will further extend to form a seventh point as $T^2$ All this is a multiplying of $k^0 = a^3 / (T^2 k) = 7$

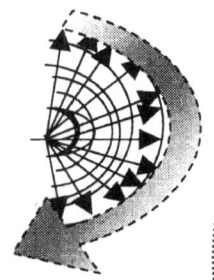

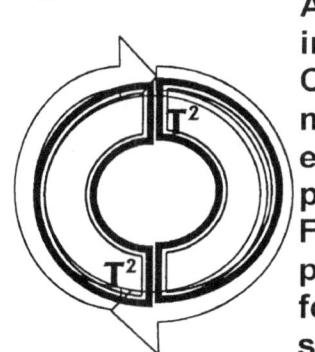

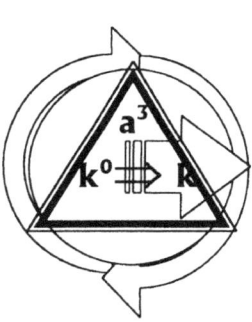

Water flowing will release from the normal line that gravity enforces and establish a link with the centre. The Coanda affect is about space in motion acknowledging a centre formed by the motion of the space in motion.

This should lead any person to investigate a centre that forms because evidently, there is a centre but that centre comes about by motion of rotating around a fixed point serving all points in motion. That leads us to the centre of everything in rotation because everything in the Universe is in rotation. As Kepler said $k^0 = a^3 / k\, T^2$ and we have to find $k^0$

At this point Newton's second law come into affect. Motion by means of the Coanda effect introduced space as motion introduced time. For the first time ever time was interrupted when motion provided time the space to interrupt. From motion by the way of the Coanda principle gravity came about as a centre formed a point where motion surrounded space, By motion space-time was established in relation to singularity

The objects are circles by dimensions and the space is also dimensions that are crossed by lines travelling through the dimension. With light being a line and the Big Bang coming from a situation that was a lot more cramped for space than at present, the correct path to follow if I wish to trace the steps of the cosmos back to the Big Bang is to reduce the straight line between the structures and find where such a line will no longer be a line. I realised I had to begin at a point where I had to find the point where any and all lines end when I reduce any and all lines which will then be the same measure where such a point will show me where the line forming the cosmos started. The same procedure will apply to the material structures all being in a sphere form in our Universe. A sphere is a lot of circles forming a unit but not repeating the space claimed by the other candidates. Such a circle also applies a straight line only known by another name but still serves the same purpose. Reducing the line will lead us to the beginning of time. The reducing of the line will once more represent the point where the sphere in its role as a multi circle will begin.

Singularity is not something hiding in Black Holes like criminal bandits. The value of singularity must be $1^0$ because that is as small as any value can get. To find singularity one therefore must trace the location of $1^0$ and there one will find singularity. What other mathematical value can singularity have that $1^0 = 1$?

Also it is true that the entire form that is the sphere is controlled from a centre within the sphere. That centre holds the sphere in form and shape. Therefore the strong form is dictated from that space fewer centres where there is no space and no form left. The natural inclining is in the form of the sphere. It is part of the roundness that the overall shape of the sphere represents and this structural strength is carrying down to the very centre. Because the circle is forever reducing that reducing which is inherently part of the form of the sphere becomes a tool in distorting of space in the sphere and is eventually removing all forms of space from within the centre of the sphere. The very centre ends up as having no space because of the reducing that continuous down to become the space less inner centre. The all roundness is the ingredient that forms the backbone of the absolute strength that the sphere has and that is the component that the sphere is so famous for. The form the sphere has allows the sphere to have a control that is coming from the centre deep inside the sphere where the space vanishes and being without space seems to keep the entire structure rigged. From the centre the sphere shape shows strength that the shape as tough as it is. How does it work in its most basic analyses?

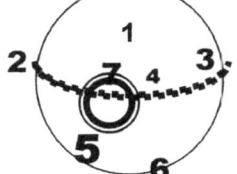

There is one more point in the sphere in the centre forming an addition in the sphere. That point holds gravity secure.

The cube has sides and the sides form a rather weak and flat surface that connects four corners. The flat surface produces a rather indifferent contact point with no special features on the surface. The corners connect to other sets of corners and those corners form a weak structure without any direct support coming from the other five sides. Without material to fill the body of the cube the cube has no direct connecting between any of the sides other than corners connecting at the edges of the sides.

Taking the vantage from the point the sphere is holding from the centre out into space there are ten points connecting to the centre. In that are the dimensions of singularity connecting to space where five connects to space in the second dimension of singularity, and five connects in the third dimension of singularity. On the other hand, the cube does show a very different characteristic, which involves only six sides (at least) connected.

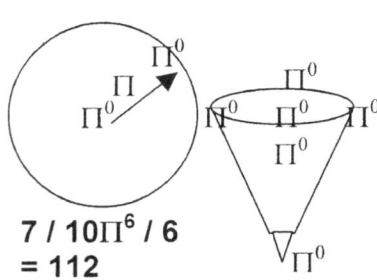

$7 / 10\Pi^6 / 6$
$= 112$

The spinning of $\Pi^0$ around the centre $\Pi^0$ establishes $\Pi$ and $\Pi$ is what produces the form gravity has. Still it is the relation or relevancy there is between the centre $\Pi^0$ and the spinning $\Pi^0$ that gives status to the form that $\Pi$ represents. In out Universe we are accustomed to and are familiar to the rules we want to place seven points holding singularity to the centre holding singularity in a relation of $7/10 \ \Pi^6 / 6 = 112$. In that Universe everything less that a duplication ability to the value of 112 protons fit but only atoms to a maximum of 112 protons fit.

At one point the line that will form the radius will go single dimensional as $r^0$ and that is equal to $1^0 = 1$. Singularity must have one value because singularity has only one dimension. Everything to the power of zero and not zero is valued at a single dimension of 1. Put what you like to the power of zero and you will get one. That then must be the mathematical value of singularity.

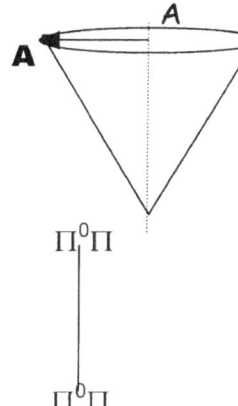

The moving of $\Pi^0$ to $\Pi$ involved relegation and not motion as we consider motion. It was $\Pi^0$ getting a side and that is all. There was no true side but only a form that came into place. Singularity (A) received singularity (**A**) and no more of anything but the shift to comply with having a relevancy forming in relation to singularity. The dots had no sides, had no length or diameter. There was not measurable space or measurable time involved. The time could have been a micro, micro second as much a trillion millennium because time had no relevance. It was eternity interrupted by infinity, as it still is the case, however the line that eternity followed was no line because there was no space to hold the line. The line was momentarily interrupted by infinity, however with no one there, there was no one to notice. The lines were not lines but relations to sides being formed.

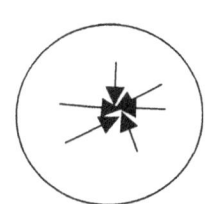

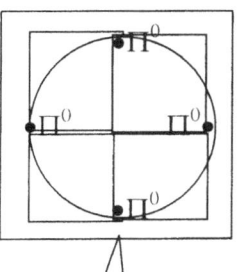

Inherent to the form the sphere offers, there is a specific location of singularity where the radius first goes single $r^0 = 1$ and then form goes into the realms of singularity $\Pi^0 r^0$. The cube also may have such a pint bur having such a point does not connect directly to six points located on the edges of the cube or any other form the is.

In relation to such a centre where $\Pi^0 r^0$ forms singularity there are always four cubes related to such a centre where the centre is part of seven points in total representing the sphere. Every cube gas lost one side to a point of the sphere where the sphere takes control of form and removes one side of the cube. In relation to the time factor that is inherently part of singularity by the extending of singularity there are five sides connecting to four points standing related to singularity by the $\Pi^0$ factor and that gives 5 X 4 = 20. That is always directly in relation to seven points singularity offers.

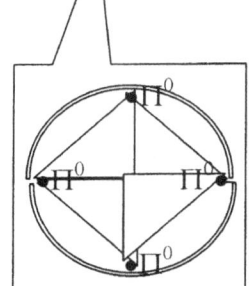

This only applies in relation to time because time is the square or then if you wish time is the flat to space being the cube. Time in the square draws space in the cube flat and that is the principle behind the effort gravity can apply to destroy space. Once gravity destroys space and time goes square then all factors preventing time to remain has gone single leaving time falling into singularity as well. This does not yet explain the

three dimensions forming the six -sided Universe we find we have.

In everything else that is not round there is no such a precise center located than there is in the sphere.
Other forms connect by eight corners attaching three lines but the lines form no other direct connection except by the corner where the three lines meet each corner.

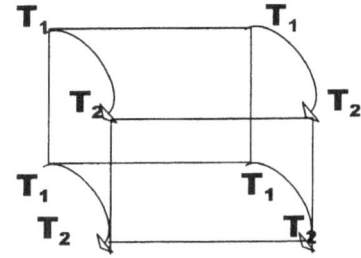

Now that we know it must be all over and we know the value we can be able to find the location with a little thought about it). In one form within the sphere we might locate singularity and we know the most common form used in the cosmos is the sphere. In truth it is hardly coincidental that science got the curvature of space time just as round as a sphere.

## The definition of singularity is as follows:

**Singularity:** a mathematical point at which certain physical quantities reach infinite values for example, according to the general relativity the curvature of space-time becomes infinite in a black hole. In the Big Bang theory the Universe was born from singularity in which the density and temperature of matter were infinite. While it probably is the greatest mind to walk the Earth that produced the spectacular in the above, a much more simple mind as the one I have noticed much more simple aspects of nature that only one with a simple mind as I have could recognise because my mind does not have the capacity for the greatness of the great minds and wonders about matters as great as the density of the Universe and how much matter goes into a flat Universe. If the Universe did start from one single point and time, matter and space flowed from that point, then that point must have a relative connecting base because such a point holding singularity must be eternal as space, matter and time link eternal. There therefore must be one point linking the entire Universe when regarding the fact of singularity. Then according to the theory off relativity there has to be one exact point holding time in a relevance notwithstanding the fact that time depart from that position and relate differently to all space-time away from such a point. Every person I have discussed facts about creation recollects images in the trend depicted in a presentation as one may find to the above. That would be the most unlikely way Creation came in place.

The recalling of pictures representing images about creation must have form, but to mathematics it had no form. From this thought the very opposite arise where Creation came from nothing but such an idea is mathematically simply not possible. The thought of nothing is just what it is, a thought of nothing and although it is in the human mind common nature to present nothing as a value in the recalling of something, nothing is a presentation of the figment in the human mind. There can be no number such as nothing and that was (possibly) Newton's biggest error. Nothing represent non-existing and that is just what nothing is, it is non-existing. In order to prove my point I wish to ask the reader to define the shortest line there can theoretically be. If such a person should answer anything but that the shortest line will be at a point where the beginning and is the very same spot he will be wrong. The shortest line that can ever be anywhere must have a start and finish holding the exact same spot. The line will be humanly impossible to create but we humans are capable of very little. When we look towards the centre of the sphere we find a point where space meets its ending point. There in the centre is a point that is there and has a mathematical value but is not there with no mathematical influence in any value. By taking the line in any direction towards the centre would leave the line eventually with one value and that must be $\Pi$. In the manner we calculate a circle we take the radius by the square and multiply that with the form, which is $\Pi$. In the calculating of a sphere we take the radius to go cube while the form in value of $\Pi$ stays connected to the radius. That puts $\Pi$ as a factor depending on the line to give the form factor meaning. This line of reasoning is actually very wrong because it is the form that stays at value even when the radius goes into single dimension $r^0 = 1$. By this reasoning I formed the idea, no at the time it was much more a suspicion that if I wanted to get to the point where the Universe started then I better detect $\Pi$ to find that location. Talking is cheap, but money buys the whiskey so where do I go from there. A line does not start at zero, but when I introduced that idea to Universities in South Africa my reasoning was called incoherent, and I still have the documentation of proof. But more about that later...

One possibility that the shortest spot can never have is having a starting point on the zero mark. If the mark of zero holds the start it must also hold the end because the end and the beginning has the same position. If the position of zero then is the beginning, the end will also be zero leaving the line without an end as well as without a beginning. The conclusion from this is that no line can start at zero because that will be a mathematical impossibility. If that line that started from zero did start from zero such a line technically would form line or spot starting at a point shorter than any possible line could and would therefore be shorter than the shortest line possible. This we see in evidence looking at a sphere. The radius of the circle forming the sphere has to start where the shortest possible line can start, but it cannot be at zero because zero would remove such a point leaving no line to grow. A line growing or extending from zero can never leave zero because of the influence of being zero disqualifies any possibility of growth. If the line then had to grow in all directions at the same pace the line must therefore be a circle. The value of the circle is Π, and that is where creation started. That gave me the clue where to start looking for singularity. One would find singularity in the value Π and the value Π will be in all things rotating in a circle. To start my explanation about my cosmic theory I wish to firstly bring some nostalgic and the relevancy will become apparent later on. Such is the importance however that I wish to place this at the very start of the prologue.

When we were boys we played with a top we called the spinning top. I cannot imagine that there is one boy in the western world that did not hold such a devise in his hand. Tying a string securely around the tapered cone started the operation and then with a jerking or pulling throw the devise is launched in a projectile manner and the big knack to success was getting the nail end firmly on the ground and by the realizing jerk the top was rotating. The champion was always the one boy that could throw his top to spin the fastest and that would create a humming sound. The louder the sound produced the bigger champion

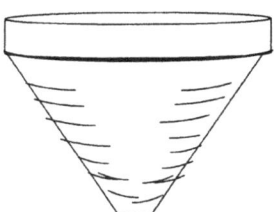

When a back braking effort produced a throw of enormity the spinning top would not only produce sound varying in pitch but also create a spin that would seem to have some instability. There are very many limitations about the spin, parameters that determine the slowest and the highest sin rate and spinning is within the parameters of such settings. The question arising is why such parameters are there in the first place?

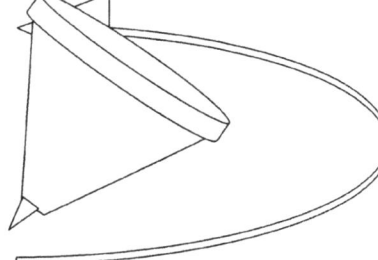

This concept explains the effect called the sound barrier in precise detail. However the condition is if only the Newtonians would look past their intellectual barriers and discard Newton in exchange for what is evident in nature. The spinning and movement of the top never stops to explain the four pillars that forms the Universe namely
1 The Coanda Effect
2 The Roche limit
3 The Titius Bode law
4 The Lagrangian Points

An enormous effort will have the top going oblong while spinning violently and as the pace reduced the top will stabilize by coming to an upright position. In the upright position it wall then spin for the remainder of the period where it will in the end start tilting to the side and in a last effort throw a few wild oblong turns and fall over.

Boys playing games will never realize scientific breakthrough explaining and grown ups do not play with toys. In this little toy played everywhere everyday by almost every one is the answer most brilliant of human Brainpower seek answers about all the cosmic riddles no one seem to understand. In the spin as such one may find two vital boundaries in the motion and the boundaries are marked by a wobble coming about as if the top is fighting some other influence. Spinning too fast pulls the centre off centre and so does spinning too slow. It is the same influence coming about at both ends of the limitation in the spin. There are influences at work, but force...no; it cannot be forces setting such boundaries. From that I started per cuing what sets such limitations because that limitation must be universal as all matter is spinning in one way or the other.

In the past these remarks made me the clown in the courtyard and no friends came to my aid because no friends were in support of my statements. A description that would be closer to is that no friend wanted to admit any friendship because such admitting may also reflect on his or her sanity.

When looking at the cosmos from whichever angle indicates the fact that the cosmos is moving. It is forever spinning and it is going to as much as it is coming from. Everything is on the move and always encircling something of greater importance. A top can spin but the parameters of its spin are limiting the motion it can apply. By not spinning the top is still spinning as the earth are doing the spinning on its behalf.

When spinning too fast the top fights something because the alignment keeping it upright starts to tarnish. The same apply when spinning too slowly but that makes sense. It is the fact that the same affect comes about when spinning too slow that triggers the questions.

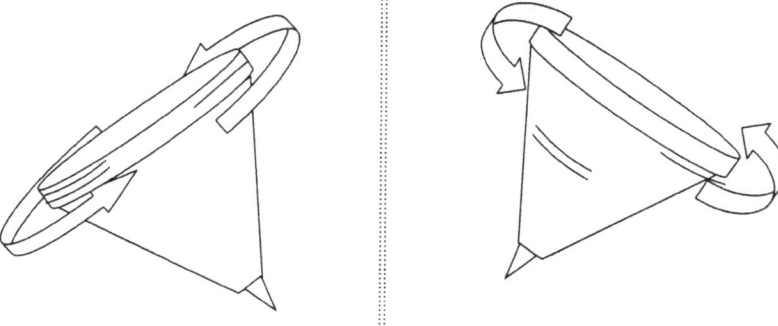

The spinning top is all the evidence any one needs to come to such a conclusion.

This thought gave me the idea that singularity might be a mathematical fact but singularity seems much more likely to be generated than being in existence as a right to birth. The thought brought me back to the top. While it is true that I can put singularity somewhere in the centre of a sphere it is more correct that where science believe singularity should be I see a lot of motion in space that should not be there if science is to be believed. I see movement of space depressing as the space speed towards the point that is not moving. I see a point holding singularity but round about the point everything is in motion. Singularity is about motion.

**If the line started at zero there was no line to start because zero multiplied by whatever results in zero as the answer. That must also be the cosmic starting point. Einstein introduced such a point and named that point singularity.**

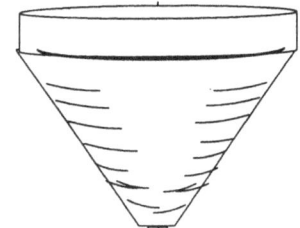

This brings us back to the 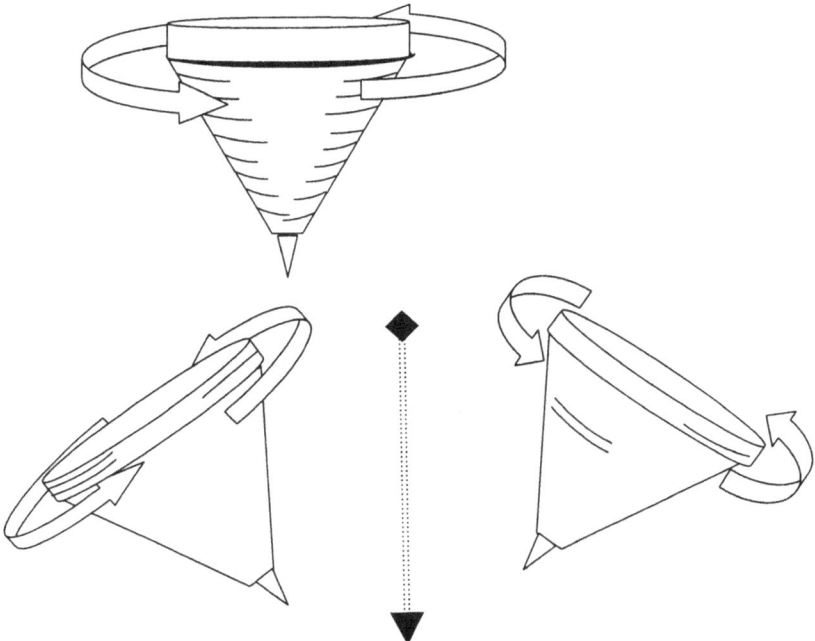 spinning top I presented at the beginning.

**I have asked as many persons as I do not care to remember why the top sinning will remain spinning around one point while turning. The answer I receive from the most educated to the schoolboy is always about momentum. That is a very simple answer and to say the least a little too simplistic by further analysis. Why would the spinning top go of centre when spinning higher than a specific velocity and lowering the velocity it would stabilize and run square to the Earth. The only slowing down significantly would it after that go oblong and then fall limping from side to side and finally after a big fight and great protesting would it land on its side and role along. I could go on about different positions bringing across different momentum of thrust but I do not wish to insult your intelligence because I am aware that you are familiar with all the law. When the top is spinning it is spinning about its own axis and when it is not spinning it still remains spinning about the earth's axis therefore when it is spinning it is also spinning about the earth's axis. Therefore the limitations applying can only result as an influence coming from the earth's axis. The second question now comes screaming across and that is in what manner could the earths axis ever affect a spinning top since the spin and the spinning top is a gross mismatch to what ever standard the Earth may introduce. It is clear that spinning objects do influence each other in contrast to Newtonian opinion.**

Every round object has a point establishing a very centre, a middle dividing one side from the other. That division determines the space from one side away from the other side. At one point there must be a point that does not fall on either side of the divide. Such a point will still be a circle, because from that side the circle divides into two sectors.

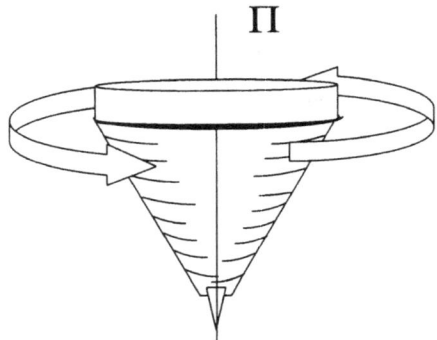

In every spinning object there is a point of infinity, a point that does not turn because it holds the dividing spin. From that point running in all directions the spin is opposing the other side. All spinning activity starts at that point diverting outwards and from that point the spin is either clockwise or anti clockwise in all directions. As I pointed out no line can start at zero because then there is no line and no rotating point can start at zero because then there is no rotation.

Calculating a square involves two aspects that we think of as sides.

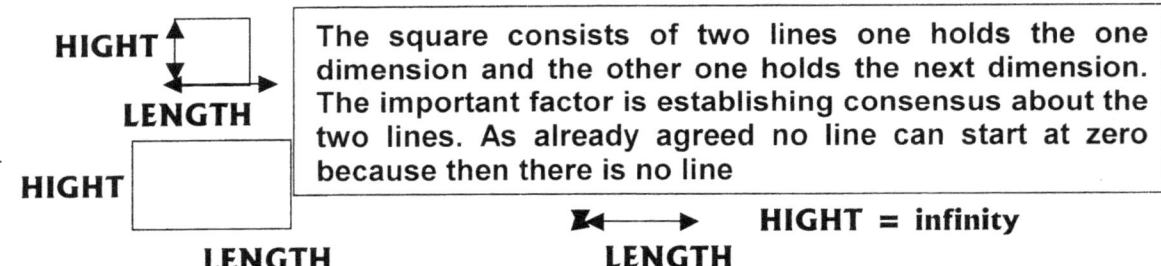

HIGHT LENGTH

HIGHT LENGTH

The square consists of two lines one holds the one dimension and the other one holds the next dimension. The important factor is establishing consensus about the two lines. As already agreed no line can start at zero because then there is no line

HIGHT = infinity
LENGTH

By reducing the one line the other line can never reach zero because then there was no such a line to begin with. That makes a straight line also inevitably always a potential square and that makes the straight line half the value of the square being 180°. At a later point I shall continue with this argument, but for the mean while I wish to come back to the circle. This same principal apply to the cube and that means everything there is and ever will be is either a square being part of a cube or a circle. With the straight line forming half the value of a square $360^0 / 2 = 180^0$ in as much as being one line and reserving one line in infinity to eternity. The straight line is just half the value of a square. In that manner the triangle is also half a square and therefore holds the same dimensional value as the straight line being also 180°

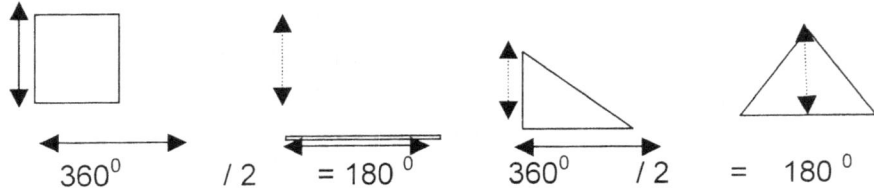

$360^0$   $/ 2$   $= 180^0$   $360^0$   $/ 2$   $=$   $180^0$

This line having the same value as the half circle as well as the triangle serves as the foundation mathematics was built on. Mathematics was not from the start but mathematics as all ells came into place at a point and before the point the basis of mathematics was in place. That is why there is something as building blocks

The circle is a square holding a round shape, as the straight line is a square holding one side to infinity. Calculating a circle involves two aspects where the one is either the radius or the diameter that is double the radius. The other is the factor $\Pi$

$\Pi \times D^2 / 4 =$ circle and $\Pi \times r^2 =$ circle

The point of singularity cannot be in space at large because space is not there and secondly what ever is there spin to slowly to have a connection with singularity directly.

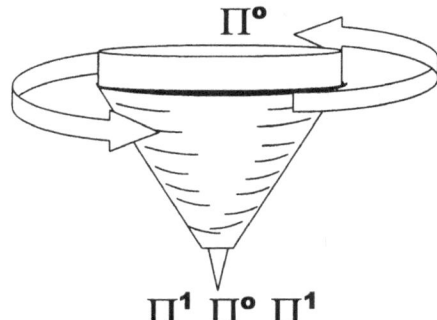

$$\Pi^1 \ \Pi^o \ \Pi^1$$

With everything in a cube or a circle or a potential of the two, brings about the implication of eternity in a form of singularity or the point of creation. Removing the radius of a circle does not remove the circle, because the circle is there, securing the ring. If the line (or imaginary line if you wish) holding the value of $\Pi^0 = 1$there has to be a point where the circle is no longer in infinity but claims existing outside the imaginary. At that point the radius may be lightly more than infinity, but to all calculating purposes it still remain as infinity.

The spin was going on for eternity because the spin does not apply, it has a value of zero and zero is another expression for eternity.

Having edges where $\Pi^0$ duplicate to present the edges singularity lost the value of $\Pi^0$ to the value of $\Pi^1$ with the same value singularity had being $\Pi^1$ to the one side and $\Pi^1$ to the other side, $\Pi^0$ must be the point splitting singularity into two parts of eternity, the eternal value of the first dimension outside eternity. It was the square of $\Pi^1$ being $\Pi^{1+1}$. That was the first dimension outside singularity $\Pi^0$ where singularity has a value of $\Pi^1$ in the form of $\Pi^{1+1=2}$. The first claim to space had a value of $\Pi^2$. This applied to both sides of the claim to space outside singularity, and the double proton became the dominant factor on matter.

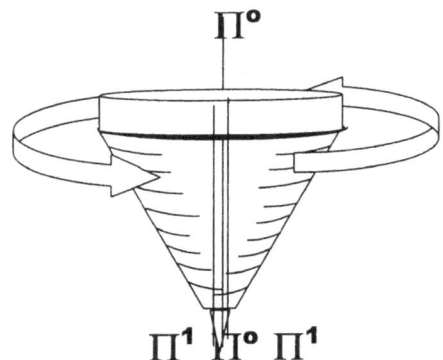

$$\Pi^1 \ \Pi^o \ \Pi^1$$

If the ultimate there is in gravity are moving that which we think of in terms of as space, then everything that gravity move must be by the moving of the space only in a lesser degree. If gravity is about moving space then gravity is about moving everything that is carried by that space that is moved by gravity. That is gravity.

**By receiving space, singularity received a value outside eternity as $\Pi^0$ received edges. Granted the fact that the edges were so small there still was no r to present a circle.**

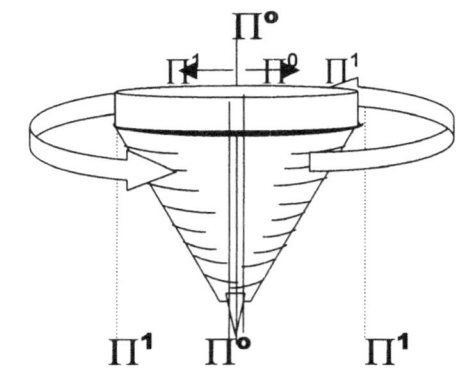

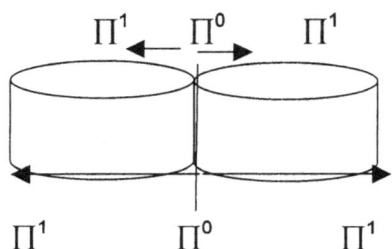

**Taken from the point of rotation the two sides are in opposition to each other in every aspect that they may contain and with all that they hold.**

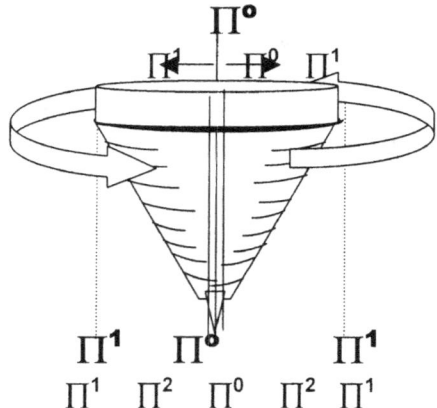

**With $\Pi^0$ little more than a figment of the imagination there is actually to values of $\Pi^1$ facing each other in a relation combining $\Pi^1$ to hold the value of $\Pi^{1+1=2}$ $=\Pi^2$ and with two sides being the very same but opposing each other there will therefore also be $\Pi^2$ to every side that holds $\Pi^1$.**

The spin charges the top in a manner that gives the top an independence, which the top very reluctantly relinquishes. The top seems to find an ability to put up a fight to the bitter end not to lose the sudden individuality above and beyond the normal structural independence that all matter has. What the top receives goes far beyond that. It has an energy making the top unique and independent of the motion of the Earth. It gives the top a seemingly good reason to fight for something as well as a will to be able to fight the fight for its life the fight of its life.

## Locating and finding the presence of singularity

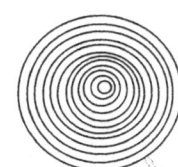

$k^0 = a^3 / T^2 k$ states that whatever is, is also spinning in order to be present.

> What is in the Universe is spinning. In the **precise middle** of all **objects in rotation** is a precise centre dividing the object in sectors that will **start the spinning initiation** from that centre point.

As I am introducing a very new idea, I wish what I try to convey.

While the toy top is, spinning one will find singularity by moving the rotating line or radius progressively to the middle by reducing the length the line has from the edge to the middle. At one point all further reducing must end but the ending cannot include zero or nothing because the rest of the line still attach the rest of the top.

**That point** albeit hypothetical, is also as much a reality none the less and is placed where that point **must be standing still** because every line **running from that point in opposing directions** is also **in opposing directional spin the other or opposing side.**

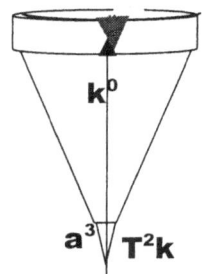

Thus, the spinning object **will have a middle point,** a very specific **centre point that does not spin** and only holds Π as a specific value because no radius can apply. But also the one value such a line **cannot have is zero** because the line **is there and holds contact** to the rest of the material bringing about that **zero does not start any** line and therefore the **value of the line must be infinite**, just as described in accordance and by **the definition of singularity**

to explain in better detail

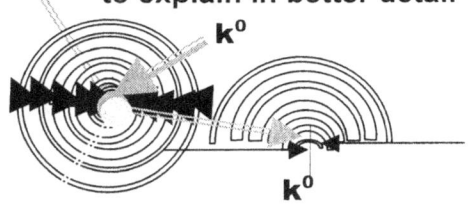

As the rotating direction moves inwards, the rings holding Π will become smaller and smaller. The reducing of the radius r will eventually end where the spin direction ends at $Π^0$. However that point where the directional spin ends is the point where the actual spin takes place. The spinning is on the precise location the point is not spinning.

However the line in singularity is not there but is only generated into existence by rotating motion. Once the motion stops the line in singularity starves and once the motion starts the line in singularity generates to life. This action we find in the top is equal to the charging of electricity, which is the equal to gravity. It is done by motion, which activates the line in singularity controlling space-time.

This matches Kepler's findings exactly. The space $a^3$ becomes equal to the motion of the space $T^2k$ and without the space $a^3$ being in motion $T^2k$ there can be no gravity. Do not think of me in too harsh terms when I say this but I have just introduced you to the centre of the Universe. Now too you know where the centre of the Universe is but I shall come around explaining this seemingly madness in a short while. Kepler said the centre can be generated by $k^0 = a^3 / T^2 k$

Singularity is a mathematical reality. Einstein may be the first to name it and Galileo (unwittingly) may have been the first to define it as Kepler was the first to formulate singularity, but in mathematical terms singularity is the most basic principle. At this point I wish to establish a fact that seems lost in all other grandeurs of cosmology. A straight line cannot begin at zero or nil it can only start at infinity. Such a statement will hardly seem appropriate but the relevancy of this fact has no limits.

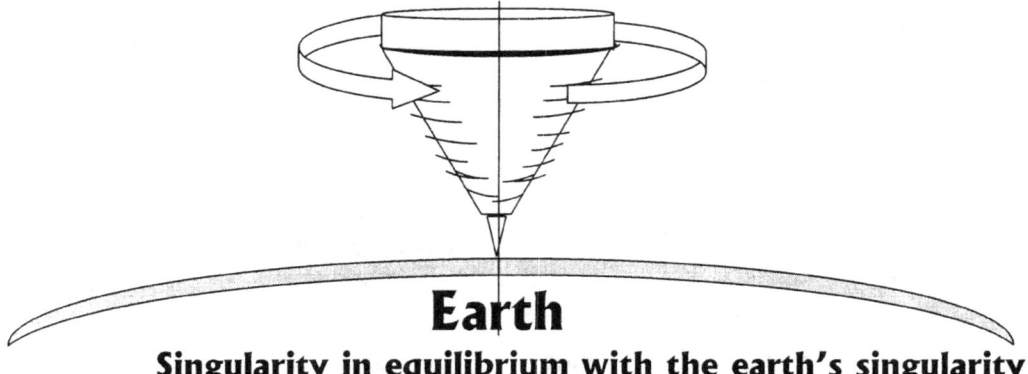

# Earth

**Singularity in equilibrium with the earth's singularity**

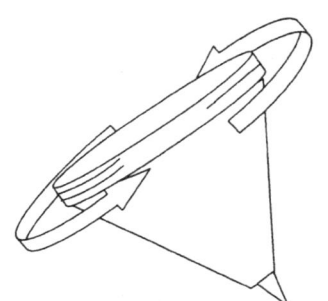
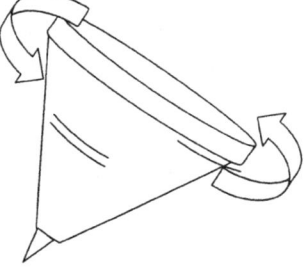

| | |
|---|---|
| **Singularity of the top singularity** | **exceeding the earth's** |

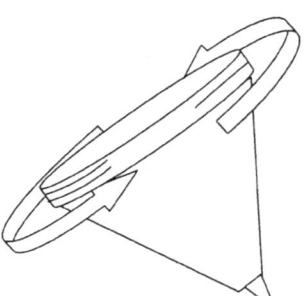
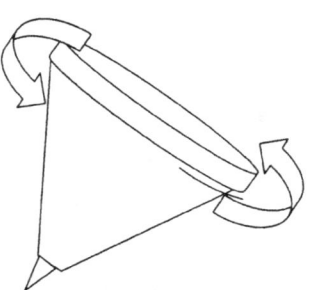

| | |
|---|---|
| **The earth's singularity the singularity top produce collapses and fall.** | **dominating and exceeding through spin as the top** |

The centre may or may not spin and the fact that it does or does not spin is all the same because that centre part never spins in any case. Therefore the boundaries set by the spinning motion does not depend on the spinning motion of the object but has to stand related to another bogy bringing about a larger spin influence.

All this was missed by science ever since Newton changed Kepler's work because Newton failed to read into Kepler's work correctly from the information he (Newton) basically discarded due to the arrogance on the part of Newton. He (Newton) concluded every factor in the whole formula of Kepler represent motion instead of the true representation it carries being space-time.

**Granted the fact that the influence the earth has on the top may be that of gravity but if that is the case then surely the sun has also influence on the earth and other rotating objects through gravity. It needs more investigation because it may bring about evidence we are not aware of.**

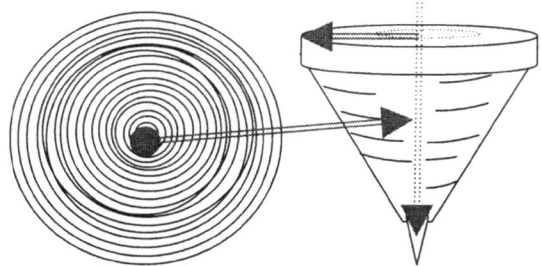

**This observation places a much bigger question mark on the statement of Newton where he proclaims no influence on two rotating cosmic structures.**

П

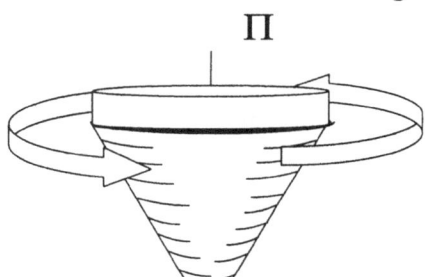

We may proceed to the wider picture that the cosmos hold. What is it the Newtonians fail to see? If an electron is orbiting around an atom, the inside of the atom must be a circle. If the atom was not a circle, it then had to be a cube. The electron cannot rotate around a cube; therefore, the inside of the atom is a circle. In a circle, there is a radius that initiates the circle. The calculation of such a circle is $\Pi \times r^2$.

The radius r runs from a circle centre of the circle. In the centre of the circle, there is a point where the radius starts. It runs outwards from that point in all directions towards the circle П. Technically, there then has to be a point where r is zero, an absolute zero. However, the circle therefore remains П. The circle does not disappear; it remains there for all to see. It is only the radius that removes.

$$\frac{\Pi r^2}{r^2} = \Pi$$

If one removes the radius from the circle, the circle remains, only holding the value of П. By removing the value of r, П becomes singularity with no place to be. Singularity is the place where there is no space to be in place. However, П remains because once r receives the slightest of space П will find space. Then the circle will grow to $\Pi r^2$ and r would determine the space. Without space, there is no r but there is a circle with the value of П. Singularity is in every single rotating object, be it the proton or the universe.

In Kepler's formula this explaining is very evident. I used Kepler's $k^0 = a^3 / T^2k$ to come to this conclusion! If singularity had a value it can only be $1^0$ and $k^0 = 1^0$ because all singularity is $1^0$.

Newton failed to study Kepler's work even though Kepler arrived at this work only when he (Kepler) received the work directly from the cosmos as Kepler was studying the cosmos. In ignorance on the part of Newton driven by arrogance he failed to notice that Kepler's work introduce for the very first time space-time as a formulised conclusion and being space-time Kepler's formula has no need of adding $\Pi$ and having the formula changed from Kepler's vision of $a^3 = T^2 k$ to $4 \pi^2 a^3 / T^2 = G (m + m_p)$.

Newton tried and failed to marry Kepler's formula with an earlier concept of him (Newton) which he (Newton) received by vision when he introduced the formula $r^2 / (M \times m)$, which was his first vision of what gravity was. With that he formalised gravity but then he overstated the grandeur of his first vision by matching his concept to the Creation at large which was introduced by the work of Kepler. In that attempt to link all creation by what he found to be condensed in the name of gravity he changed his first formula from $r^2 / (M \times m)$ to $F = G (M_s \times m_p) / r^2$ which Newton then afterwards introduced accompanying his term as gravity. This was one of the most unfortunate and corrupt misleading in science that ever took place. Consider the seriousness of misguiding that came from the case of the Piltdown "Ape man" Hoax and by comparing the influence such diversion inspired the Piltdown "Ape man" incident becomes an innocent little party prank compared to the implications that sprouted from the Newton diverting of science. But it is not only Newton that is charged with the deliberate misleading in science because Newtonians are aware of the so many missing answers to questions Newtonians never dared to ask in their attempt to hide all the shared blame.

By merely diagnosing gravity being naturally pulling from the centre is rather avoiding the question with simplicity because the question arising from this answer is where then is the centre of the Universe to where the final pulling is heading? By using the gravity formula $F = G (M_s \times m_p) r^2$ the centre is the vital issue directing as well as controlling every aspect involved in the formula. When this formula is mathematically translated to English the formula says there is a pulling of structures $M_s \times m_p$ towards a centre r from both sides $r^2$. There is the centre $r^2$ that is the Sun. There is the centre $r^2$ that is the Milky Way. But that is not where the Universe stops. The Milky Way centre is not an individual end-of-the-finality-of-the-Universe in terms of coming from as we are going too. Mainstream Science declared that all the pulling is toward a centre $r^2$. What placed r where and what was r be to achieve that much pulling power? We see that the Universe is somehow always coherent and disciplined because the rotating of structures indicate the presence of gravity.

The Newtonian formula insists on a centre pulling particles. In the formula $F = G (M_s \times m_p) r^2$ such a centre is most important. Such a centre is as demanding as it is commanding and yet, Mainstream Science never came to pinpoint the centre of the Universe. Kepler on the other hand showed the precise location of the centre of the Universe. The pulling of all mass in the Universe must be towards a centre and because of Newton's introducing of a Gravitational Constant (G) in the Universe; this then demands a centre to form G. Such a factor with gravitational powers must point all gravity forces towards one centre since gravity is undeniably located in a centre of all objects. When the realising of this all out important centre arrived and Newtonians became able in recognising this factor Mainstream Science should have then at the time have been working towards identifying some centre before proclaiming the serious implications arriving from the presence of such a centre. Never once was there too this tome some launching of an all out search to locate this centre.

There was a silly attempt designated to Einstein because of his superior mathematical abilities to calculate the entire mass of the entire Universe but what prominence would the mass have in pulling without a precise indication of a specific location where the pulling of the entire mass is heading. Without the centre all other factors loose their validity. Where then will such a centre of the Universe be? This is what makes cosmology the shambles it is. Kepler gave us the answer centuries ago but no one ever tried to take notice of what Kepler said without Newton's interfering. Kepler gave us the ability to see what lies beyond the limitations of the visual as well as the very obvious.

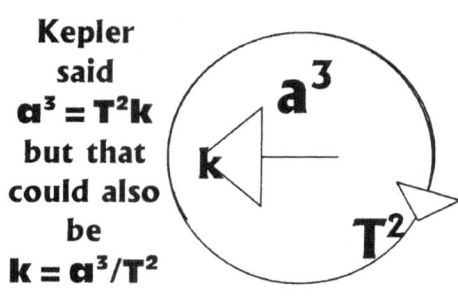

**Kepler said $a^3 = T^2k$ but that could also be $k = a^3/T^2$**

**When translating Kepler's mathematical expression into English we can see what Kepler said also read as $k = a^3 / T^2$ where $k$ is one point from a centre point that is space $a^3$ relating to time $T^2$. From a centre comes space-time. The centre $k$ brings space $a^3$ in ratio to time $T^2$, which is space / time $a^3 / T^2$. Reading this correctly cannot bring any dispute…yet it does…and it has been doing it for centuries on end!**

Others like Newton and Einstein came much after Kepler and coined the phrases but Kepler formulated the concepts of gravity as well as space-time. They (Newton, Einstein and many others) named Kepler's innovations. That is very clear but only on the condition that Kepler is read correctly and Newton gossip about what Kepler is saying is ignored. What Kepler said in mathematics all the brilliant Mathematicians through so many centuries were unable to read although the coded language was written in mathematics and as it is their field of supreme speciality they have captured as theirs, therefore they should have the ability that would enable them to decode what they are the masters of which they claim to be able to decipher. It is the mathematics, which the cosmos use to allow the cosmos such communicating with humans! The cosmos said that all space stands excluded form all other space which then forms an all including unit.

That is why the Sun is so cold and outer space is so hot (and no… calling the Sun cold and outer space hot is not a mistake on my part and neither is it due to a printing error but the Sun is as cold as it gets and the outer space is as hot as it get. We'll get to that one later on…). The two share a cosmos in which they are both apart while forming one unit. The separating part in the unit is motion driven by heat and the motion as well as the heat carrying the motion that sets the two apart while both are in the unit. I mentioned previously that there is an undeniable connection between heat and gravity. Let us do some investigating and try to establish answers are read correctly and Newton gossip about what Kepler is saying is ignored. What Kepler said in mathematics all the brilliant Mathematicians through so many centuries were unable to read although the coded language was written in mathematics and as it is their field of supreme speciality they have captured as theirs, therefore they should have the ability that would enable them to decode what they are the masters of which they claim to be able to decipher. It is the mathematics, which the cosmos use to allow the cosmos such communicating with humans!

The cosmos said that all space stands excluded form all other space which then forms an all including unit. That is why the Sun is so cold and outer space is so hot (and no… calling the Sun cold and outer space hot is not a mistake on my part and neither is it due to a printing error but the Sun is as cold as it gets and the outer space is as hot as it get.

We'll get to that one later on...). The two share a cosmos in which they are both apart while forming one unit. The separating part in the unit is motion driven by heat and the motion as well as the heat carrying the motion that sets the two apart while both are in the unit. I mentioned previously that there is an undeniable connection between heat and gravity. Let us do some investigating and try to establish answers.

From since the time that man discovered intelligence man (if he ever did) has been with the presumption that the Sun is the hottest centre in the solar system. Later on in the more present time it came to someone's attention that the Sun also holds the solar system in gravity.

The Earth by its standard and dominating its sphere of which it can control with influence is the hottest centre in the space of its domain and it holds the moon centred to the Earth. The gas planets are the hottest centres in the relation with the most heat and they all hold their satellites captured by a hot centre. All space structures hold in every centre there is that is confirming their independence at that point of securing independence the centralizing of the most heat it is able to concentrate and from that centre holds all material captured or controlled in the domain of what that forms the independence of the structure. I can go on and on but heat in the centre couples gravity to space-time, just like Kepler said before he was spoken for on his behalf and without his permission or his agreeing to it.

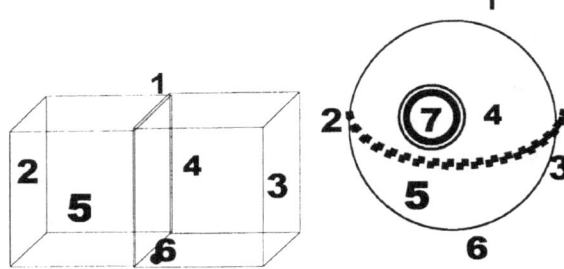

**Kepler's formula also indicates that a sphere is within a cube that is holding**

Taking the outlook from the point the sphere is holding from the centre out into space there are ten points connecting to the centre. In that are the dimensions of singularity connecting to space where five connects to space in the second dimension of singularity, and five connects in the third dimension of singularity. On the other hand, the cube does show a very different characteristic, which involves only six sides (at least) connected.

$a^3 = (T^2 k) = a^{3+2+1} = 6$ with the sphere presuming the position of singularity as part the of $k^0 = 1 =$ **singularity**. Einstein proved that at the point where space reduces and such reducing reaches a point where space as a factor in the third dimension disappears into the single dimension (space going flat) gravity is overwhelming. Einstein interpreted this, as the complete Universe going flat while the Universe going flat, that can only be within singularity since singularity represents the Universe as flat as it can get.

Humans (including Einstein) interpretation of the Universe is faulty but the faulty aspect does not include the fact that the Universe is going flat but only which is the flat going Universe referred too. According to Einstein he proved that the Universe is alternating between going flat and holding space but his lack of studying Kepler lead to his spontaneous misinterpretation collected from our culture and his incorrect interpreting of what the Universe actually is. We all have a faulty perception of the Universe because not only he (Einstein) as an individual Scientist but all humans throughout has also never asked the Universe what the Universe is. Kepler did and the Universe answered using the mathematical equation $k^0 = a^3 / T^2k$, which when interpreted means singularity placing space-time is the Universe.

No one ever thought about this statement in sincerity because from a Newtonian aspect it seems silly. But rethink the silliness presented by the Newtonian Universal centre and compare that thinking about what the Universe told Kepler then decide what is silly. Newton's never acquiring the effort to do a study of Kepler's work withheld him (Einstein) from reading his very own mathematic translation accurately because apart from Newton Einstein must be considered the second most important Newtonian ever. What Einstein saw was that space disappear and he then jumped to the conclusion that the space he saw in his mathematical equations was outer space referring to the space falling outside the parameters of the material occupied space secluded by dimensional borders. In the sphere placing the borders that the sphere holds there are deliberate and very distinctly placed edges or points forming a specific distance from the centre. The centre is also proven beyond any debating.

The centre of any sphere has to be at the very point where space completely falls away. It is at the point where all the points of line centres meet by the crossing the centre of their individual connection coming in to contact as a group. In that way one may assume that the lines connecting the controlling points on the other end is crossing on a centre point that all that is participating in the constructing of the sphere is democratically electing such a centre. Please note this conclusion very well because this forms the heart of the Coanda principle. That will put that position where the lines cross which in itself is centralising all space in the sphere at that point, such crossing point will become very distinct and controlling where that point forms in the single dimension and singularity is the single dimension. But Kepler also solves another riddle that truly got Newtonians unstuck. This too which I now refer, is what is referring to when they refer to the Hubble constant.

The growth we see in the Universe is an adding of space in every cycle completed by every cycle, which all the protons complete. The adding is the smallest addition that can come about in the shortest period of repeating by cycle rotation there can ever be. This growth of space-time next to singularity confirms the growth of singularity as singularity recall the space it uses to grow in the time it grows. The margin of growth will be by the extension of **k** in the formula $k = a^3 / T^2$. Every cycle completed in the relation to space by the initial value of **k.** $k = a^3 / T^2$ leaves ultimately $a^1$ extending as space or as Kepler chose to indicate it as $k^1$. But that two has to be compensated by the duration of time reducing the time aspect by the margin that the space expands. This confirms what is evident in the Hubble Constant. The further one look at time the more time seems to race because time has the invert properties we give to space.

The Hubble Constant is the most incorrect study made in science. The only fact proven is that there is shift in space but measuring it brings total and widely different values. In fact, not one study ever proved to have the same results. In fact the study has such wide margins of difference that they stopped the publishing of all further outcome results on the Hubble Constant. It is because of these differences that every reading is different. And yet with not one reading being equal, the brave Newtonians still have the audacity to determine the age of the Universe in accordance with this misinterpretation of fact.

Now comes a perceptual question: with the Universe being all there is and all there can be, then how can it expand and whereto can it expand. I do explain this oddity but then I put the movement of gravity in a completely new line of thought.

The entire concept works on the density shifting from gas to liquid and then to solids and whereas solids increase in density constantly gas loses density in favour of liquids reducing in density. This is the role that time plays in the Universe by time determining the precise density applying at the time in the space forming the very spot.

When one observe the cosmos one observe the night sky as one big black hole that is forming. From what we observe the night sky also fills with tiny lights here and there and in between, and the better the lenses one uses the more lights are here and there filling up everywhere. But in all the blackness and all the vastness and all the sparingly filled spaces there are three relevancies, which result in gravity. Let us acknowledge what we see from the controlling centre. There is a position that is in motion that is forming the very edge of the outside. To be in motion the position must be in relation to a point from a centre. From the centre there must be a specific allocated space ending at the object in motion and starting from a centre that has no dimensions. The object in motion determines the one limit and the centre with no sides and no space, which is standing still in singularity, determines the other limit. By that we can see there are only one way of looking at what we can observe and that is from the outside in.

This is the one perspective. There are the others. From the outside there is a centre orbiting an unfilled space with an inner centre. The centre orbiting has to have an allocated centre with no sides because that centre secures motion that is independent from the other space surrounding but which is including the independent space forming the border.

centre

$\frac{7}{3}$ +

The orbiting object also secures an individual independent and own but from the orbiting object the limit it holds ends at the edge it forms. Being a sphere the orbiting object secures seven positions and the larger containing sphere is ten of which seven is within the singularity dimensions within the centre, which we not observe. Immediately following that as part of that relevancy comes the containing sphere that holds space-time and another tree positions. The three positions puts a relevancy of three to the holding space that already caries seven. There are fore ever another centre that secures seven positions just because singularity chose the sphere at the value of $\Pi$ and in the sphere 7 positions is made up of six sides that hold relevance to a precise centre. There are $a^3$ but then there are $T^2$ putting $a^3$ at a value. Then there is a relevancy named $k$, which puts $T^2$ at a relevancy. None of these are fixed markers because the relevancy can and does swap sides placing importance as alliances changes. When $T^2$ focus on another $a^3$ the relevancy about $k$ changes and amplifies the importance of yet another space. When one applies the Coanda effect one would see just how easily new alliances come in place and secure new centres that charge new relevancies between newly established points. Going either "bigger" or "smaller" is only shifting focus on another relevancy. What started out as so small it was one spot is now so big it houses the entirety. We must not view the Universe from any perspective we have because in the Universe large and small is not a matter of size but it is a matter of time. The relation big and small, hot and cold, young and old has with other parts of the Universe is merely a matter on the time line of development but it does not bring inferiority to what ever is in focus. It is our human conception that big is better that makes us look at the Universe in such a dismal way.

I suspect this cosmic growth of all material is equal to the growth of a human hair or a human nail the presents as the duplication of cells because life takes command of what is made available by the Universe and then manipulate space-time to claim such growth by taking charge of the opportunity to use the growth to the benefit of life. But this growth constitutes multiplication from the very centre of the most inner part of the where **k** = infinity plus one.

Mass is the result of applying gravity by reducing space. Protons are the only diminishing devises of space-time there are in nature but the protons remove space as it concentrates space-time to furnish material with growth of material. The more protons there are allocated to a specific space the more space the larger number of protons will diminish. Gravity is not the result of mass. The belief of mass brought about by large numbers of protons confined in little spaces is not always true. In some cases the mass produced by a large number of protons does no result in a heavy confined element since there are those with high number of active protons which should enforce a large gravity, the opposite is true as the element show as an elaborate anti gravity by being an airborne element or a gas. There are those holding protons in clusters with numbers matching heavy metals but that is categorised as gasses and gasses produce high ratios in antigravity.

Kepler thus gave us the answer about what Hubble found what was happening in the Universe centuries ago and centuries before Edwin Hubble's discovery. **From Kepler's formula one can see that time and gravity is the same because as gravity weakens so does time reduce and as space expands so does the influence of gravitational reduce because gravity has less time per unit to control;** $k^0 = a^3 / T^2 k$ **more space per unit. Gravity is $T^2 = a^3 / k$ since the object cannot depart at any further distance between the centre and the object and is captured at that distance. In addition, gravity is $k = a^3 / T^2$ for the very same reason. The circular bonding $T^2$ of space $a^3$ is enforcing an orbit $T^2$ to gravitationally circle around a specific centre k, which indicates the gravity $T^2 = a^3 / k$ in relation to the other gravity component $k = a^3 / T^2$, and it means $T^2$ is a circle of gravity and k is the straight-line distance of gravity applying motion. Still Mainstream Academics ignore my statements that gravity is space in motion and motion of space is time: precisely as Kepler said. Any area to the cube is space $a^3$.**

Gravity is the result of the motion of the number of protons that through such motion creates the reducing of space in the space less centre. Gravity dismisses space and by doing that the stronger gravity is where more particles can have fitting into less space occupied where that reducing of the space is bringing about extensive mass increases into the volumetric occupied area confining more material into less space. Gravity is space measured over time. Gravity is the space in comparison with the time affecting the space. It is the motion of space relating to the time of motion of space. Gravity is the moving away of what fills space by extending singularity where singularity responds by bringing about space between the structure and a centre being within another and larger structure. The larger structure in consider to fill the role of the governing singularity will hold more material in less space which is then more material that is confined to less space. The centre structure is reducing more space in the time factor between the moving structure and the centre structure. Gravity is the increase of heat occupied by the reducing of space in a spherical unit. If Newton only tried less to deny Kepler any recognition and gave Kepler more deserving accrediting about Kepler's input in the total work, he Newton would then have seen that that was what Kepler formulated gravity to be. Kepler formulated gravity as space $a^3$ over time $T^2$ in relation to a centre $k$. It is the space that relates to time in relation to a centre just as Kepler introduced gravity to be. Kepler said space $a^3$ standing is over time $T^2$ in motion $k$. $a^3 = T^2 k$ and to all those who tries to give space-time some godly appearance with mystic properties can lose the séance- like attribute they wish to connect to space-time. Every bit of space however insignificant or however demanding forms a relation with time, which is what separates the different space from one another. It is the separation coming from time differences that distinguish space from one another.

If particles were that close and yet they expanded by growing apart, that then is proof that gravity is not about particles pulling each other closer. If the pulling closer was gravity then in that case the Big Bang provided the ultimate opportunity to unite what there was instead of expanding what there is. I came to realise that gravity is about removing the space between the body and building on that which the body already holds as well as the space surrounding the material that serves as unoccupied space whereas material holds occupied space. A fan drawing air into the blades has the same sucking or pulling that one experiences with gravity but in the case of the fan we know it as air that is in motion and the motion extends to affect those objects placed in

**Fan contracting air by producing flow of space-time and not just air.**

the line of the air flowing. Gravity comes about as space $a^3$ applies motion $T^2$ and from singularity $k$, gravity is as much part of space as the motion of space is part of gravity.

Gravity is working on a principle of indicators pointing dimensional integration and separation of space through heat densities applying different grades of space intensity. That means the space does not mingle, but forms layers. This is unlike one would expect from the advocating by Mainstream science about the characteristics of space. By gravity acting space becomes denser and therefore space can become a liquid and as all liquids do, space then depends on specific densities being in specific positions. With the specific densities borders come about in space. It is as Kepler stated gravity to be even before Newton came up with an idea that there was such influencing going on and named the influence gravity. Gravity is $a^3 = T^2k$, which is the space $a^3$, that forms through the moving $T^2k$ thereof giving the space $a^3$ independence as the independence comes about of speed differences which is motion in relevancy which is $T^2k$. It is distancing $a^3$ from $k$ by applying $T^2$ in the surrounding space and this is done by $a^3$ duplicating in motion when applying $T^2$.

## The Oxford dictionary of Astronomy defines gravitation as follows:

Gravitation is the force of attraction that operates between all bodies. The size of the attraction depends on the masses of the bodies and the distance between them; gravitational force diminishes by the square of the distance apart according to the inverse square law. Gravitation is the weakest of the four fundamental forces in nature. I. Newton formulated the laws of gravitational attraction and showed that a body behaves as though all its mass were concentrated at its centre of gravity. Hence the gravitational force acts along a joining the centres of gravity of the two masses. In the general theory of relativity gravitation is interpreted as the distortion of space. Gravitational forces are significant between large masses such as stars planets and satellites and it is this force, which is responsible for holding together the major components of the Universe. However on the atomic scale the gravitational force is about $10^{40}$ times weaker than the force of electromagnetic attraction.

## Gravitation Constant
**The constant that appears in Newton's law of gravitation. It is the attraction between two bodies of unit mass at unit distance apart. Its value is 6.672 x $10^{-11}$ N $m^2/kg^2$ when the distance is expressed in metres and the masses are in kilograms. Although it is described as a constant, in some models of the Universe G decreases with time as the Universe expands (see Brans-Dicke theory), but there is no evidence for this.**

## Gravitation Mass
**A measure of the quantity of matter in the body. It is measured in kilograms. Mass determines the strength of the gravitational force exerted by an object.**

Now even three hundred and fifty years on, science still came no closer to explaining what gravity is in contrast to the fact that Mainstream science established even more forces than the one that Newton declared at the time. If Newton only was less presumptuous about his genius and took more notice of Kepler's work he (Newton) could have seen just what Kepler said what the cosmos told Kepler mathematically about what gravity is. That effort would have saved so much misconception. But even almost four hundred years on Newtonian disciples will not recognise my personal effort to indicate too the world what Kepler said what gravity is. I have been trying to indicate this thinking by using academic channels but on grounds not related to my effort I was dismissed so many agonising times by so many academics in charge of Official policy protection. Although it is most apparent (to me at least) that I can tell what Kepler saw and tell them that, still they the Newtonian priesthood silences me just like Newton silenced Kepler. Newtonians should have realised centuries ago that Newton and Kepler did not have the same mathematics in mind. Consider the following and then decide

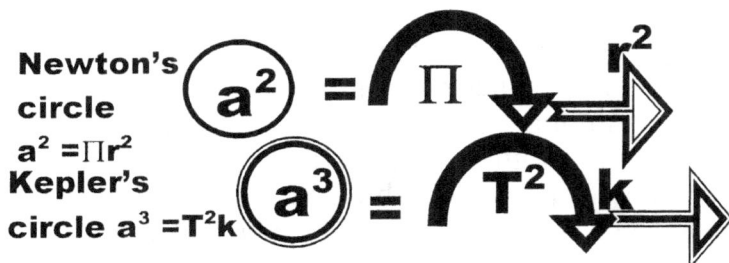

Newton's circle
$a^2 = \Pi r^2$
Kepler's circle $a^3 = T^2 k$

In the manner the two measured, the circle Kepler used a different means of measure of the circle than Newton did. In the case of Newton the radius, r goes square while the circle $\Pi$ indicates the single factor. Kepler deliberately ignored the factor $\Pi$ for reasons I explain elsewhere.

Newton whished to see a mathematical equated circle and the formula therefore was in need of revision. Kepler intentionally stated the very opposite of Newton because the circle indicator $T^2$ goes square and the diameter indicator **k** remains single. That is how the cosmos relayed the given information too Kepler and that is how Kepler correctly interpreted the given information but Newton thought himself as being brilliant enough to change all that in favour of his ego. While Kepler formulated his ideas according to cosmic information Newton saw this effort of Kepler formulating the cosmic numbers

Kepler measured as mathematically being incorrect. The question Newtonians failed to ask is: How can the cosmos give mathematically incorrect information about the cosmos?

**Newton's mathematical vision was the way to calculate the space by using a mathematical formula used as**

$$a^3 = 4\pi \text{ X } (r^3/3)$$

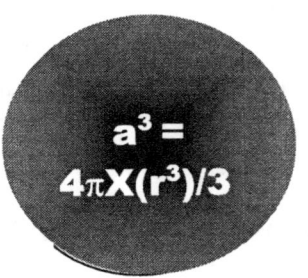

**Kepler's cosmic vision was that in the formula** $a^3 = k \ T^2$ **the space** $a^3$ **is equal to the movement** $T^2k$ **of the space, which comes about as time** $T^2$ **in relation to a distance** $k$

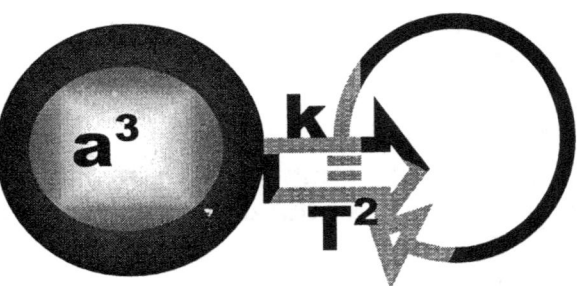

**Newton whished to see a mathematical equated circle and the formula therefore was in need of revision.**

The object in motion is displacing the space that the object occupies in relation to the surrounding space and space being between the object and the centre. The object is dismissing space by motion as it is in motion through space. But there is another factor just as relative that is also applying motion. The applying of motion by the departing object only holds any relevance to the centre from which it is applying a distancing or an escaping attempt by increasing the distance between the centre and the escaping object. On the side of the escapee there is another space active, trying desperately to escape by enlarging the influencing field the escaping singularity is fighting to establish. Seen from the escapees point of view in relation to what the escapee tries to bring in place what is important to the escapee is that the only motion there is the escaping object trying to escape. However much the escapee is contained the escapee is only aware of its motion it finds as it sees all motion being only progress in an continuous charge of escaping.

Just as Kepler stated gravity are three relevancies acting in opposing as well as sustaining motion where the one action works by interlinking with the other without being aware of the other. We have to recognise the relevancies applying. It is motion performing contrasting relevancies and without all three independent actions of motion contributing as one not one act of gravity is possible. There are three relevancies applying motion in the formula Kepler left us. There is the space in which the formula presents the dimensions. The centre provides the space the dimension to move through the space as the object is trying it's utmost to depart from the centre as far and as fast as it could. It is displacing the space in which it is by it is in using a time as short as a period as it can manage. It is running of into the distance from the centre. In this motion the smaller object tries to repel from the larger object claiming to have the centre of space while it is still part of the bigger space in which the smaller space part of. The Universe is the atom.

Let's look at gravity while we use our common sense.

There are three factors forming gravity. In the one scenario, gravity is all about the motion of one part in space applying relevancy to another motion by dismissing space through the effort of moving through the space in a common factor. The lesser-developed point holding singularity is in an all out effort to depart from the centre spot where the gravity is vested as the strongest applying influence. The effort behind this is speed or if you wish I can use the term velocity but it will still mean it is the comparing of different motion each holding a different value but still requires

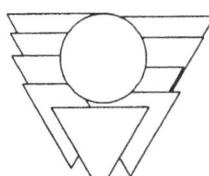

While the orbiting structure has the effort of displacing the occupied space, the material holds in an effort to produce a cooling effect on the material in the occupied space it holds. There is another centre in another object that has to be much larger than the roving orbiter and as such the centre of the centre object is removing space towards such a centre in an effort to secure the centre of remaining cool and thereby preventing destruction by overheating. From the centre object there is no escaping object but only the space it is harvesting in an effort to sustain singularity in the centre of the centre object and where the strongest point of gravity is located.

The atom forms the Universe in all its dimensions because the atom forms the star. The star is what the atom is producing the star to be. The galactica is many stars and every star incubating or fully developed is a combination of what the total number of atoms form. When we look at what to find in the Universe we better find it in the atom or we will not find it at all. The atom has a containing centre, which we are of the opinion that gravity is centred in that location. The proton is the containing or the preserving dynamic but there are also an evenly matching and equal dynamic other partner.

There is the electron and the electron defines the boundary of the atom. The electron is what determines the volumetric space that confines all space we think to be as the space claimed as the atom. The electron acts as if in defiance of the proton and for that there is a considerable good reason. While the smaller but independent space is busy with the great escape effort by putting a distance **k** between the space in motion and the centre it fights to secure an independence from that centre. The larger space is accumulating as much space as it is contracting all space towards the centre. The independent and escaping object is moving through the space surrounding the object in this effort to put distance between the object and the centre. If it did not do that in motion it would hurry towards the centre instead of circling around the centre year after year. From the smaller objects vantage the smaller object is carrying through space displacing the space surrounding the object by motion in space. From the point the smaller object has is the space is standing still while the object is applying motion to get away from the centre. The object is applying motion by dismissing the space it is moving through.$a^3=T^2 k$

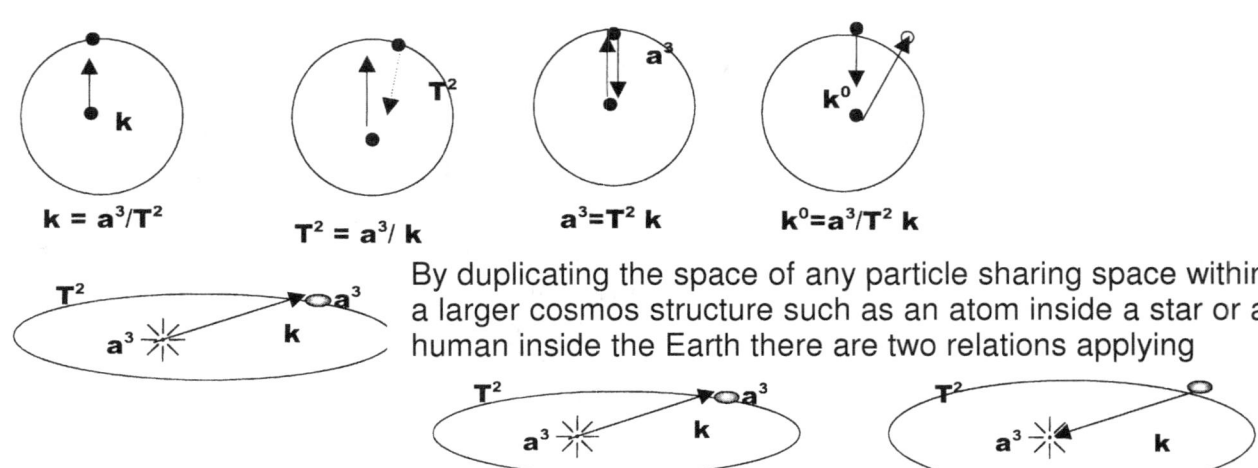

$$k = a^3/T^2$$

$$T^2 = a^3/k$$

$$a^3 = T^2 k$$

$$k^0 = a^3/T^2 k$$

By duplicating the space of any particle sharing space within a larger cosmos structure such as an atom inside a star or a human inside the Earth there are two relations applying

The factor **k** represents as much space as **a³** as a factor represent motion. But **k** also represents as much space as it represents motion because it represents motion. This is the Titius Bode law where each planet adhere to an individual and specific space –time or **a³=T² k.**

I argue that if it is the correct practise to use $F = G \dfrac{M_1 M_2}{r^2}$ to calculate gravity then the radius holding the gravitational constant must lead one to the centre of the Universe. As I confront science dogma and principles, nobody is willing to publish my work. I have to walk the road alone and fight the battle by my private effort without any support anywhere. If Newton is the problem one have to go pre Newton to find the problem.

The problem is that when looking at Kepler's table then if there is **T²÷a³** according to the table matching a column, then mathematically **T²÷a³** must be **k⁻¹** and where **k⁻¹** goes negative it shows space reduces time. It shows space in volume goes single by movement of space and not objects.

| Planet | Mass per Earth unit | k⁻¹ Movement | a³ of space volume | T² During time units |
|--------|--------|--------|--------|--------|
| Mercury | 0.06 | T²÷a³ =0.983 | (a³)= 0.059 | (T²)= 0.058 |
| Venus | 0.82 | T²÷a³ =0.992 | (a³)= 0.381 | (T²)= 0.378 |
| Earth | 1.000 | T²÷a³ =1.000 | (a³)= 1.000 | (T²)= 1.000 |
| Mars | 0.11 | T²÷a³ =1.000 | (a³)= 3.54 | (T²)= 3.54 |
| Jupiter | 317.89 | T²÷a³ =1.000 | (a³)= 140.6 | (T²)= 140.66 |
| Saturn | 95.17 | T²÷a³ =0.999 | (a³)= 868.25 | (T²)= 67.9 |
| Uranus | 14.53 | T²÷a³ =1.000 | (a³)= 7067 | (T²)= 7069 |
| Neptune | 17.14 | T²÷a³ =0.999 | (a³)= 27189 | (T²)= 27159 |
| Pluto | 0.0025 | T²÷a³ =1.004 | (a³)= 61443 | (T²)= 61703 |

If you are a student in physics then you should read the following information with care and with much consideration because your mental health might be at steak here. One could think of another name for physics and that would be Newton's mythology. It is about the subject of gravity and is most important. The "Newton's mythology" comes from the fact that students have to learn what the professors claim to be true and what was never was proven. Students have to repeat in examinations that the formula $F = G \dfrac{M_1 M_2}{r^2}$ is truthful and viable while it was never proven. Do you realise that it is an accepted practise that all students that are studying physics on all levels are subjected to the most intense brainwashing and thought control found any where on Earth? This must be some sort of a joke you may think but thinking that way in disbelief is just what those practising the mind control wish you to think! According to the tables, all movement is according to some other value than mass. They never prove Newton's philosophy on gravity but those persons conducting teaching in the subject of physics force all physics students to learn Newton's gravitational concepts and accept the facts as if it has been proven beyond all other facts. Students have to believe that Newton is correct or academics will see to it that they fail their examination. The condition of being accepted in physics is to accept Newton without questioning the proof that is never supplied.

Let those academics now prove precisely how mass brings about gravity and then afterwards test you on how Newton is proven correct and not on you repeating their facts blindly about what they say is true about what Newton said, which they say is true. The manner they present Newton is completely hearsay and that method may not be used in any court of law. Let your professors now prove how it is that Newton's teachings are correct and then examine you on the process they use to prove Newton's concepts rather than test the state of brainwashing you have submitted to. At present they say Newton is correct and then they test you on your ability in repeating that Newton is correct without ever proving to you that Newton is correct. That is not testing your knowledge but it is testing the mind control you have submitted to. Let those physics professors now prove Newton and their ability to prove Newton correct and then test you on the manner they use to prove Newton to be correct. The truth beyond all other truth is that Newton's gravity has never been proven (because try as you may it is not possible to prove Newton's formula forming gravity mathematically) and because academics know that, academics require the blind acceptance of Newton by students. This unconditional acceptance of Newton's correctness relies only on the pre-conditioning of students' mind set and academics depend only on the student trusting the academic "say so" about the institutionalised correctness of Newton. That Newton is correct nevertheless and notwithstanding that there is no founding proof about this matter, is what students should be accepting blindly. I'll bet you they are more surprised than you about me accusing them of systematically mind altering the student's physic and ability to think than you are.

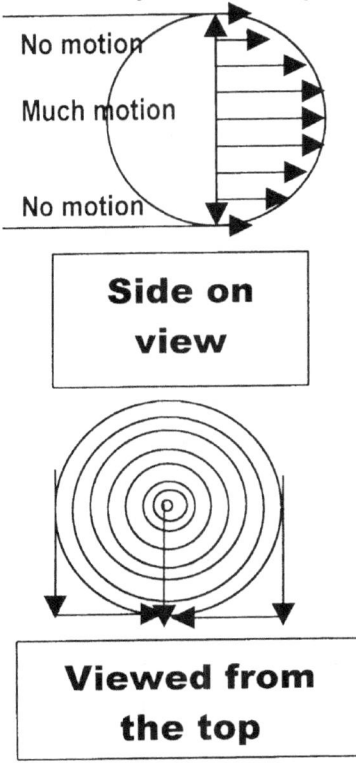

**Side on view**

**Viewed from the top**

**There is a simple yet specific reason why the Universe uses the sphere as the form in which to develop stars. The form of the sphere provides the sphere with such form that in the center of the sphere there is little room of motion going down to no providing of space to move. In the center of the star which is a sphere, the center is about contracting and bringing security while the outside of the star is about a space providing motion. The two factors we find evidence of in the Kepler formula $a^3 = T^2 k$ where both the line k that is enforcing contraction as well the motion $T^2$ that secures duplication or motion attach to the other part of gravity to the space there is in the star. The outer edges are filled with motion advancing elements while the center is filled with contraction advancing elements.**

This means that all space in motion is resisted by space in motion contradicting each other. In applying motion to space there is never just acting but always counteracting, which is precisely what Newton's third law on motion states. Because there is a motion in one direction there will be a counter motion performing a balance to the reaction on the motion and establishing a motion be a reactive motion. But also it proves Newton's first law on motion by proving that equilibrium existing of the action of space in the counteraction in space brings about equilibrium and being in motion that is in countermotion provides for the same effort as being in rest or in a stable state equal to being in rest after forming a circle eventually. Since there is equilibrium between the motion of expanding and the motion of contracting any additional motion adding to the "stalemate" existing will bring in Newton's second law into action in the manner Newton described.

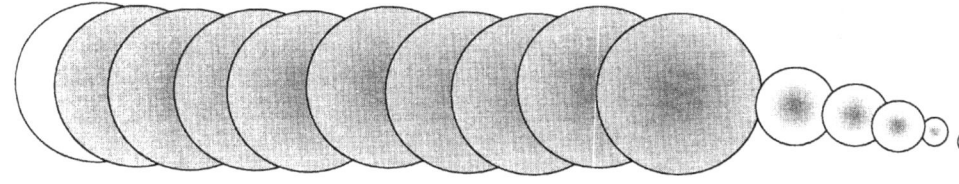

The containing structure represents duplication and in that it applies a norm set on conditions applied by the governing singularity of the space- time, which is supporting and maintaining the point in singularity. By containing the subordinate singularity with the ultimate goal to initiate the inevitable uniting which all singularity will do in the ultimate end of all development when all singularity will once again re-unite. But in the mean time all singularity is fighting the fight of their life to remain independent, as much as possible and by what measure it may have to produce.

When the two are mostly on equal motion the gravity each produce will bring about motion of maintaining independent by creating space through duplication. When the motion equality falters the space reducing of the dominating space will relinquish space by contracting space faster than the dominated space can accomplish by using the duplication of space. When the dominated space falls prey to the incorporating of such a smaller space into the domain of a larger space the fight for independence is carried on but the terms of conduct changes. As the dominating space adheres to the containing of all space within the domain of the larger space the lesser space will be reduced to the point it may have to relinquish independence totally.

**The space we find is not just space but the space we find is motion of space as the space reduces to a centre by duplicating or by motion of repositioning in a rotating relation by doubling what there is in a repeat thereof. This motion, to which I am referring, is extremely apparent in the illustrated imaging of the functioning of a Black Hole where one can witness the total collapse of space toward and around a mathematical centre point forming where space and motion ends.**

By not agreeing to such terms the lesser independent singularity continues the fight for independence by an all out resisting relinquishing all individual form and to join and unite singularity. Loosing independence will amount to relinquishing all forms of independent space being independent because of the independent motion it contributes too the unit of space –time within the unit of space–time. The resisting is performed by establishing individuality through the maintaining of individual space and form and in such maintaining of individual construction it confirms independence in a mode of self protection.

The self-protection is a resisting and the resisting the joining of the larger space where such resisting of relinquishing independent space brings about mass. The establishing of a mass factor is part of the resisting losing independence in a self-protection drive from early uniting with the domineering singularity. As the dominated singularity fights to remain independent it is forming resisting and when the resisting is not capitulating such resistance to capitulate its independent form, that is what we then see as mass. That will demand that the mass is brought about by all that the space contains, which is where all the atoms are individually and as a group and as a structure formed by the group where all that are fighting forms a unit and as a unit it is fighting for individual independence

What this formula, which Kepler introduced tells more than any other fact is that there is no space if space is not in motion and all space there is must be about motion or else there is no space. That is establishing the fact that can only be by motion and motion is what space does to move from one point in time to another point in time by the square of such motion. If there is space the space is duplicating space by projecting space from a past through a present to a future and that means the space is duplicating what was in the past towards the future through a present. But if the space and all space there is, is not duplicating by motion it does not qualify to be in the Universe. The fact of space $a^3$ is about the equal $=$ motion $T^2 k$ thereof.

In the commanding space $a^3$ forming the superior part of the space $a^3$ that holds lesser space $a^3$ in the unit as an independent part of the unit, which is included as a part in the same unit that normally occupies space-time as particles or at least by particles with independent space-time The particles hold some space $a^3$. The space $a^3$ the particles hold is directly in relation to the particles the containing structure that has to in duplicating in. Every particle in the unit has to fight every other particle in the unit as well as the unit as a whole for space within the time the group puts on the unit. The more space $a^3$ there is being relevant to the structure in comparison to the structure as a whole and as a unit that forms other relations with other units, which is duplicating in the sense of being a unit once again is relevant to the space $a^3$ that the structure destroys as a unit. The bigger the space is that the smaller space hold in relation to the size of all the space combined where the combination is part of the complete unit forming individual space-time as a group but still occupying space-time in the group as independent particles while in the unit.

In this unit where the smaller individual space remains part of the unit but having to match to the capabilities of other space also in the unit being as part of the unit the less the smaller space will find it able to match the duplication effort compared in ratio to the entire group but still will find it able to focus a duplication standard that holds relevance to individuals in the group. Two sectors emerge in the star where the outer sector advances duplication as the main focus of the star and the inner circle is the sector that place gravity on dismissing space-time. In the inner part the particles focus much more on the dismissing of space-time in the sector as the main focus of the entire unit. The outer part is overshadowed by space-time duplication. Space-time dismissing overshadows the inner part. The bigger the space is the more the unit as a whole will favour that particular independent to group where the individual will identify its preference in the star by selecting the group that will compliment its achievement. The space will tend to associate with that group that promoted its goals. The larger space will tend to form alliances with those large enough to favour duplication and those smaller but being more solid will sink down towards the dismissing sector in order to select a suitable position it can locate in matching the smaller space needs.

Gravity in whatever form comes as the result of movement by the interaction between material showing a density difference between confined space and non-confined space.

The spin of the circle forms 7° on the one side of rotation as well as 7° on the other side of the centre of rotation.

That puts the centre turning by 7° x 7° = 7² = 49. Then on the other side of the divide the same repeats where there too forms by turning 7° x 7° = 7² = 49. Then by using the law of Pythagoras and incorporating Π° on both sides 10 forms as a space value outside the circle.

A double 10 plus a developing singularity value of 1.9991 represents singularity, which represents Π turning around Π° to form Π. This means the one side of the circle holds Π = 21.991 ÷7 and on the inside connecting to the circle centre holding singularity at Π°= 1 the value of Π is 3.1416 ÷ 1

Where gravity as the movement follows the curve, the space holding 7 at a double square in relation to the circle forming 7 has gravity at Π = 21.991÷7 and on the inside gravity is Π =3.1416 ÷ 1. Anything that relates to singularity in terms of the outside of the circle as Π = 21.991÷7 will form as liquid and everything relating to singularity forming part of the circle is solid and holds a relation of Π =3.1416 ÷ 1. When there is contact between the object and the earth the relevancy forms gravity at Π =3.1416 ÷ 1 and mass as a factor forms while when in space without forming contact the relevancy changes to space forming a value of Π = 21.991÷7

$$\Pi = \frac{3.1416}{1} \qquad \Pi = \frac{21.991}{7}$$

From this reference we find gravity forming in relation to Π that includes what part is outside the circle and what part is inside the circle. By having a solid value of Π =3.1416 ÷ 1 forming a solid and holding mass or forming part of space that has a value of Π = 21.991÷7 distinguishes the alliance of the object. That makes Newton wrong that gravity forms by mass.

"GRAVITY IS DIVIDED IN TWO FACTORS, BEING <u>LINEAR DISPLACEMENT</u> (Π / Π°) WHICH IS WHAT <u>NEWTON'S GRAVITY</u> IS AND,

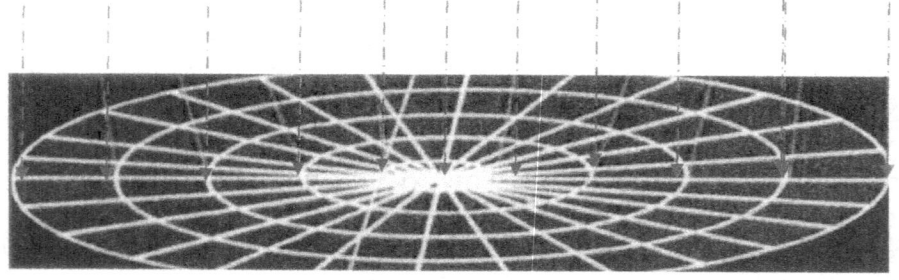

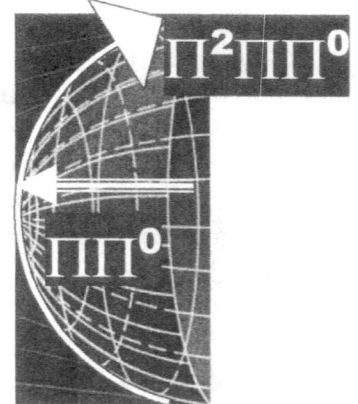

CIRCULAR DISPLACEMENT (Π²/ Π)

WHICH IS THE "GRAVITY" EINSTEIN RECOGNIZED

As Kepler proved there are two lines formed when gravity forms. Te one line is linear going straight and the other is circular as it revolves in circles and the one form always flows into the other form as gravity grows or shrinks.

The containing structure represents duplication and in that, it applies a norm set on conditions applied by the governing singularity of the space-time, which is supporting and maintaining the point in singularity. By containing the subordinate singularity, the ultimate goal is to initiate the inevitable and eventual uniting of the two positions where singularity will again form one unity. The space-time first has to dismiss while the space-time fights to gain and duplicate as the one does which all singularity will do in the presence of the other that destroy space-time.

**The duplicating sector of the star... is the outer layers**

**The inside part is dismissing sector of the star that contacts**

**The duplicating sector of the star is the part providing motion through duplication**

end of all re- fight as are the bring by motion reducing of the

This will continue to the ultimate development when all singularity will once again unite. In the meantime, all singularity is fighting the of their life to remain independent, as much possible and by what measure it may have to produce. When the two mostly on equal motion gravity each produce will about motion of maintaining independence creating space through duplication. When the equality falters, the space dominating space will relinquish space by contracting space faster than the dominated space can accomplish by using the duplication of space. When the dominated space falls prey to the incorporating of such a smaller space into the domain of a larger space, then the fight for independence by the lesser partner carries on but the terms of conduct changes. As the dominating space adheres to the containing of all space within the domain of the larger space the lesser space will reduce to the point where it may have to relinquish independence totally. By not agreeing to such terms, the lesser independent singularity continues the fight for independence by an all out resisting relinquishing all individual form and to join and unite singularity. Losing independence will amount to relinquishing all forms of independent space being independent because of the independent motion it contributes to the unit of space –time within the unit of space–time. The resisting is performed by establishing individuality through the maintaining of individual space and form and in such maintaining of individual construction it confirms independence in a mode of self-protection. The self-protection is a resisting and the resisting the joining of the larger space where such resisting of relinquishing independent space brings about mass. The establishing of a mass factor is part of the resisting losing independence in a self-protection drive from early uniting with the domineering singularity. As the dominated singularity fights to remain independent it is forming resisting and when the resisting is not capitulating such resistance to capitulate its independent form, that is what we then see as mass.

The more space $a^3$ the particle claims as an independent particle in relation to the space $a^3$ the container hold as a unit formed by the whole lot as one inclusive unit that relates to the space $a^3$ the container duplicate is relative to the space $a^3$ the containing structure destroy as a whole unit representing all other space that is combining in the group as representing the group where then the group effort of dismissing space-time in each individual capacity and to the performance what each individual may achieve where such a combined dismissing manifest in a precise centre of the groups space-time.

This is the point focussing the dismissing part where the groups dismissing of space-time coming from every atom would gather and present the unit effort, which is in presentation to the duplicating in motion all the atoms form as a group and in all this there must form a balance to gain gravity. As individual occupying space $a^3$ the atom is an individual container by own dimensions and as such duplicate space $a^3$ in this regard. The atom resists the dismissing of space- time in which the individual space the atom holds is also included by the confirming the structural form to the atoms relevancy being $k^0$ in singularity that is bringing on an independent value that fits the particular $k$ in the atom relating to $a^3 = T^2k$, which is part of the whole unit represented by the combined value of $a^3 = T^2k.$

This relevance means that without a specified container producing individual particles that control the specifies duplication and destroying of space in relation to the outer space such a container will apply a diminishing relevancy of space-time displacing that will equal the displacing of a number of 112 protons and no more which are all working as a unit within a confined unit we call outer space and in conjunction what is dismissing in the unoccupied space- time where the unoccupied space-time can withstand. We know that is a theory because the atoms in space can sustain much less than 112. In short the value of the walls serving our three dimensional Universe can sustain no more than that what a possible 112 protons gathered in one atomic cluster may displace.

Inside side this outer container, which we see as outer space there is inner containers being stars that bear the direct relevancy which singularity is applying being inside stars, puts much more strain on the surviving abilities of atoms. In outer space the atom has an own relevancy of seven and the space demand on the atom is only three that the atom must maintain in order to duplicate. But in stars the containing star places a demand of the containing seven plus the space creating three in relation with the time applying inside a star, which is four. Let's put what was said just now in conjunction to the Earth. Since the Earth has no singularity demand that is that much better developed than outer space is insisting on as being a limit in the Universe needs to afford space-time self sustaining of all there is on Earth by a relevancy of $k$ to $a^3 = T^2k$ is adequate in that which the atom normally can sustain and leaving lots and lots of space to spare. We call this stat of affairs inside the atom to be quantum physics where we directly associate the concept of quantum with none quantifiable volumes of unaccountable space in disuse. But in bigger units the space-time displacing relating to space duplication presents with much more demand on atomic structures. As the demand of singularity in such units grow some relevancies within the atom come into play and I developed a formula to place such a demand in relation to singularity where the ultimate demand sets the standards. As I stated there is within the star that shift as the star progress through development. At first the star leans heavily towards duplication. Then as the star develops the star moves across a broad range of specifically identifiable stages coming from one extreme where our Gas "planets" (which is stars in the making) is to where a tiny star such as the Sun is growing towards sizable monsters and then on to cosmic destroyers.

Through out the variety of development there is a balance unfolding that at first support motion by duplication and leaves dismissing for a senior partner to commit while growing the presence of a superior partner. Then the space-time development allow the smaller star to drift away from the domineering partner as space –time develop amidst all the partners involved including the material giant and not so much giant. The growth confirms the security of individual singularity in the presence of other singularity united under an elected unifying centre singularity. From the motion that points to the stage where atoms dominate as they bring about motion with much less dismissing the stages come and go but the direction of development is always in the direction of centre singularity committing unifying in the extreme by dismissing space-time. In the end the star is totally committed to dismissing space-time by being absolutely unmovable and as it is so solidly stationary it places all motion outside the star into outer space. The aim of this development is to secure all singularity which was at first vested in every atom in independence to a shift towards and eventually including all the atomic singularity into one controlling singularity where the purpose of atoms in independence is taken over by one all including and controlling governing singularity in the centre of the star. All the singularity that was present before is then included in a centre spot that is not even a dot any longer but then it returned to the spot. The black hole is a star as all other stars but the Black hole completed the journey by taking all the atoms in the unit and unified all singularity into one position that replaced all atoms and secured their singularity into one spot.

At first when the star is in a duplicating prone state, the atoms in a range of elements control and produce the drive of the star. But gradually the protons in the atomic cluster fuse into larger units and the larger proton cluster units starts to challenge the drive and eventual destiny of the star by fusing together elements with much less protons in clusters. The working process is immensely more integrate and complex than the process I describe here but this is the shortest introduction I am in a position to give. In short at first in the duplicating stages the atoms take charge of the driving of the star but as progress in development soldier on the centre singularity takes command and the star dismisses space-time while it also fuse together elements in the dismissing process.

At first the singularity sustaining the atom has to group together to sustain the star unit but in time the unit becomes string enough to sustain the star without depending on atomic independence to overcome overheating. The main picture arising from the explaining is one of a balance that controls the star through out its development. In the Sun for instance which is a minuscule small star a relevancy in the outer region might be all–favouring duplication, which is favouring duplication relating to singularity and with the atom having a sustaining displacement of favouring the electron position there is no actual danger of the atom demising. The star is still all about its duplicating mission. On Earth as in outer space the atomic difference in mass between the proton and the electron is 1836 to one. The electron is displacing the same space-time but compressing the space-time displaced 1836 times less than the proton does.

The figure mentioned is a specific predetermined ratio brought about by the dismissing of space-time in conjunction with duplicating of space-time showing discrepancies between the duplicating factor and the dismissing factor. That means the "outside" is 1836 times away from producing what the "inside" within the atom core does reduce space-time. In the Sun this ratio will be much less because the relative mass of the electron will form a relevancy number of 27 where it is on Earth only 3. That is one factor I placed on the ratio whereby the density intensity of the star space-time increases from as the star atmosphere (if you will) of the star such as the Sun grows denser and gravity makes the space-time more compact. The density shifted towards the intensity the electron has and as the intensity of the density progress by becoming more compact through gravity the space between particles will match the electron at $(\Pi^2/2)(4(\Pi^2+\Pi^2)/7)=$ **55.66**. This displacement value produces gravity and it produces electricity by intensifying what is in outer space at a value of $7/10(\Pi^6)/6= $ **112.162** to what limit space-time displacement may endure while remaining three- dimensional. The number of 55.6 personifies the maximum proton displacement ability while remaining in form within the sphere. After that the sphere begins to show miss forming. In is no coincidence that that all gravity driven structures has to have a molten iron core and electricity can only be generated by an iron armature in the presence of a magnetic inducting copper chargers producing a field. In spite of the proclamation of science about the Sun being only a hydrogen star the concept is nonsense because not hydrogen n nor any other element except iron is able in our Universe to produce through motion the effect of space-time reducing which we either call electricity or gravity. Electricity and gravity is the same ting.

The electron position or in other words the favouring of the duplicating tendency within Suns ability on creating a working environment within the Sun as that will have has a diminishing factor going as low as 27. This means the atom can reduce the electron space-time occupation by 27 times whereas the atom can sustain as much as 1836 times further than the electron can and the electron matches the speed of light. As the relation in the atom within the Sun degenerated by 27, which means it loses to find compensation in the time duration extending by the ratio, which the space reduces in the atom is left with a sustaining value of the electron plus the neutron applying space-time displacement without involving any of the neutron at all.

By declining of personal atomic space in order to avoid deforming and losing individual identity which will lead to accelerating early uniting through absolute space-time demising such diminishing that will bring about the relinquishing of individual independence carried by having independent gravity securing the atomic identity which brings over the result that produce mass. That is the mass that the electron will consume in the space reducing and producing an enhancing of mass within the star. When the composition of elements within the star is such that the combined effort of all the atoms reducing space and thereby improving their individual space-time dismissing as they relinquish the same factor in duplicating space-time by motion. By moving the control of gravity from the individual gravity vested in the motion of the individual atoms towards and in control of a centre elected spot holding all invested gravity secured such increase of time by the reducing of space will produce the favouring the dismissing trend much more than a duplicating by motion provided by the compliment of electrons spinning about. When the central governing singularity takes charge of gravity within the star the star goes dark as there is then less electrons in the star and it will reduce the amount of photons flowing a way from the star because what we think of as light coming from a star is the ejecting of excessive electrons not needed by the spinning atoms and such ejecting of an overflow of electron production will have the star shining at night as a bright little boy shining by dismissing pebbles of light-photons into space. It is when the star gets dark that we can know the real monster woke up rise and dismiss in earnest. The star going dark will happen when the centre gravity will request more motion through dismissing of space-time that would light photons have the ability to escape. The proton then takes command from the electron and changes commitment within the star from duplicating in motion to dismissing by excluding all motion from the star. When a demand of space-time displacement to the value of 56.6 protons becomes the norm the star will seize having space-time concentrated to a liquid by concentration as the star by that time exclude all electron functions and stop shining. This comes into affect when the demand on space-time duplication and reducing the reduced due to mass overload the atom to space without a heat envelope. Only the nucleus will be able to sustain the further diminishing while the reducing of space is directly coupled to the increasing of time.

The atom would shrink to such little space it will have space within the star that only the centre nucleus of the atom will take up to fit. More reducing by applying motion in creating space differentiation will leave a star with so little space the space will be insufficient to secure a position for the neutrons and the star will then have the name of being a neutron star. Going even further will find the proton rejected from the star. That is how gravity applies because it is a matter of relevancies applying between space holding and demanding conditions and space reducing in relation to insufficient motion bringing about much less space duplicated. The space duplicated brings about mass as a result. We shall again return to this topic in the new suggested theory later on after much more exchanging of information and arguing about the introduced information has taken place.

In the previous explanation it becomes clear that there is two forms of gravity applying through out the cosmos and not just one form. Saying this I first have to reconfirm what Kepler proved that space could only be when space is in motion and the motion is in relevance to a specific controlling centre. The accepting and the understanding of these principles are absolutely vital to our understanding of the cosmos. This brings across the truth about the expression that one must not think of the heavens in the same terms as we think of the Earth.

Laws applying in heaven and laws controlling the Earth by nature are one Universe apart. The Earth serves life while the rest of all the heavens are hostile to life. It even seems to us that the rest of whatever fills the Universe is meant to destroy life. Except that what is on the Earth, the rest of all created has one purpose and that is the destruction of life. If life cannot find any means of supporting life in surviving in a natural state anywhere in the rest of the cosmos out there we have to adapt our thinking about nature in considering the cosmos as totally different from what we find on Earth. We have to accept what we invaded and infested on Earth, but that invasion will be the only part of the cosmos we are likely to invade and infest.

Newton saw what physics applied on the Earth and Kepler saw what physics applied in the cosmos. Kepler saw space is in maintaining space by the motion thereof. The accepting and understanding of this is absolutely vital to our understanding of cosmology. This brings the truth about the way we have to regard cosmology. What is applying on Earth is almost definitely not applying in the cosmos at large. We may never think of the heavens the way we think of the Earth. Heavenly concept stands widely apart form the Earth because the Earth came about to support life whereas the rest of all the heavens do not even know about life existing and is hostile to life. In the cosmos Kepler's gravity overshadows the gravity Newton saw.

Kepler saw space is in maintaining space by the motion thereof. In this statement there is a balance maintaining equilibrium of space specifically duplicating by motion in precise equal duplications of the previous space that is repeated by the duplicated space by precisely copying as the following bisect of the previous copy of space to perfect in precision. I once again at this point have to remind that such duplication by bisecting is within the space less surroundings of the proton. It is not in the Universe we see when looking at the night sky.

This is what the formula $a^3 = T^2k$ translates to when turning the written mathematical code to the verbally pronounceable English. There is a balance forming equilibrium on both sides of the divide by producing $T^2 = a^3 / k$ and when barriers are broken and lines are crossed the defining ratio change to $T^{-2} = k / a^3$ where the singularity distance in relation reduces by the time component going negative progressively. What this brings to light is that there is two points forming relevancy which indicate a separation of space although both is sharing in one space with both in the position of the identified space having motion that is balancing gravity by motion. In the previous explanation it becomes apparent that there is two forms of gravity applying through out the cosmos and not just one form. Saying this I first have to reconfirm what Kepler proved that space could only be when space is in motion and the motion is in relevance to a specific controlling centre. In this statement there is a balance maintaining equilibrium of space specifically duplicating by motion equal to the previous space that repeated a duplicated space by precisely copying as the following double of the previous copy of space to perfect precision.

This is what the formula $a^3 = T^2k$ translates to when turning the written mathematical code to the verbally pronounceable English. There is a balance forming equilibrium on both sides of the divide by producing $T^2 = a^3 / k$ and when barriers are broken and lines are crossed the defining ratio change to $T^{-2} = k / a^3$ where the singularity distance in relation reduces by the time component going negative progressively. What this brings to light is that there is two points forming relevancy which indicate a separation of space although both is sharing in one space with both in the position of the identified space having motion that is balancing gravity by motion.

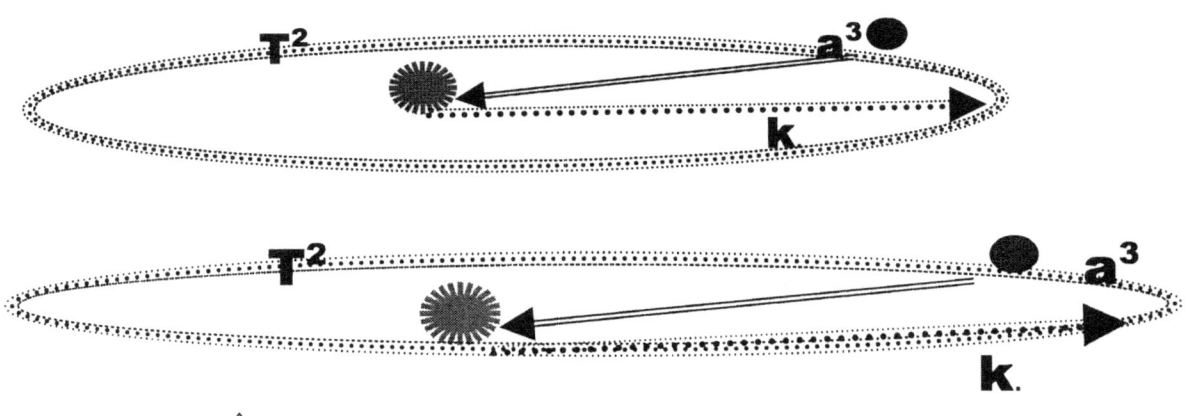

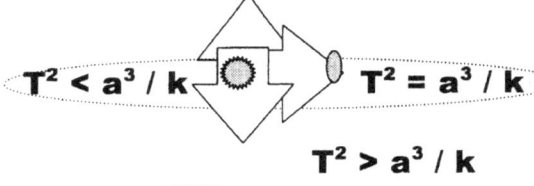

$T^2 < a^3 / k$     $T^2 = a^3 / k$

$T^2 > a^3 / k$

Let us see what gravity is in reality when two objects perform a commune with gravity applying. Gravity applies between space occupying structures in space not occupied and a centre

The space **_a_**$^3$ that is dominating the space **_a_**$^3$ from the centre

The space **_a_**$^3$ that is dominated by the space **_a_**$^3$ from the centre

The smaller space **a**$^3$ is distinctly distinguishing the larger space **a**$^3$ as the larger space **a**$^3$ is housing the smaller space **a**$^3$ where the factor **k** is as much indicating by length the larger space **a**$^3$ as much as it is indicating the end of the length of the larger space **a**$^3$ at the location of the smaller space **a**$^3$ by directly pointing at the position the smaller space **a**$^3$ holds. Where the larger space **a**$^3$ ends the smaller space **a**$^3$ is. The two remain as an inseparable single unit in double motion where the motion identifies the unit as much as distinguishing the separateness in the unity and always remain in absolute relevancy.

**a**$^3$ is space occupied by material seeking independence from the centre but that is not all because space **a**$^3$ is space holding an identifiable position all the way through the length of the line indicated as **k**. The space **a³** refers to a space **a³** within the space **a³** which all depends on where the motion draws the attention and the space **a³** will only find relevancy when motion sets whatever space **a³** one refer to apart from the other space **a³** referred to. But by identifying one both finds identification because the one is not identifiable if the other is not prominent too. The prominence comes from distinguishing both sharing a joint position.

**T²** is space in motion towards the centre of the space holding the space **a³** that is in motion, which is the space **a³** that is validating the space in motion towards the centre.

**k** Indicating the distance of the motion of space and in space in relation to a very specific centre. By indicating the point which **k** indicates **k** also indicate the space **a³** becoming the unit of all the space **a³** being in contact with the centre and being in space from within that centre from where **k** indicates space **a³** which through motion is distinctly not the dominating space **a³** that is in motion towards the centre but the space **a³** in motion that is differentiating by distinction separating the space **a³** that is dominating from the centre the space **a³** that is dominated by the space **a³** from the centre and through this motion relevancy the relevancy is holding all space **a³** connected to the centre.

If space were zero or nothing as Mainstream science so affectively teach us then Kepler's principle formula would need the changes Newton brought about. But it is true and stands tested like no other research ever coming either before or after Brahe and Kepler's work.

From the implementing of $a^3 = T^2k$ we can see that

$$
\begin{array}{llll}
\left(\begin{array}{l} k = k^{3-2} = k^1 \\ a^3 = a^{2+1} = a^3 \\ T^2 = T^{3-1=2} \end{array}\right) &
\begin{array}{l} k = a^3 / T^2 \\ k = a^{3-2}\,(T^2) \\ k = a^{3-2} = k^1 \\ k = k^{3-2} = k^1 \end{array} &
\begin{array}{l} a^3 = T^2\,k \\ a^3 = T^2\,k^1 \\ a^3 = T^{2+1}\,(k^1) \\ a^3 = a^{2+1} = a^3 \end{array} &
\begin{array}{l} T^2 = a^3 / k \\ T^2 = a^3 / k^1 \\ T^2 = a^{3-1} = T^2 \\ T^2 = T^{3-1=2} \end{array}
\end{array}
$$

**is the same as  is the same as It is all the same**

$k = k^{3-2} = k^1$ **is in direct relation to** $a^3 = a^{2+1}$ **and that is, is in direct relation to the formula** $a^3 = T^2 = T^{3-1=2}$.

**With this information staring mainstream science in the face and scream pleading at them to recognise the information they turn around and ask why can man not fly off to other galactica at the speed of light**

$$k = k^{3-2} = k^1 \longrightarrow$$

**It takes time for space to fill k in the distance. In fact, it takes the distance that k developed since the Big Bang** $k = k^{3-2} = k^1$ **to fill the distance.**

**It also takes time** $T^2 = T^{3-1=2}$ **to produce the distance forming** $k^2$

**It takes space** $a^3 = a^{2+1} = a^3$ **to form** $k^3$ **since coming from the Big Bang**

We find that manmade structures in orbit in outer space have a relative very short life and the corrosion up there destroys the material considerably in a relative short period. This is most apparent when comparing such corrosion material decay in Antarctica. In the South

Pole articles remain seemingly destruction free for centuries whereas in the desert the heat quite literally dissolves material and even more so is the case in outer space. The heat n the desert as the heat in outer space corrodes material many times more that what is the case in outer space. That means it is not merely **k** but that what forms the concentration forming **k** that also has a strong influence.

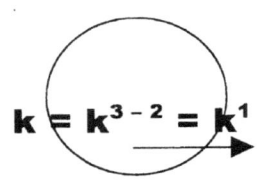

$k = k^{3-2} = k^1$

**It takes time for space to fill k in the distance. In fact it takes the distance that k developed since the Big Bang $k = k^{3-2} = k^1$ to fill the distance.**

**It also takes time $T^2 = T^{3-1=2}$ to produce the distance forming $k^2$**

**It takes space $a^3 = a^{2+1} = a^3$ to form $k^3$ since coming from the Big Bang**

When the astronaut is departing from space on Earth or filling Earth space it will take the departing astronaut $k^2$ time to reach $k^1$ and fill out $k^3$. At present and in this moment our most impressive astronautic engineers will devise an engine that would cut $k^1$ by say half. This achievement will come as they increase the power output say for argument sake to double what it is at present.

There are always two singularities in relevancy. The motion of $T^2$ seen from the centre in contraction uses the $T^2$ coming about as the **k** factor for the lesser space $a^3$ applying motion. Therefore where $T^2$ is representing motion to the larger **k** it is taking $T^2$ as the figure that represents **k** as a motion indicator to the smaller $a^3$. It means that $k = a^3 / T^2$ and $T^2$ to the smaller $a^3$ is the **k** factor of the smaller space soldering from point **T** to point **T** which then is the relocation of $a^3$ by the distance of k. $k = a^3 / T^2$ means $a^3$ was moved the distance of **k** in the time $T^2$.

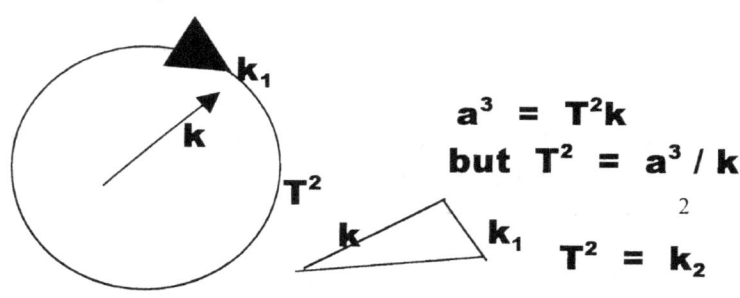

$$a^3 = T^2 k$$
$$\text{but } T^2 = a^3 / k$$
$$T^2 = k_2$$

It is conducive to remember that there is another part of the two relevancies applying where one is $a^3$ that is relevant to **k** but also there is the point where **k** has a duty to place a relation to $a^3$

When the astronaut is departing from space on Earth or filling Earth space it will take the departing astronaut $k^2$ time to reach $k^1$ and fill out $k^3$. At the point $a^3$ then serves the new k will relate as much as it has to adhere to the $T^2$ time it takes to keep $a^3$ attending the new orbit $T^2$. At present and in this moment our most impressive astronautic engineers are devising an engine that wills double $T^2$. This achievement will come as they increase the power output say for argument sake to double what it is at present. What we see happening in space with objects in motion translates equally to what happens to objects entering the Earth's atmosphere. There is a smaller projecting of space that changes **k** because of an altered **k**.

Mainstream science promotes the idea about particles coming into contact with the fuselage of spacecraft when they enter the Earth atmosphere and thus cause friction that entertain heat which then rises as a result of such friction occurring. I know well I had this argument before but I cannot underline the incorrectness more of the way that mainstream science views the principle. By acknowledging this very incorrect way they see what applies when the heat blankets the incoming aircraft will disallow in further accepting of the understanding how and what will apply in such conditions because those condition express cosmology in detail.. There is no friction between particles of the craft and of the atmosphere that is destroying the frame of the craft because it also is true that in outer space there are not enough particles there to bring about such a structural decomposing in outer space or in the atmosphere of the Earth an altitude to do it. What happens is that relevancies reapply and we know science acknowledge that material (they say some but in fact it is all particles) that material reduce the space it occupies when coming from outer space into the Earth atmosphere. In the atmosphere the reducing of space $a^3$ comes about because $T^2$ increase upsetting the ratio in $a^3 = T^2k$. When the space in the atmosphere became too small to allow the time it takes to enter because the distance $k$ decreased faster than the space $a^3$ could compromise with the time $T^2$ changing from what is present in outer space comparing that to the time in to atmospheric space. The space shrink and that push time back into the pas when the heat surrounding objects were much hotter than t is now. Time moves back as space decrease towards the time the Big Bang was present, as we know the Hubble concept suggested. With the information in hand for a period of almost a hundred years and where the information forms the basis of modern cosmology since the information formulated gravity and not merely produced a name for gravity as our English friend did it is amazing that such accidents can happen and it is more amazing that no one in Mainstream physics has not the slightest idea or inclination as to why this is taking place! To top the cake with a red-hot cherry we know that our most impressive astronautic engineers are assembling a machine that will scramble the ratio Kepler introduced to a level in outer space where the ratio will be more than what the ratio in the Sun is. Surprisingly they are not in the least surprised that not one object in outer space is using an excessive velocity.

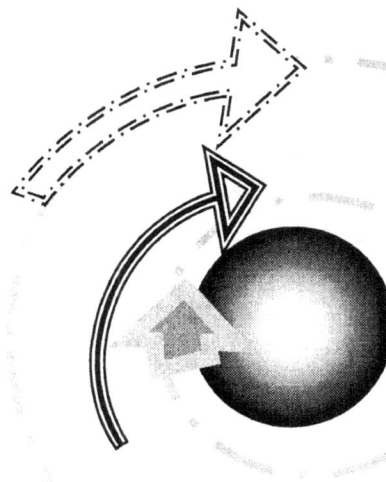

**An object can rotate in outer space as long as it can maintain a speed that will keep the object rotating in that orbit. The speed requires that the distance from the centre of the Earth to the centre of that object rotating must remain even at all times. <u>That is the gravity applying up there in outer space</u>**

When the motion decreases and a lesser motion differentiation sets in the object can rotate in the atmosphere as long as the motion will last and it can maintain a speed that will keep the object rotating in that orbit position. Nevertheless, the difference is that the speed required orbiting declines. <u>**That is still the gravity applying up there in outer space**</u>

## Gravity is the maintaining of speed in relation to other motion that is either contributing to the object in motion or the influencing of such a motion.

Travelling is about bringing space in motion. Motion is combating heat and heat brings about expansion. Expansion is producing material as a substance that accumulates by growing into more material and producing material is about duplicating through filling vacant space. To move from one point to another point the material must release from the space it filled and fill the space it is moving into by which measure it then produce material. The lesser the material is, which duplicates in the way of being an individual unit is by taking up space in a larger space unit where the space that is taking up space forms a part of the larger and containing space. The method of filling space as forming a part of the containing space being as such the unit of such a larger space unit. But in as much as filling the larger space the smaller space still hold individuality be motion pronouncing the independence as well as the inter dependence of both individual spaces in the unit. It is the time it takes to duplicate such a large unit when comparing the two individual space units in relation of time each has taken to duplicate, that the ratio between the two duplicating will prove to be of less time duration or "slower" in time duration that the larger motion will take to complete the duplication of the larger space in comparing to the time it takes to duplicate the smaller space. On the other hand the more the space is that is in the process of duplicating the longer period in time such a process will take to perform such duplicating of space per time unit. To duplicate space per measure of space size which is having more **k** that holds time further from singularity by extending the space the "slower" the larger space duplication will be because of the bigger effort it takes to duplicate more space although both uses the same time frame to duplicate. It is not that complicated because a lorry duplicates by motion as does a bicycle duplicate by motion but to get the truck up to eighty km per hour takes a hell of a lot bigger effort than to duplicate a bicycle at eighty km per hour. The motion $T^2$ requires effort to reposition space $a^3$ as space $a^3$ duplicates using time $T^2$ to shift space $a^3$ across a distance **k**. The more $a^3$ there is to extend by the increasing of **k** the more $T^2$ will be required to complete the task. The faster the duplication is the further the distance is from the centre of singularity and the longer the rotation will be in relevance. To duplicate space using the same time and duplicating a smaller space will bring about a much reduced distance from the centre using a shorter **k** in $k^{-1} = T^2 / a^3$.

We know that science do acknowledge the fact that material (they say some but in fact it is all particles) reduce when the space it occupies is re-entering the atmosphere as it is coming from outer space into the Earth's atmosphere. In the atmosphere the reducing of space $a^3$ comes about because $T^2$ increases, resetting the ratio in $a^3 = T^2 k$. When entering the space in the atmosphere the occupying volume of space becomes too small to allow the time it takes to enter because the distance **k** decreased faster than the space $a^3$ could compromise with the time $T^2$ changing to bring effective re-aligning. This still is all to do with the Roche principle as space turns to heat to align with the new limit $T^2$ applying from what is present in outer space comparing that to the time in to atmospheric space. At entering the atmosphere such entering takes the space shrink back to the size and the heat that applied at the time the Earth was released and formed an acceptable barrier with outer space. The space differentiation that started then we now call the atmosphere. By the object entering the atmosphere of the Earth that entering pushes time back into the past. It takes the time in concern, which is the atmosphere of the Earth back to the time when the heat surrounding objects were much hotter than it is now, even in outer space. It refers to a time when outer space had the space-time density we now associate with the atmosphere. By entering the Earth atmosphere it shifts time that then moves back to an era when everything was much more denser than what it is at present. By entering the space-time increases as space decrease shifting the factor **k** towards the position it had when everything was much closer to the time at the time the Big Bang was in full swing and with all characteristics being present then as we know the Hubble concept suggested. With the information in hand for a period of almost four hundred years science completely ignored it all. Where the information should form the basis of modern cosmology it shines in its absence since the time information was formulated even before gravity was named this information was available to everyone that showed the slightest interest. But the interest asked for more clarification than merely producing a name for gravity (which our English friend did). It is amazing that such accidents can happen in science and it is more amazing that no one in Mainstream physics never showed the slightest appetite to research the idea that Kepler introduced or inclination as to why this is taking place! To top the cake with a red-hot cherry we know that our most impressive astronautic engineers are assembling a machine that will scramble the ratio Kepler introduced to a level in outer space where the ratio will be more than what the ratio in the Sun is. Surprisingly they are not in the least surprised that not one object found in nature in outer space is using an excessive velocity like moving at the speed of light Oh, I know they have everything confused in the red-shift and the blue shift because again no object can travel even close to the speed of light. The Red and Blue shifts are all about lenses swapping relevancies and that I explain to a certain detail in "**A Cosmic Birth ... Dismissing Nothing**".

The concept behind travelling is about bringing space in motion. Motion is combating heat and heat brings about expansion, which is one more concept of motion. Expansion is producing material as a substance that accumulates by growing into more material and producing material is about duplicating through filling vacant space. To move from one point to another point the material must release from the space it filled and fill the space it is moving into by which measure it then produces material by means of duplication. The lesser the material is, which duplicates and by duplicating forms an individual unit, such duplication is increasing the space in duplication by reducing the space through duplication. By taking up space in a larger space unit where the space that is taking up space forms a part of the larger and containing space the relevance of producing more space by duplication which at the same time is reducing the space by halving the duplicating space changes the relevancy the smaller space shows in relation to what it holds in the larger space. The method of filling space is forming space and filling more space as a part of the containing space where the small and the large space is being as such one unit of such a larger space unit. But in as much as filling the larger space the smaller space still hold individuality by separate motion introducing a different $a^3 = T^2k$ in motion and that motion is pronouncing the independence as well as the inter dependence of both individual spaces in the unit. It is the time it takes to duplicate such a large unit when comparing the two individual space units in relation of time each has taken to duplicate, that forms the ratio between the two sharing a unit but not the duplicating in the unit.

Even in a unit the individuality one finds will prove to be of less in time duration respectively or "slower" in time duration that the larger motion will take to complete the duplication of the larger space in comparing to the time it takes to duplicate the smaller space. On the other hand the more the space there is that is in the process of duplicating the longer period in time such a process will take to perform and complete such duplicating of space per time unit. As explained the $k^2$ of the larger space becomes the $T^2$ of the smaller space and the more k the smaller space produce the more inclining it would have to reduce the $T^2$ factor of the larger space because of interdependency. The relevancy each one shows becomes a factor indicating the duplication of interdependence of the other. In order to duplicate space per measure of space size is to be having more **k** that holds time further from singularity by extending the space of the "slower" or the larger space will use time to duplication.

This will be because of the bigger effort it takes to duplicate more space although both uses the same time frame to duplicate as both holds motion in contrast or in different directions but still find equilibrium the unit that has to synchronise the time aspect in order to hold the unit together. It is not that complicated because a lorry duplicates by motion as does a bicycle duplicate by motion but to get the truck up to eighty km per hour takes a hell of a lot bigger effort than to duplicate a bicycle at eighty km per hour. The motion $T^2$ is seeing to the duplicating by motion but that requires an effort to reposition space $a^3$ as space $a^3$ duplicates using time $T^2$ to shift space $a^3$ across a distance **k**. The time is the equilibrium factor and the space is larger and smaller respectively therefore the **k** is different, but also at the same time the **k** implicates the occupying time factor $T^2$ and where each $a^3$ duplicates differently the **k** it duplicates will be longer or shorter. The more $a^3$ there is to extend by the increasing of **k** the more the $T^2$ of the occupier will be required standing in as **k** to the lesser space to complete the task. The faster the duplication is the further the distance is from the centre of singularity and the longer the rotation will be in relevance. To duplicate space using the same time and duplicating a smaller space will bring about a much reduced distance from point of duplication to the next point of duplication in relation to the centre performing the control using a shorter **k** in $k^{-1} = T^2 / a^3$. Where the two objects sharing relevancy are also using the same time the further development brought about space-time having more space in less time.

Singularity provides space-time but singularity is without space and therefore being without motion that takes up time to complete the duplication of space. Singularity starts at eternity and from eternity all space-time develop. The less the space is the faster the motion will be in duplicating the space because the smaller the space will be in need of duplication. But also the faster the motion is the closer such motion will be in relation to the centre as far as relative duplication goes because the bigger the extending is of the **k** in distance by measure of duplicating it applies and the less space it occupies from duplication to duplication.

Gravity is the strongest where space is the least and therefore the time that it takes to fill the space by motion will also be the most in time duration. A Black hole is altogether singularity and a Black hole is all about reducing more space into less space by faster motion dragging time to eventual eternity when space in singularity within the Black hole reaches infinity. The motion is so fast the motion reduces the space into infinity but also drag the time to eternity by the same measure. The time factor slows down so much that the light is unable to duplicate enough space in which time will allow to escape in the space that the light in the space has available. By only having the atom dismissing space-time, as is what happens to most stars in our Universe such reduced dismissing will lead to more reduced contraction. That means less relative motion. In the end when the Universe will draw the final curtain the final gravity will produce a speed so fast the motion will extend the time duration into eternity as it stretches the time beyond Universal limits and too achieve that it reduces the space, by collapsing all space into infinity. I refer to this action as being in the Black hole but one must remember that the Black hole is the ultimate unifying that all atoms within one certain unit can reach forming a single Unified structure. The atom's final stance is the Black hole that became a massive single atom. In our Universe however having the atomic dimensional qualities, this process of dismissing such space is found only in the atom, which achieves it by applying gravity. Somewhere down the reducing line one find the proton is reducing space into the oblivious by increasing motion to the ultimate, but that is the proton and the star is only all the proton's accumulated efforts.

By only having the atom dismissing space-time, as is what happens to most stars in our Universe such reduced dismissing will lead to more reduced contraction. That means less relative motion. The lesser relative motion will contribute to a smaller ratio in the space (not more compact but just less space used to fill) in need of duplicating. With that a shorter duration or period of time will be required to allow the duplicating to come about. The more motion that is required the more in space in the process of duplicating will come about and the further the relative duplicating will be in terms of duplication in ratio to the rest of the surrounding space. This only applies because the relative duration prolongs as the space reduce to comply with the bigger volume of space in need of duplicating.

Speeding up the motion will extend the terms of duplication produced by the motion as the space reduces to extend the time duration. In short: going faster will take longer in time because space reduces by motion duplicating more space per time unit. $a^3 = T^2k$ – this comes down to $T^2 = a^3/k$ and that means extending $k$ which brings about faster motion that will prolong the time duration as much as it reduces the space in motion in relation to the space holding the motion of the space in motion. Every time space halves, each it will take with it the same time and therefore the time doubles through the space that duplicates. The fact that the space duplicates halves the space as much as it doubles the time within the process in duplication. As the space halves each space has an individual alliance with the time therefore as the space reduce it will prolong the time when a quicker or faster motion comes about.

Motion is gravity and gravity is strongest where space is least. If an atom is being confined in a smaller reduced space the circle of the atom will have the electron circle growing smaller which will have the electron rotate around the centre core of such an atom. The time is a fixed factor set by the occupying space but with a smaller circle to complete. The same atom will use a lesser confining space allowing the atom more space to be within. The duration the electron has to complete one cycle is the same but when the atom is bigger. The electron travels faster to encircle a bigger circle as it does take to encircle the smaller circle in the same time period. The duration of the spin that the electron will take to complete a cycle will be in the same period as when the circle was smaller therefore the pace the one electron will move about will be much different from the next electron cycle of the other atom in the other lesser confinement. Duplicating space at a faster interval will mean taking space-time back in time which will increase the direction of time to a time where singularity was starting to provide space with time, that is taking space-time back to $k^0 = a^3 / T^2$ $k = 1$ but going in such a direction involves the reducing by measure of $k^{-1} = T^2 / a^3$ to the point where $T^{-2} = k / a^3$. At a point round about singularity the gravity that the space acquire will crush the space the object claims back to the size it would have had, when the Universe was condensed to round about singularity. This cannot happen because long before it happens all space will become heat and the heat will dissolve material into photons. This is the direction we, whom are captured by the Earth on the Earth are heading if not for mass forming to secure our atomic individuality firstly as an atom and then as an atom in a larger unit that is forming a group. Let's carefully look at the general use of gravity as is mostly applied between objects in consent of remaining individually separated by space and with respect honouring each other's independence.

While the smaller (planet) is in a wholesale effort to escape and secure sovereign independence there is the larger partner that is providing the centre from which the smaller object is running. The centre contracts the space it claims and from the centre the object in escaping is as much a part of space-time than the rest of the claimed space-time being the occupied and the holding part of the space unit. Both are relevant as both have a part of space in the unit forming the unit. The centre partner is providing the retraction of motion of the departing object in containing the departing. The second and centre object is retracting the space surrounding the centre object in an effort to supply the object with space the centre reduces through gravity. In relation to the centre the centre is applying motion that is reducing space and the more the space reduces the more heat surrounds the centre point where the space disappear. The space containing the heat disappears but by the space disappearing there is much heat left in the rest of the space as concentrated heat. As the space reduces towards the centre the heat level in that space rises.

The centre object is applying motion by dismissing space towards the centre as the centre applies gravity. $k = a^3/T^2$. Then there is the third factor, which is the space itself that is in motion as well as providing motion. This is $T^2 = a^3/k$. As much as the smaller object is running away from the centre, the centre is contracting all the space it claims to be space-time by diminishing the space from the centre. The centre forms a larger space $a^3$ that provides a flow of space $T^2$ which produces the time aspect that is being concentrated by establishing $k$ being the flow towards the centre $k^0$ as all the space-time moves the length of $k$ from $k$ inwards to the centre point $k^0$. From the vantage space holds it finds all space-time equal that it is moving towards the centre of the Universe. The Universe I am referring to is the pivotal position as the Sun is in the case of the solar system. This we see with light coming towards the observer locating the observer as being in and being the centre of the Universe. I explain this statement in much better detail later on because that statement defines our improper view with which we approach the cosmos. As much as the runaway is running away the centre is contracting the space-time and as far as the centre is concerned there is no special thought going to the runaway because the runaway is all part of the space-time centre but a part which is not that much successfully contracting. The centre is tidying the flow of the runaway but not containing the flow of the runaway. The contracting is successful it is fighting off overheating in a coming together and this that we see we see it to be gravity. The third factor is the space reducing as it is moving and as it is moving and reducing the space by the same margin it is increasing the heat towards the centre by gravity's ability to decrease levels within the decreasing space moving the space towards the centre. $k^{-1} = T^2/a^3$.

We have to accept that rules apply where singularity stands in regard to other singularity. Of the two one is a domineering dominator seeking control as a dominated subordinate fighting off the control by seeking independence. If no working relation is yet formed there is an ongoing fight for position between the two whereby the one will compete to destroy the other lesser developed into submission and the other will put up a relentless fight to flee and secure its independence. I was asked on occasion about my ability proving this statement. Well, we all have eyes and we all have minds so we better use it therefore we all can think about what we can see . We can see there is some dominating going hand in hand with some flight to prevent full submissiveness or a fight to destroy or achieve one of the two relevancies.

In every case it is space-time in motion flowing towards the centre holding a centre spot valid as the space-time is flowing towards the centre and that is providing the motion that is affecting the others sharing the relevancies. The difference (I suppose) between space and space-time is that space is just another meaningless human concept while space-time is having a flow or a motion of a valid substance and such flow is validating the particles or objects and the space-time holding them.   Seen from outer space that motion of a fluid substance is the factor that is bringing about the gravity.

Gravity is the relevance of motion of a smaller space putting a movement in relevancy of another moving space within the same space but acting as the larger space while sharing space as one unit with the smaller space and being in the same space. If any reader is in doubt about my statement then tell yourself in all honesty what a force is…but be honest while you explain to yourself what a force is…(the force Newton suggested is keeping the Universe glued) and then go and scientifically differentiate in mathematical detail what the difference is between the powers of a Pagan god and a force. From my personal view a force is just motion applying and that is what Kepler said gravity is. Even if you wish to maintain the silly idea about gravity being material pulling each other all over, such pulling demands motion to initiate the pulling or the tendency to apply motion when given a chance to do so.  The pulling starts and ends with motion. The answer about what gravity then is can translate directly from and in relation to the findings Kepler's work produced. In relation to the space surrounding the orbiting structure as well as the space between the centre and the orbiting structure the structure in motion is steady and motionless in concerning the motion of the space, which the centre of the largest sphere is dismissing space towards that centre. The orbiting sphere has a lesser capability to draw space towards such a centre and in that the smaller sphere applies motion in order to secure the maintaining of the lesser singularity in the effort of combating the overheating of the lesser structure.

The smaller object is applying motion by getting through space while in the larger object centre is applying motion to space-time and the space-time is providing the motion linking the relevant object to find equilibrium in motion applied.  It is only in this book that I ever refer to space because there can be no such a thing as space in cosmology.

The one object tries to put space in between the centre and the object in a specific time and the centre removes the space between the centre and the object in that same period. That is making the space there is, space-time.  That forms a circle and the size of the circle depends on the space relation with the period in time, which produces speed or velocity. There is space through which the occupying material moves. It is at a specific space volume during a specific time period. It is velocity or speed. If the space part is too little comparing to the time part then the time part will contribute more to the ratio and the object will decrease the distance between the centre of the circle holding the gravity applying spot and the object. It is then moving faster than the space is moving towards the centre and the space the object occupy will extend the distance that is between the centre and the object.

**What Newton saw, Kepler describes best.  $T^2 < k < a^3$** means that the object is falling out of the sky because the time it takes to complete a circle requires much more duplication of space within the space available to the object by the motion for the purpose of duplicating and the space available is not able to provide a large enough period of time to counteract the centres retaining the rotation by restricting such a rotation with an equal contacting.

$T^2 > k > a^3$ says that if the departing object seeking independence shows a greater motion than the distance **k** can provide therefore it will have to increase the space orbit in the time period by establishing the space increase in adding orbital space to establish more space to orbit. By increasing the space $a^3$ within the new $T^2$ such extending will force **k** to grow bigger and in so doing provide more space in which to orbit. This is mostly artificial such as one would find in the way rockets are launched but it ring true (although by the tiniest of margins) where the Universe develops by means of extending the **k** factor. It echoes that which we see in a normal fashion as the Hubble constant and that law describes such expanding. Even comets adhere to time old routes with cycles that are well established and as old as the solar system is. By launching the rocket straight up into the sky following the $7^0$ inclination that forms a sphere $T^2$ becomes $k$ and $k$ becomes many times the value of $a^3$ that finally reach outer space. It is a case of this radical increase resulting in more space and with that in the result thereof we find that when the time of the cyclic relation provided by an extended $T^2$ is to slow and a larger cycle is required because of a velocity ratio that favours the object in rotation, the space between the object and the centre will increase and so will the radius between the centre and the object increase. $T^2 < k < a^3$.

When the space $a^3$ does not have the ability to produce the required motion $T^2$ or the increase in speed that is required to accelerate the speed value and the level the Earth centre demands from such an orbiting object to remain in that orbit it will seize to provide the opportunity to the orbiting object to remain in the orbit. The slowing down of relevancy in speed hampers all further progress of extending $k$, which is enabling the satellite the opportunity the Earth, provides to allow such escaping to continue. The shortfall will come from $k$ as the length of $k$ is reduced and the deducted is in place to compensate for the short falling in the rotating motion that such an inadequate $k$ will provide. Then the formula is $k^{-1} = T^2 / a^3$ If the rotating time is smaller than the space the centre provides from the centre to the centre end the distance $k$ will reduce and provide a smaller space $a^3$ in which to rotate $T^2$ as to establish the required equilibrium needed to secure harmony in gravity $a^3 = T^2k$. This is what we Earthlings experience as gravity, but which is not gravity because it is a bi-product or half the result of the full compliment of gravity. It resulted when some balance went imbalance that crossed the limits of harmony. When the time factor is equal to the space cycle the orbiting structure is rotating within and hold its own in the company of the contracting motion. The lesser space orbiting should claim as much space as the centre is disposing to have equilibrium in space-time. In that the motion providing the escaping has to be the same as the motion providing the equal contracting motion. The motion of leaving is equal to the motion of staying with the circle. It is because of this that comets orbit the way they do and any thought of inter cosmic travel is completely ridiculous. To leave the Sun the structure that tries to leave the solar system must beat (not only meet) but beat the gravity coming from within the very infinite of the Sun's inner core where the diminishing of space-time provides the space less ness and timelessness needed for fusion between atoms. In order to leave the solar system the craft and all that the craft contains will have to fuse into one atom or dissolve into liquid heat.

By the eliminating of the motion where such elimination is coming from the centre all the space, which the smaller space is within is part of the diminishing and that includes the lesser space that is applying motion. Therefore it is the task of the smaller space to capture space and identify the captures space. When the rotation speed cannot keep up with the dismissing of space and the space dismissing is then overpowering the orbiting space, then the rebalance of gravity steps in where it will try to dismiss the smaller space in total. The orbiting structure will start its descent under such conditions and the orbiting structure will then begin "to fall". If the object is moving more rapidly than the space is depleting towards the Earth the orbiting object will "lift off'. But it is all a relevancy of speeds applying placing space in relation to time. It is $a^3 = T^2k$, just as Kepler said. Where the departing speed of the orbiting structure equals the diminishing speed of space in contraction that the centre produce, an orbit of $a^3 = T^2k$ at that point holding time will come about and gravity is in equilibrium. Then gravity in equilibrium departs just as fast as the space holding the departing object diminishes and the departing velocity is the same as the diminishing velocity

When examining the illustration seen at the bottom left the motion in the top illustration indicates as one can see that the motion does encourage the seeking of independence from a centre by a lesser independent singularity. We may take the controlling object as representing the Earth gravity that is securing the object forming the role of the satellite onto a centre that could be the Earth. The reason behind this effort of the lesser-developed independent singularity is seeking independence is because it is threatened by singularity that is the better developed and more controlling. I explain this statement much better a little further on in this book when I get into the Roche principle. From this securing and the breaking of such gravity securing comes the sound barrier and when such securing border is broken the sound barriers represent a control that becomes invalid. The reason the centre holds the most gravity is by now very well discussed and argued. But the centre is where space disappears and where the Universe goes flat because where the centre is where one will find singularity being the singularity that becomes the governing singularity, which is forming, is forming the centre of the Universe. In relation between the three factors there is the one that is in relation by applying motion within an occupying space. In the space between the two other participating factors there is the occupying space.

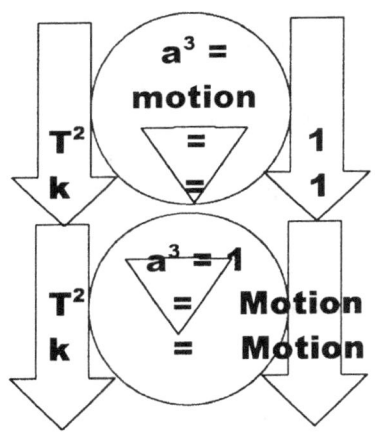

The balance that Kepler noted in his formula of $a^3 = T^2 k$ is two motions applying in relevancy to each other. These two motions must form a balance to produce the balance of space being equal to the time in which the space is.

The linear motion that the orbiting structure applies is a fight to secure independence of the lesser-developed singularity where the motion is contradicting the space surrounding the motion of the second object's centre. In the centre of the dominant structure is all the gravity domination secured just as Einstein said. The lesser singularity is seeking independence from the unit forming the containing object and therefore tries to depart from the centre.

That space then forms a relation with the space that is roving performing the duty as the surrounding space that serves as the roving occupied space. From any of the centres the whole picture seen from the position end of the line the entire space will be filled by the motion of the space contracting.

Observing such space it is apparent that the space in its role as the super container of all proportions will seem to be very motionless. We have to contemplate that the space we regard, as outer space is as big as space can get and the larger the quantity of the volume in motion gets, the more motionless it will seem. By staring at the biggest there can ever be, it will seem to us as being motionless if then only of the shear magnitude that is involved. From the centre nothing is departing but the centre is sustaining the centre by providing dismissing of space that is flowing towards the centre where in the centre the motionless and space less singularity kills off space in motion while the space flowing towards the singularity at the centre is replenishing the centre of space, which is being dismissed by the centre. It is space in motion as Kepler realised $a^3 = T^2 k$. Motion is independently coming about holding three positions independent in equilibrium.

This independence drive to secure independence is part of any structure having the potential to produce and apply heat that will bring about that the second and lesser singularity will search to bring independence to the lesser singularity. It is how the cosmos react to heat coming about as reaction of heat increase.

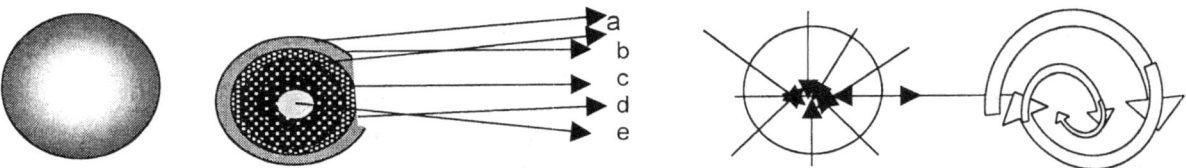

**The gravity we recognise and which we consider being the "force" influencing our very existence is the reducing of space by concentrating space to form more dense heat in different layers. The gravity we find in this position is the one sector that Kepler introduced as $T^2$**

The second form of gravity is the phenomenon, which Mainstream Science chose to name momentum but in fact is the second leg of gravity. It still holds gravity as motion but it has a directional change. This factor we normally think of as the one that Kepler introduced as k. In the centre is singularity, which at the time may seem inactive but as soon as conditions re-apply the centre can form singularity that can come alive by motion that introduces the position thereof. By electing a centre point singularity dominating can be anywhere in the cosmos at any given point selected by the motion of all the atoms in a liquid or a solid state or a gas state with the only condition is the asking of relevant motion forming a centre in the application of the Coanda principle guarding singularity as $k^0 = a^3 / T^2 k$ anywhere. This phenomenon we named after Henri Coanda. This statement I shall return too to prove little later on. The singularity stands related to other singularity and all though it forms singularity that group together electing such a centre such a relation can only exist when relevancies to other singularity comes about as space –time or to use another term which is space in motion in relation to the individual singularity grouping to select a centre that will apply to all atomic focusing on a centre gathered by all atoms concerned.

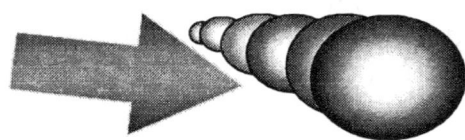

**Gravity is the maintaining of motion and motion produces speed or velocity in relation to other particles also maintaining a speed or a velocity but moreover gravity is a balance that is forming or the striving to unite singularity or aim to achieve an eventual uniting of singularity.**
**Gravity is not a tug of war and neither is it a magical force coming from nowhere. Gravity is about half circles forming lines that produce half squares in the format of Pythagoras.**

In Newton's gravity formula Newton placed the relation of the two objects in gravity in a square to each other by the moving closer of the particles. But Newton brought in mass as the principle factor the one that is responsible for achieving gravity as a contracting force and according to they're thinking, which is suggested by those most learned in science results in gravity forming. That cannot be the case because in that case there cannot be anything such as micro gravity. Micro gravity is what comes about when mass floats about . It is thought that since the floating object is out of reach of the Earths mass the question is what will then cause the pulling of mass and in what direction will such pulling be heading. So far we have seen that gravity is motion and motion enlarges or reduces mass but mass cannot produce gravity because mass receives gravity. Since we know that which is keeping the objects afloat above the Earth is motion differentiation coming g about as the two crafts harmonise but no equalise their comparing motion or speed. However we find ourselves very much unable to explain the existing of the so-called micro gravity Consider the affect of gravity in micro gravity. It is not micro gravity we meet in outer space but it is micro mass. The pulling of whatever is up there in outer space is less than down where we normally are on Earth.

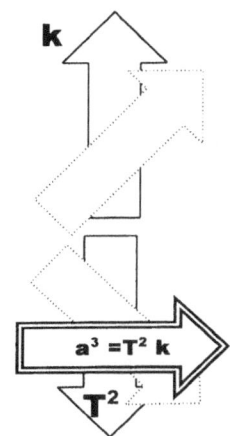

Gravity is the maintaining of motion and motion produces speed or velocity in relation to other particles also maintaining a speed or a velocity but moreover gravity is a balance that is forming or the striving to unite singularity or aim to achieve an eventual uniting of singularity. Gravity is not a tug of war and neither is it a magical force coming from nowhere. Gravity is about half circles forming lines that produce half squares in the format of Pythagoras.

Every relation formed about singularity is about surviving overheating and protecting singularity as much as preventing the loss of individual singularity by destruction. This loss of individual singularity is in the worst scenario forming the Roche limit and in the least scenario securing matter onto the earth.

If it is the mass, which is responsible for the pulling of what we think of as gravity, then the gravity can resemble a form of possibly being micro because the mass that is producing the gravity must be the guilty party of going micro. In order to produce micro gravity one must insist on micro mass to produce such micro gravity. Since this argument is childish nonsense we have to realise the gravity up there is coming about from motion where motion produces gravity. A higher motion will increase the space between the centre and the structure while a slower spin will result in the decreasing of the space between the structure and the orbiting structure. The mass remains the same but there is a specific border that an incoming object shall not cross or the crossing will change micro gravity into gravity, which produce micro mass becoming mass.  When being up there or down with us the object holds the same composition of material, which we think has as mass in the normal flow of conversation. But the mass becomes a factor not when the specific border is crossed bring the object under the control and in the command of the Earth centre singularity. If it did Galileo is wrong and if it does not Newton is wrong. All objects fall at the same rate notwithstanding what mass it has or has not. On route down to the ground there is no mass because all mass fall the same rate. Only when being on the Earth with direct or near direct contact with the Earth mass becomes a differentiating deference.

Again I stress that mass as an applying factor cannot contribute to the balance of gravity except for dragging along when being at the bottom and lying on the Earth with the motion of the Earth causing friction between the object and the Earth causing what we humans wish to distinguish as mass. Newtonians, you can go on bluff yourself as much as you wish but Galileo insisted that all objects fall at the same rate notwithstanding whatever that mass of the falling objects might be or not be because when falling mass and all differences associated with mass differences are compromised and that discounts mass as a factor in falling.

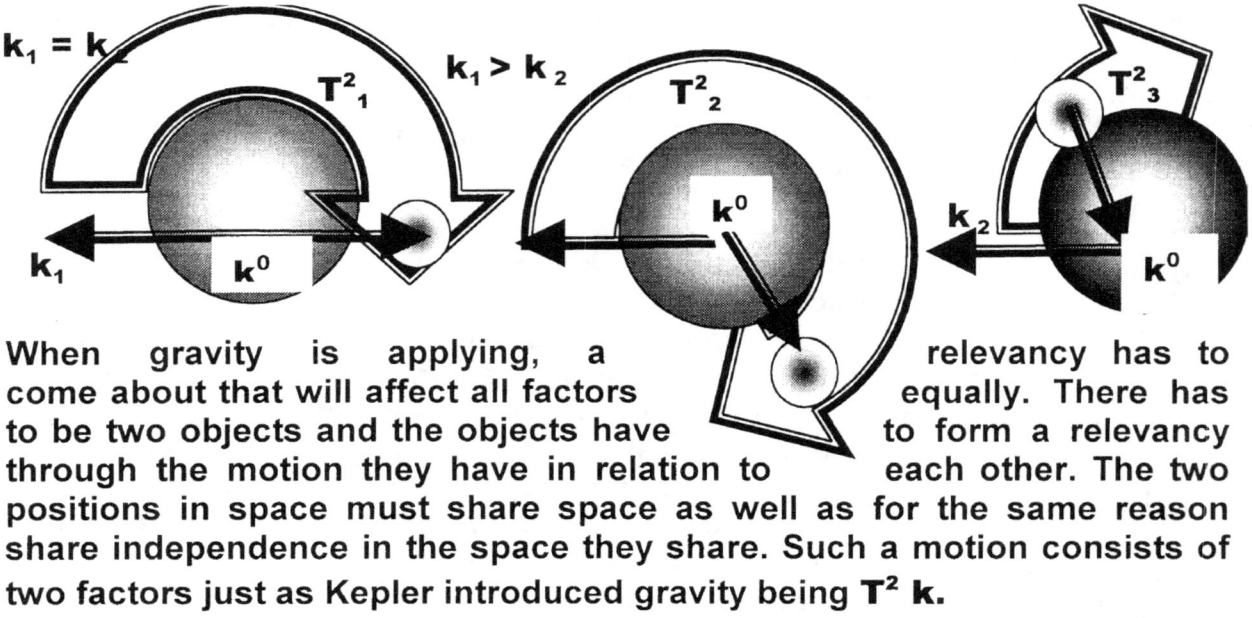

When gravity is applying, a relevancy has to come about that will affect all factors equally. There has to be two objects and the objects have to form a relevancy through the motion they have in relation to each other. The two positions in space must share space as well as for the same reason share independence in the space they share. Such a motion consists of two factors just as Kepler introduced gravity being $T^2 k$.

That once again confirms my view that mass is friction caused by the lack of motion contributing to duplication and therefore mass is truly only resistance. According to this explaining mass in contributing to gravity has no role to play as we all can clearly gather from evidence about performance of the objects in the increasing of gravity. Neither has mass any function in this process. It is motion that is creating space. It is about motion $T^2 k$ producing space $a^3$. With the increase of motion $T^2$ the factor $k$ would be affected and that will affect factor $a^3$

In Kepler's formula there is on the one side space in the cube but at the same time there is a larger measure of space on the other side of the relation. The one side holds the space implicated in the relevancy to a cube but in the case of the bigger space only a line indicate the distance through which the space will measure. The distance is indicating a space that runs from the centre where at that centre there can be no motion as space runs into the contracting end where the motion stops. The space, which is indicated by the running from $k^0$ to the point cared by $a^3$ at the end of the line $k$ can only be if such a space is in motion. Kepler said it...Kepler said space couldn't be if it is not being in motion. In that space presented by $k$ the space $a^3$ hold direct relevance to the time $T^2$ that the motion takes. That means every particle no matter how minute it may be but if the particle complies to the independence it has in protecting independent individual singularity it contributes with having gravity by the motion of $k^0 = a^3 / k T^2$. They have space-time because that is space-time. That is the space located between $k^0$ and $k$ distinctly pronounced by the applicable $a^3$ to the tune of $T^2$. Having those principles has what fills the space qualifying as having space-time.

There at the end is a space $a^3$ identifying the space that is covering the area running from a specific given centre hat is covered by the length of **k.** There are within one space a relevancy created by motion of space created by two factors sharing a space. The one shows a negative tendency and performs in an effort in moving away from a centre. Then there is a line that is connecting arch to the centre, which is securing the object by containing the first objects effort to bring about individuality. This produces two points and the $T^2$ factor accommodates the square relevancy that comes about through motion applying. If the one **k** comes to soon following the second **k** the $T^2$ factor would be too small to accommodate the two $a^3$ sharing the spot by motional duplication and the object will fall to the centre because of **k** reducing. When the two points forming **k** is too large the $T^2$ will force the object into a larger orbit. If the motion in $T^2$ is not enough to provide **k** the required distance **k** will reduce the space holding the moving of $a^3$ in place. The circle $T^2$ in which the space $a^3$ moves will then reduce in size by placing the distance indicator **k** into a negative state or a declining measure.

That is the gravity we experience but that is a small part of gravity only brought about by motion discrepancy. The object applying true cosmic gravity does not show any tendency toward mass or indicated that mass will affect the falling of the object and with the falling of the object all sizes will fall equally if all sizes and masses in that spot has the same velocity discrepancies effecting all falling objects equally. It proves Galileo correct but it also proves Newton incorrect. We see this with so many satellites and even space stations that plummet towards the Earth. The object did not get more massive and did not through adding mass to either the orbiting object or the Earth began its descent as it fall to the Earth. Neither did the object become less massive and flew away from the Earth. When the speed of the object goes into imbalance such diverting of the balance occur. The relevancy of the speed balance between the motion of the Earth and the motion of the rotating object changed to accommodate both $a^3$ in the $T^2$ that **k** would allow. That is gravity. That is what Kepler said gravity is when he said $a^3 = T^2\,k$. What no one ever took notice of is that gravity acts precisely in the manner Kepler stated. If the motion increase the space increases and if the motion decreases the space decreases as gravity applying is motion in space forming space in motion. The more the motion is the more space is produced and the more space is affected by the increase or decrease of the motion of space. Newton's $4\pi^2 a^3 / T^2 = G\,(m + m_p)$ has no part to play and it is only Kepler's $a^3 / T^2 = k$ that comes into the equation since **k** is $G\,(m + m_p)$ in any case.

When **k** increase all factors has to increase to compromise for the extending of **k.** If **k** extends space has to reduce because **k** is in direct but inverse proportionate relation. In this following argument we find two opposing forms of space where each plays its part in order to maintain a compromise done by both with mutual respect. The one we consider as the lesser trying to escape from the domineering, which is more developed and tries to contain the escapee. With **k** extending the lesser escaping space $a^3$ remains just as big as it is but forms a smaller part in the bigger space $a^3$ being the one in retention of the escaping space $a^3$. By the bigger space having a bigger area in retention the smaller space is confined to a bigger space while remaining the same space and therefore in relevance by application is reduced in the whole relevancy where it now has a smaller part in the overall enlarged larger space $a^3$. But in the time aspect the completing of one cycle by the smaller space $a^3$ within the improved bigger space $a^3$ that is much bigger and is holding the much longer outer circle of the larger and more of the containing space $a^3$.

The time it takes to circle about a longer space rim will bring about the circling around a bigger space in total using the same time that is taking longer in duration. The roving space can claim more space that will then fall into the space to be concentrated from the centre by the centre as more motion applying to the independent captured roving space will introduce that increase of space into the accumulated space shared by the factors. By introducing more space into the equation it provides a new balance that will suit all the factors in achieving the maintaining balance required. The time component will travel a wider space using the same time component but stretching the duration therefore increasing the time the time used per space unit gained as space holding becomes more but the length of each unit becomes shorter than previously. As the circle increase the time will be adversely affected in duration of space-time. What this implies is that one cannot have space-time and where space increases have time that is not affected by the change in the space.

We gave this forming of separate gravity coming about by means of performing individual motion a name being the Coanda effect to mention one amongst many others. The Coanda effect depends on singularity being a circle and motion establishing an independent singularity. Then the singularity cuts the Kepler formula in two parts. Evidence about this has been with us since the time of the great Leonardo da Vinci whom was the first person to see the potential manipulation of space-time by changing singularity direction by motion.

A low-tech mechanical human device might teach us something about the most basic rules about gravity if we pay attention to the rules as they apply. When a bicycle is motionless and free from support that keeps it erect it will fall down going straight downward towards the Earths centre of gravity. It tips over on a side as it falls onto one or the other side. As soon as independent motion other than that of the Earth comes about in a controlled manner the controlled motion alters the gravity as the motion brings about a balance that establish another form of gravity and is in a way redirecting or channelling the motion from downward spiralling to side ways moving. If the bicycle comes into controlled motion it will redirect the gravity controlling the bicycle. However we should never forget that the bicycle as well as the way the bicycle acts. The action performed by the bicycle is as artificial to the cosmos as life is artificial to the cosmos. It may be a coincidence that two bicycle builders were the first flyers in the air but it just might not be that big a coincidence. The bicycle represents the first phase required to fly, which is the part just before the part where the object must get off the ground. That is what the bicycle is doping: it is firstly getting the balance of gravity off the ground. The stability gained from motion is much more than what we humans read into it and it has even less to do with human skills

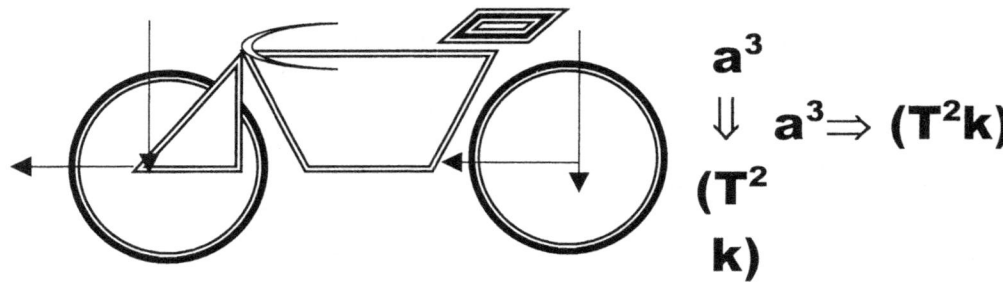

$$a^3$$
$$\Downarrow \quad a^3 \Rightarrow (T^2 k)$$
$$(T^2$$
$$k)$$

A person that acquired the skills of peddling a bicycle has achieved the method of rearranging gravity within singularity. Without motion the bicycle falls on the spot it holds. When the bicycle is put in motion the bicycle can maintain the upright stance as long as the motion applies. When the motion stops the bicycle drops. To introduce motion to the bicycle the motion brings about a stable unsupported upright stance where balance can result from the motion the Earth enforces to the balance coming about by the bicycle using independence gained from motion of the space holding the bicycle because the gravity effecting the redirecting of the Earth gravity response comes about as the result of additional motion that is introduced to the bicycle, This is the very same process that the aircraft need to get air born because it replaces or repositions the singularity the Earth holds to the singularity the bicycle develop in motion. The aircraft only takes the change in direction of what the gravity is insisting on through changing direction in motion through phase one and into phase two. It all is still part of the Coanda effect. With more motion contributing to acceleration the bicycle will become airborne on condition that it is also given the advantage of a set of wings to increase the effect of creating space-time to the advantage of the motion requiring the change in singularity direction.

I specifically chose to use a bicycle in my explaining because the bicycle is the object that relies the most on singularity achieving the required balance in which to operate. It is singularity in balance that keeps the bicycle up right while it is singularity that allows the bicycle to move or duplicate the material that forms the composite we know as the bicycle by relocation the combination of atoms through time. It is an act of balancing singularity that gets the bicycle as a machine working properly. It is also the next best thing to illustrate how singularity by motion provides gravity in addition to that which the Earth already produces. There is no possibility to ride a bicycle by any stretch of the imagination in outer space and it is the close connection to singularity control that is behind the effort.

**This is where motion through duplicating changes the dimension in equation**

**In normally applying gravity, we find contracting lines running vertically as the lines connect with the Earth centre. This line form $7^0$ with the centre of the Earth it**

**The motion of the bicycle not only extends the vertical connecting lines and not only changes the direction of the vertical connecting lines, but does both. The value added and the change in direction contributed is what brings about flying and moreover is the cause of the sound barrier.**

**When the bicycle is motionless, the bicycle is part of the Earth by gravity applied.**

**As soon as life steps in and brings about separate and artificial motion but still uses the support of the motion that the Earth provides it will inevitably do better than the Earth as long as the motion that life provides is not in conflict with the motion the Earth provides the bicycle becomes an object with the ability to transform the direction of the Earths domineering motion by redirecting gravity there in find the ability to change the direction of gravity.**

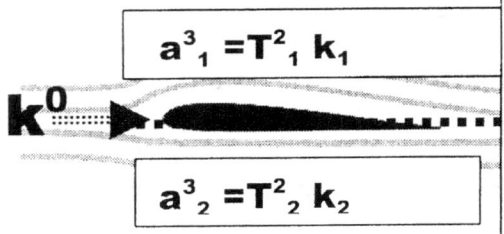

$$a^3_1 = T^2_1 k_1$$

$$a^3_2 = T^2_2 k_2$$

By moving linear the wing becomes the k in Kepler's formula and the earth is $T^2$. That is the reason why the aircraft can fly. By creating an uneven $k^0 = a^3$ on top of the wing compared to the $k^0 = a^3$ below the wing can lift or descend. It is a balance going unbalanced.

The wing is representing that which is holding singularity and while maintaining singularity the singularity in maintaining puts the object in a Universe apart from other Universes. The aircraft becomes a separate identity maintaining a singularity and as all singularity does, such a singularity insist on two factors. The one factor is duplicating by motion the production of space-time while the other is the dismissing the space-time by controlling of space-time. The contact with space-time allows the motion to present the wing (and the aircraft) as being much bigger than in reality because it is not only the size of the space-time that maintains the singularity but it is also the contact or relevance which holds the dimensional area of control which stands as a controlling factor. These factors combined make the aircraft become bigger as the capability of motion suddenly allows the relevant size of the aircraft to grow in stature.

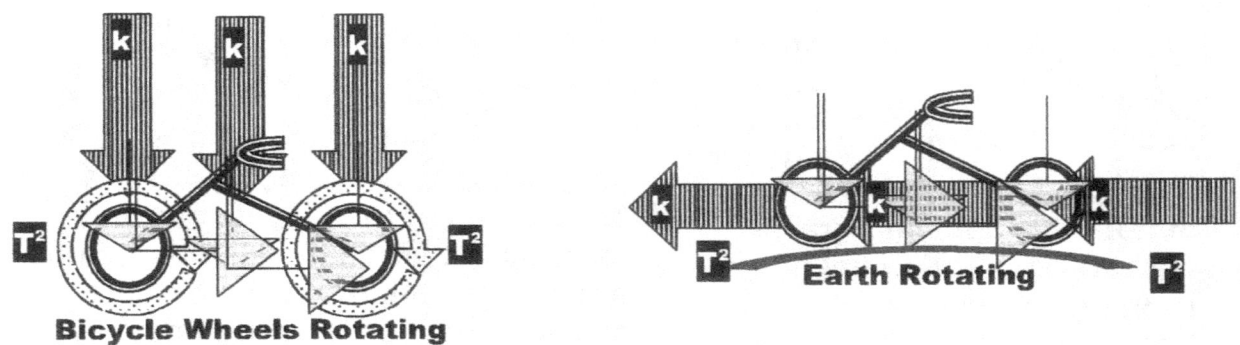

**Bicycle Wheels Rotating**

**Earth Rotating**

The contact that the air or space has on one side which is more than the contact that the wing has on the other side where airflow is restricted, makes the side having more airflow larger by motion than the other side has in size by restricting air motion. The motion enlarges or restriction reduces the contact and therefore the size per time unit. This alters the size as much as it redefines the balance and that makes the craft fly. By making contact to more air makes the wing on the one side bigger than the other side. In that the one part is flying faster than the other part being restricted.

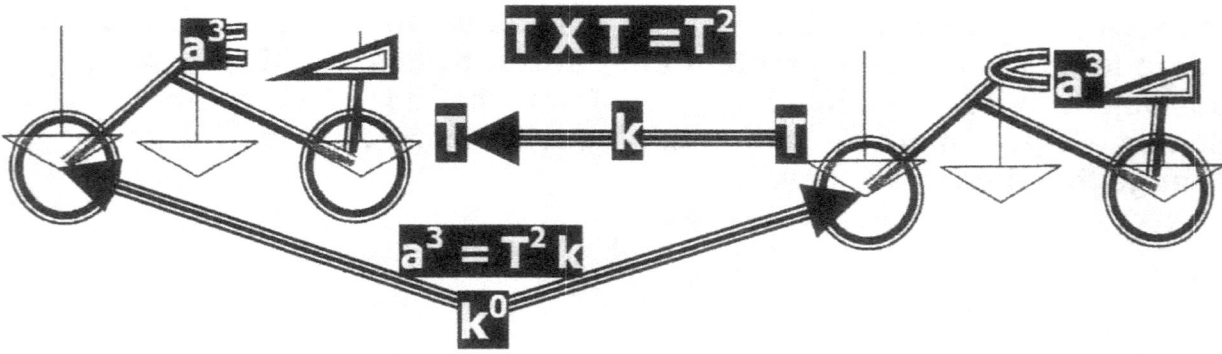

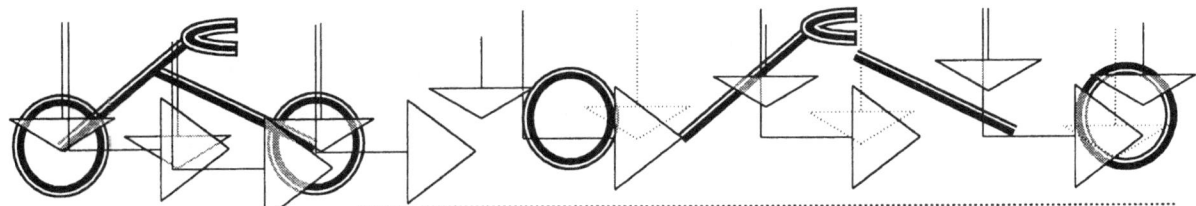

In normally applying gravity, we find contracting lines running vertically as the lines connect with the Earth centre. Motion provides extending of the $7^0$ establishing the centre connecting points to the Earth to which it connects.

The Earth takes the position that was before the motion of the independent object came about previously and held by the sun. By establishing the directional motion in accordance with the $k^0$, which the Earth then provides instead of as previously provided by the sun, relevancies replace previous ones.

When the motion of the bicycle accelerates such points forming, connections extend to match to motion. The motion then contributes by increasing the space factor to keep the commitment with gravity valid. The bicycle breaks its form but because it is structurally bonded. Other aspects concerning gravity have to commit to the breaking of space. When expressed extremely crudely it is put as follows but is very bluntly stated. Yet, it still is the best way to explain the basics of the sound barrier. The bicycle is the compiling of the independent space within the atmosphere space that is holding the motion in duplication where the motion is continuing from a facet going to the next facet by duplication. While the bicycle is in the space, in motion that is part of the space holding all aspects within by the atmosphere and the atmosphere is holding all in it together in the atmosphere of the Earth. Because the conflict the gravity experiences by having motion within motion, gravity first tries to break the object that is in independent motion. As the motion continuous, as motion extends the reflex to the situation is then to contain by the breaking to the connecting devices such as the sound waves in the adjoining space. The atmosphere does the breaking of the relevant **k** on behalf of the object in motion since the moving space holding the object in motion as a unit shows much stronger bonding in structure unifying. We experience such breaking of space as the breaking of sound, which is showing motion or gravity differentiation.

When the bicycle is motionless the bicycle is part of the Earth by gravity applied. As soon as life steps in and brings about separate and artificial motion but that still uses the support of the motion that the Earth provides, it will inevitably do better than the Earth as long as the motion that life provides is not in conflict with the motion the Earth provides. The bicycle becomes an object with the ability to transform the direction of the Earth's domineering motion by redirecting gravity. This redirection is not an extension of gravity but gravity continuing and in that continuance we find the ability, which allows the changing of the flow direction of gravity.

> With the motion of the bicycle increasing the bicycle is duplicating more than what the bicycle did when the bicycle only depended on the Earth duplication to allow motion-duplication. Since the bicycle is propelled at a faster pace than it was when it only and totally relied on the duplication of the Earth the relevant space the bicycle now holds in relation to what it did when it was not in individual motion increases because of duplicating the space and it does so considerably as the square of time insists the cube of space has to comply with such an duplicating space increase.

In the normal relation that the bicycle and the Earth has when the bicycle is not in motion or with the bicycle (as we might think) being motionless the space the bicycle occupies increased in relevancy to what the Earth holds. With the extra duplicating of the bicycle being in motion the ratio of space held in time in comparing that to the Earth space holds and with the Earth not increasing space by motion increase it brings about that the bicycle is larger than it was before it started it's individual

It is again Kepler's $a^3 = T^2k$ that changes the gravity relations

The bicycle now physically got larger because it is duplicating space per time unit more than it was duplicating when it only depended on motion coming from the Earth. In relation the space the bicycle holds grew more, or the space that surround the bicycle, which is actually time became smaller. Since the bicycle holds more space the bicycle should claim more time. But since the Bicycle does not even present a morsel of time or atmospheric space the Earth will dominate and even to a point where the earth will liquefy the bicycle. So depending on how you look at it, the space the bicycle claims got more while the space that holds the bicycle got less. But since the space that holds the bicycle and through which the bicycle may move is time and that time became less the space or time moves back closer to what was applying during the Big Bang and therefore it gets hotter. The space holding the time in which the bicycle duplicate became less because the space by which the bicycle duplicates became more.

It is not only the space the material claims but it is also the space the electron maintains as space that maintains the electron that is part of the issue. The more space the movement requires the more space will the electron hold in preparation for the movement. It is the space surrounding the material that is in motion as much as the material itself. That part forming the wind drag is a time concern.

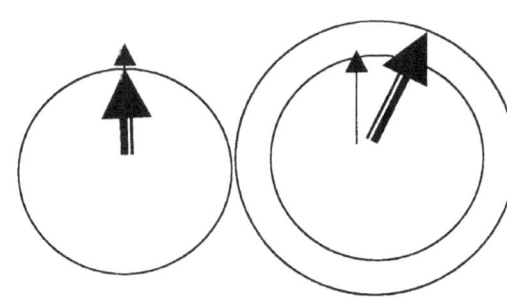

**The independent object serves as the outer relevancy while the contact the Earth has with its centre forms the contact with the containing singularity. When the independent object starts moving the atmosphere allows flexibility, however this flexibility is only limited to a point. As the speed increase the limits on the space that retains the object in motion starts to stretch and such stretching has definite limits.**

The line of the Earths containing gravity is redirected from going vertically downwards to horizontally sideways, which then becomes separate from the vertical line running to the centre of the Earth. The bicycle holds a change in gravity flow because of motion interfering with the duplicating of the lines running along the gravity line. The line running according to gravity is in a conflict with the extension that the bicycle motion brings in as the line of the bicycle that implicates the gravity extend in the direction coming about from the introduced motion changing the bicycle gravity line from one vertical running line to a horizontal running line and that changes the line of gravity as it amplifies the line of motion.

The line indicating the change of direction of the motion of the space holding the bicycle that is bringing about such motion will then being implicated by the Earths gravity completely redirect the direction of motion into a square in an opposing direction. From going straight down by the bicycles lack in motion when standing still these gravity lines introduce new directions that the line in gravity flowing holds. That then changes from coming straight down to going straight ahead in a horizontal direction forming a new link in relation to the normal straight down vertical and the one committed by independent motion. The relevancies about gravity changed. Every factor receives a new aspect and complete change comes about. By moving space $a^3$ which the bicycle holds within space $a^3$ which the Earth holds $k$ changes to take the place that $T^2$ held as a new $k$ connects $k$ not the Sun directly anymore but it now connects to the Earth centre on this occasion while the bicycle is in motion in the manner it previously connected the Earth that is performing in the motion in line with the Sun centre.

The Earth takes the position previously held by the Sun by establishing the directional motion in accordance with the $k^0$, which the Earth then provides instead of as previously provided by the Sun and while providing $k$ it brings about the space moving When establishing independence that the bicycle holds in space $a^3$ by providing motion other and above that motion of what the Earth provides in its relation to the Sun the object then takes on a directional change in motion. The change in direction also implicate other relevancies as the new motion removes the Suns direct contribution as a direct factor to the role of performing as a secondary factor leaving the pivotal contribution which is the major contributing factor then to fall to the Earth in performing from then on until motion of the bicycle stops again as the major singularity centre and gives the pivotal control over to the Earth.

With the Earth established as the domineering centre controlling the pivotal motion of the bicycle the Earth now presumes in the position the Sun had before the bicycle created independent motion and the bicycle now applying motion that produces independence fall into the role which the Earth held on behalf of the Earth and the stationary bicycle when all considered the Earth and the bicycle was one unit. It still is a unit but holds independence in the unit and even more so than before. Before the motion contributed to independence the bicycle still had independence though that independence came from atomic motion that separated the structure of the bicycle composition from that of the Earth. With the individual motion coming about such motion secures the independence to a next level that will be one step away from advancing to semi self control independence when it starts flying.

The bicycle motion will eventually end as $T^2$ but the relevancy where $T^2$ now will end is the drastic change that came about. Because the direct control shifter from bringing the Sun in as the pivoting factor to the Earth holding the pivoting factor the gravity that is upholding the cosmic equality sense is still upholding the cosmic sense as it still applies since individual space providing motion brings on changes in the factors but not the factors implicating the cosmic law while the relevancies that is required to reposition the structure factors is still in place. The Earth now forms the centre whereas before when the bicycle was motionless this duty fell upon the Sun centre. Circling about a fixed and secure centre now bringing about motion that takes the position the Earth has before the bicycle. This is a shift of one position in the gravity relevancies of $a^3 = T^2k$.

This too is the prerequisite to flying by first establishing individual space through establishing separate motion in relation tot hat of the Earth while it is creating motion because the motion establishes antigravity. While the aircraft is motionless on the Earth the Earth motion create gravity and that motion also applies to the craft structure albeit that the aircraft seems motionless to us. The mass of the Earth establishes motion of space spiralling down in time.

With the space being in motion the object resist surrendering the form its construction has and with that refuse the accepting of the joining with the Earth by uniting with the solid structure. This resisting action is delivering that what we believe is mass. It is where $k^0 = a^3/(T^2k)$ unconditionally. My argument about gravity being motion becomes prominent when we consider the motion versus space occupied by airplanes using wings to fly. Without motion the aircraft and all that it carries have mass or weight. As soon as the motion overcome the space restriction the Earth enforces on the airplane the airplane loses mass and become airborne. It is only when applying the illogic science use to explain mass where there is no sign of mass present or that mass may contribute to the forming of gravity in any way possible.

By trying to correct the incorrect holds mass and weight, as a differentiating factor but that does not make sense in any case. You can try and argue till your tongue feels numb but in arguing about the matter no one will ever be truly convinced about that one can believe as being totally convinced about mass still being a factor in outer space. The protons form the centre of the atom and by spinning 1836 times more than the electron the protons dismiss 1836 times the space that the electron does. By having so many more protons in any unit dismissing space a total effort will contribute to displacing more space-time than the effort of a lesser number of protons can achieve. There is a way to detect gravity and that is in looking how we fight gravity or eliminate the effect of gravity securing objects on the ground. We know that $a^3 = (T^2k)$ locks objects in position and flying eliminates our being locked onto the ground. Let's us look at why we fly.

Not being able to match the retainer, the object has to reduce the distance from the centre to secure a more relevant position with a suitable distance to relate to, with the duplication of space within the retaining space. This forces the object to relocate to another point where the retainer will match the volumetric space claimed by the retained object and where the containing space is by volumetric size matching the claimed space. As the object moves down to find the correct position in density level as to relocate, the object is restricted in space duplication to match that of a solid structure we call Earth or soil. Gravity is all about density and motion, where mass has no role to play.

The Earth applies motion in the atmosphere of rotation but from the human position we have on Earth we have to follow the Earth by mimicking motion the motion of the Earth. From our stance we are moving in a continuing straight line and the straight line we receive from the Earth is giving us the impression that we are allowing a straight line there is that we can see. Before investigating the principles of flying we first once again must define why the mass is keeping the object secured on the soil. The space the object claims within the space the larger object retains is not enough to duplicate using the time it holds in relation to the volumetric space retained.

The solid structure provides a restricting boundary through which the relocating or repositioning object cannot break. The effort of the relocating the object in missing the required frequency has to follow such a search as to find the position of relocation that will match in relation where the specific duplication will match the required density to duplicate the space in the prescribed harmony to the restriction of the retaining object. Because the retaining space is obligated to reproduce a much larger extending **k** from singularity that is required of the smaller space in the same situation, therefore the smaller space is in search of a position all along the line that produces the larger space in an effort to locate a comparing location to match equal duplication resulting in matching factors of space and distance in the same time experienced between the large and small spaces sharing one unit. Such a position will allow the smaller space to have a comparing spot in order to comply with equal duplication when using equal time duration as the time will allow the space then to commandeer a position in the **k** that will serve both sharing the space in the unit. By not being able to find the equal spot in the **k** that will match the space to the duration that the Earth insist on this will force the space to tend to locate while pushing against the Earth and because the soil is solid the soil; will not allow unrestricted entry for the space to go in search of identifying the prefect spot of choice. This relocating and searching for an equivalent position that will match the **k** equal to fit the smaller space in comparing to a point matching the needs of both the **k** factors where the smaller space will fit into the duplication that both objects will have in space in time providing the space that will provide the mass as a factor of control. It is a difference in speed bringing about mass and not mass bringing about the restriction. In accordance with our view we receive because of our position that we hold on Earth our motion is a straight line although the Earth shows a curve but in our minds that curve concerns us little people less. The Earth has motion and by all standards that life apply such motion can never exceed the motion the Earth displays. Our motion serves as an addition to the motion that the Earth provides but we cannot substitute or out perform what the Earth achieve. While flying and duplicating the concept serving mass does not enter any equation at this point. Then later on man found the magic he always were in search of real power in the form of converting heat to space and with this remarkable achievement man found means to break free from the establishment the Earth enforces. Man had more power available to use than just what he could harvest from human and animal muscle power.

There is very little difference between the bicycle in motion and the flying aircraft except for the motion intensity and by which the duplication of space contact will allow the aircraft to fly and leaving the other object which is the bicycle with much less space contact being without wings unable to fly. The only difference is that the aircraft produces a higher relation with space in possible motion that is contributing to space contact and space duplicating by employing a greater surface to service. In each case the aircraft wings in space contact is providing more space between the two in relevancy where the Earth forms one the object in motion being the space craft or the bicycle. Of the two in motion the aircraft service much more space than does the bicycle. By providing a bigger motion $T^2$ time factor will extend further providing the space $a^3$ more opportunity to duplicate then, which the bicycle provides. That will increase the $k$ factoring the case of the aircraft because the more motion will fill more space $a^3$ and with more space $a^3$ filling the totality of space $a^3$ will be consisting of the larger space $a^3$ to entertain more space-time that is in occupation than that of the smaller space $a^3$ has space-time to serve this relation as steady as it seems will prove to gain a bigger relation without with out providing more space –time contact by the wings the motion in itself is adding to space in motion without any additional space added in real terms in neither of the two cases only because the motion and the improved wing capacity leading on a larger contact of space $a^3$ with more of the same space $a^3$ notwithstanding the fact that the quantity of space $a^3$ in either case remained the same. This means the other factors in the Kepler formula being in relevancies of $T^2$ and $k$ will also have to increase to comply with the balance ratio. On the other hand when the motion goes into the extreme and $k$ proves to increase considerably the space $a^3$ as a factor would have to compromise visibly by reducing the space $a^3$ it claims. When the contact distances increase the other factors has to compensate to produce the sustained equilibrium. This comes about as a result of the wings that produce more space duplication received through increased space contact with more space because of the velocity increasing and with more motion as well as a bigger contact area more space becomes involved, which is promoted by the speed factor by discriminating in favour of the flying device to get the aircraft flying. The space ratio by duplication increases in ratio to the dismissing ratio by the combining effort of all protons within all atoms that is forming the flying machine and the machines cargo. However the fact that motion is in place as a cosmic event is totally artificial by cosmic standards of motion. In both examples the motion is artificially applied as a result of life's' extending life influencing. We must realise that the motion we find in the action of the flying machine is not cosmic driven although it is an interpretation of a normal cosmic occurrence that will take place but under much different circumstances than the manner life enforces the motion at the time the action that is brought about as a result of life's obtaining the manipulating abilities to translate to human achievement and as far as life is finding a means to manipulate a cosmic law to increase the benefit of life accept that the action is totally artificial by cosmic standards. There can be no such natural motion where a rock starts flying because it has received from some U.F.O. out side source a set of perfect fitting wings. Judge responsively what belongs to the cosmos without life and what life can reproduce in spite of the unnatural state of such duplication might be. It is most important to realise before classifying and grouping this normal physics action inspired by the intervention of life that there can be no such motion coming about on a planet without the presence of life bringing about artificial motion. One can inspect the moon all you like with the best telescope available to man but one will not see any flying object zig zaging the moon surface. No bicycle can by own initiative come into an upright position and start moving on two wheels. The motion is not cosmic inspired and only by seeing the difference can one have the mindset to venture into the activities applying within stars. Let's see what is artificial about life in relation to how the cosmos relate to life. Mainstream science hold the opinion that life in the cosmos comes at a dozen a penny with change repaid. What a lot of crap his idea is and I mean dirty crap. Life is alien to the cosmos while the cosmos is fiercely hostile to life being an alien in the Universe. Even on our planet life has to obey certain and very specific conditions in some cases otherwise the Earth as friendly and nursing as it is will bring about life's demise. One should try and live a thousand meters below sea level or ten thousand meters above sea level and watch your personal demise comes to you.

Let's venture slightly away from the cosmos by trying to define the role of life as the only force found in the cosmos.  Life according to my personal defining is absolute managed heat within specifically designed cosmic fibre having the ability to apply forces of a wide variety giving life power or a valid force to manipulate space-time by manipulating some motion or rules applying on motion thereof. When the body holding life is not hot or with very low intensity heat it is not with life. If the body has no motion of any sorts it is not with life. If life lost the ability to manipulate gravity in the form of low electricity life has lost living. Life can create motion by manipulating space-time it occupies or which it can control or manage. The only place in the Universe known to man, who is not absolutely in all respects completely hostile to life, is this blue dot we waste for gaining money and profits. We should never confuse life's ability to accomplish with that we associate with cosmic events because an apple falling from a tree is life's manipulating motion because it needed the intervention of life to get into a position to fall from the position it took being in the tree and that is not a cosmic event. If the apple came from the outer space it would have been fried charcoal before it reached the Earth and that result is a cosmic event. That part is the part everyone in science including Newton and Newtonian disciples ignores or chooses to ignore.

We by which my referring includes most forms of life that has the ability to stand independent on Earth and from the Earth stand on Earth above the very top layer of soil holding our space in the space of the Earth. We cannot have independent excluded space if we do not fill the space we have on Earth. While being on Earth my position is $a^3 = T^2 k$ where $k$ is because of the mass in movement standing in for $k^0$ by being $k^{-1}$ Since my body duplicating is less than that which the Earth has to duplicate but is confined to the time on Earth all the same such a body will forcedly find that the distance from the Earth controlling singularity such a distance is corrected constantly as to fit into and apply with the standards that the Earth standards insist on.  However with me calling the Earth space-time a force whereas it is the normal gravity flow the force comes from counteracting the flow of gravity. Being $k^{-1}$ we are also $T^2 / a^3$ which is reducing us in the space we hold and that is only our mass that comes into affect as the Earth repeatedly insist that we try to reduce $a^3$ further to comply with the $T^2$ the Earth is applying and which we have to use without any further options given by the Earth.  If we wish to confirm our independence of the space we have within the space of the Earth which contains the space we hold by moving through the larger then we have produce a larger $k$ factor by extending the normal $k$ the normal $k$ factor we receive from the Earth to the order of at least $k^1$ to find the ability to move from $k^1_1$ to $k^1_2$ which will allow us to enforce our own gravitational force in spite of the Earth's much stronger natural flow of space-time because we use $T^2$ to move from $k^1_1$ to $k^1_2$. So we have to improve both our independent position $T^2$ as well as $k$ to accomplish motion. But that puts Kepler's formula in question. Using $a^3 = T^2 k$ and producing a larger $T^2 k$ it means $a^3$ must also improve. That it does by doubling the space it use during the motion. The space $a^3$ becomes the next space $a^3$ because the motion $T^2 k$ is providing the way that will bring about the matching duplication the motion contributes.  This is not that uncommon physics.

A car holds the space $a^3$ and is moving by $T^2$ through the distance of **k**. When the car speeds up to a higher velocity the gravity will increase on the part of the car because the distance **k** will increase. With the mass or space in motion that remains is not able to remain even with no increase to the actual material used to move nevertheless the potential mass is increasing by the square of time where that is the gravity or the time by the square that increases. The increase in space is the producing of more of the same space by duplicating the same space more in the same virtual time. It is the motion providing the material a duplicate value of its mass (duplicating because of the square used by time) that then forms the increase in the material mass, which our human instincts of sensationalising prefer to call the momentum of the object. But that motion is that what gravity is. The way gravity is applying is acting in the same manner everywhere but man has subdivided the concept under so many names given to misrepresent each fragment of the entire concept unity we divided that we cannot even find the basic principle any more.

Gravity is not a force as Newton suggested but a motion that is formed by a natural flow of space-time between space occupied and space waiting to be filled and when filling it's forming a relevancy and this applies throughout the Universe. The only force there is can only be found on Earth in the form of life. Life is the only force and only found on Earth. In spite of all absolute madness that most of the important persons in science whish to propagate in they're apparently attempting to promote an even more mindless concept, which is atheism. They try so hard to pretend that life is a natural flow of normal cosmos that they go as far as to show how mindless ideas they truly come to conclude. Without a God it means that life which is a God linking factor, must be in abundance and if life is that plenty everywhere it is as common as star dust and then it has to be so commonly found we will trip all over other life through out the cosmos. Until proven otherwise we find life on Earth and nowhere else and that fact are written in rock in spite of all idiotic atheistic gibber. In life we find a force different to the book in the minutes detail to any other factor there is in the cosmos at large.

Only on Earth there is life being a force with the ability to manipulate space-time, which is placed under life's control under its control by providing motion other than and above the motion the cosmos does provide in order to maintain and sustain space-time. We use the Universe as if the Universe was meant for us to use. We increase what the cosmos gave us to use as if the cosmos was created deliberately for our purpose and for us to use which is just as corrupting madness as is the jabbering nonsense promoted as the religion called atheism. It is precisely in such a manner that light use to accomplish moving ability to travel in from singularity to singularity. Because singularity in space is space in darkness we consider the space we see as night as dark and therefore invisible. The darkness is light outperforming visible light by duplicating much faster than can visible light. Darkness is light which breaks down and rejuvenate space much faster than light frequency can the photons find a way to escape from the gravity applied by specific singularity points. Being in another frequency of duplication can the photon manage to secure its escapes. With the escaping that the photon does the photon can release and join the next singularity in the period being in a position where singularity takes charge and survive by galvanising a small portion of the heat forming the photon and by singularity releasing some part of the overall heat by removing space-time and forming the motion from the previous to the next infinite position in singularity in rejuvenating the point which is representing by producing space-time, which then will include the photon reassembling with the next singularity forming the space-time of the next singularity. Looking at the issue in this way we can begin to appreciate that light is the duplication of the photon by the singularity charged by the motion that provides the singularity by charging the intensity. The flow of light is about duplicating more than dismissing although dismissing does form part when the photon changes singularity. In that way the light loses intensity to the singularity that releases the light when the singularity releases the light. This process reduces the intensity of travelling light as it travels and is recharged by the singularity on route to somewhere in the future.

In contrast to the duplicating of light is the duplicating of material is more intense and more profound. The duplication of space filled with material is the use of heat compacted in space selected from the surrounding space in the atom forming a unit, which provide the material the ability to confirm the space they hold onto the space they move into without conforming or giving up ground that is filled atomic space, which is much more than just singularity. It is singularity that is sustaining more heat than the singularity will ever require and much more than what particle ever will require. Singularity empty of material which can take charge of light can generate motion to duplicate the photon whereas material use the heat the photon provide when the proton is clashing with the atom.

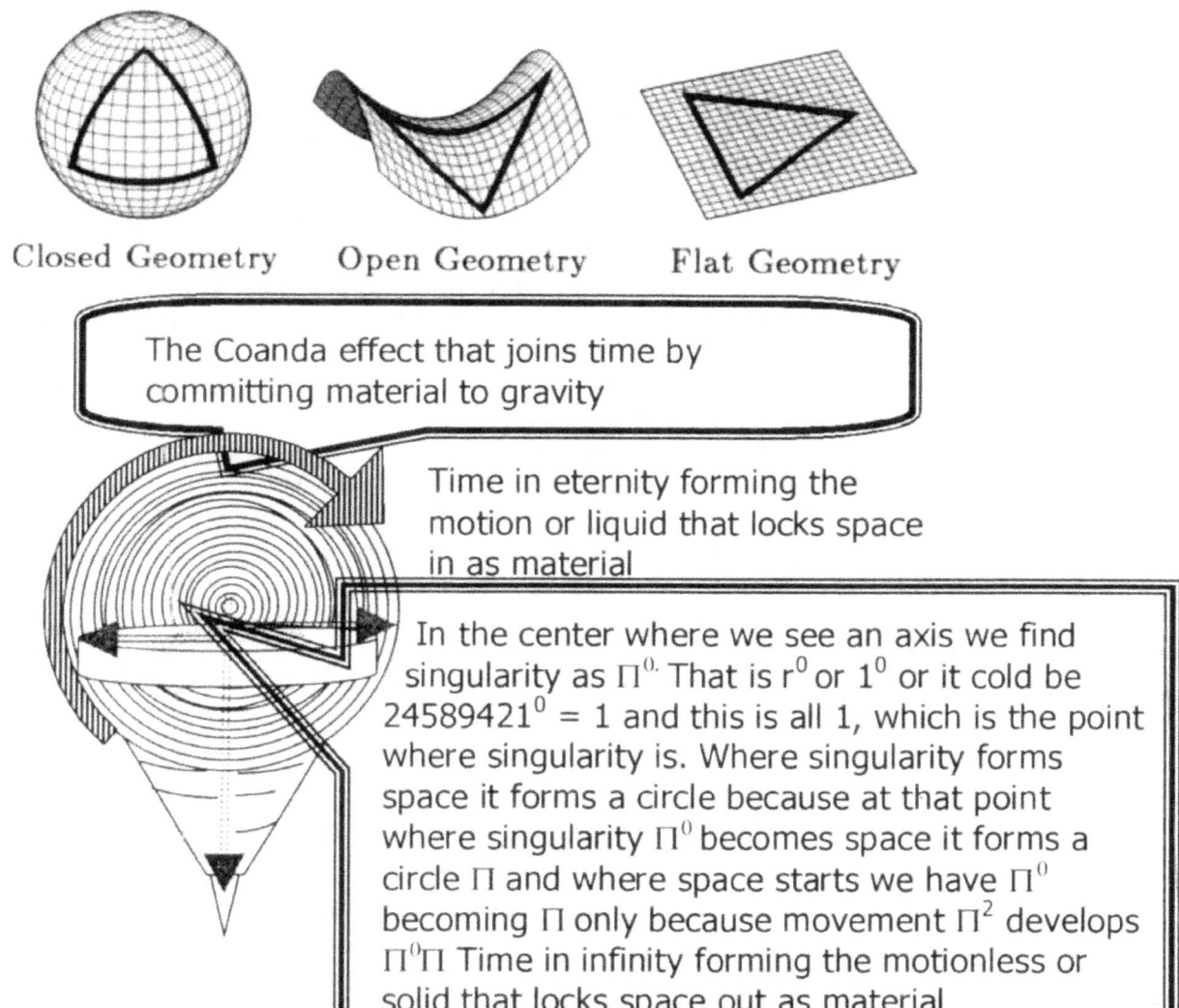

Closed Geometry    Open Geometry    Flat Geometry

The Coanda effect that joins time by committing material to gravity

Time in eternity forming the motion or liquid that locks space in as material

In the center where we see an axis we find singularity as $\Pi^0$. That is $r^0$ or $1^0$ or it cold be $24589421^0 = 1$ and this is all 1, which is the point where singularity is. Where singularity forms space it forms a circle because at that point where singularity $\Pi^0$ becomes space it forms a circle $\Pi$ and where space starts we have $\Pi^0$ becoming $\Pi$ only because movement $\Pi^2$ develops $\Pi^0\Pi$ Time in infinity forming the motionless or solid that locks space out as material

The heat provided by the photon is only a part of the total heat that is required by the atom to replace the dismissed space-time that the atom needs for duplication as well as dismissing of space-time at the centre. The singularity placed in charge and inspired to dismiss has the task to make space-time redundant whereas the electron is in charge of the factor that is making the duplication of space-time, supported by space-time protected by motion in contact with unfilled space-time then even requires more than what the photon can deliver because that is why there is shadows forming the dark side. Looking at this in a clear and sober perspective we once again find a reason to believe that heat and light is the antimatter that matter ate all up and still wanted more. Material is still eating away at light as it did when space began and material still craves for more light, which is heat. We once again find a reason to believe that heat and light is the antimatter that matter ate up and wanted more. Material is still eating up all the light and then still wants more.

**Man could create motion but at first such motion was far less than that motion which the Earth provides. The motion of man's ability was vested in what his muscle power could provide. But a very short while ago man grew wise to machines and the fact that machines can provide more motion much faster than could animal muscle bring about motion. By supplying machine motion it gave man extra ability whereby man extended the relation between what the object has when in normal contact with space and when extended by extra motion allowing more space to apply to the surface of the object, thus enlarging the object surface in the relevancy brought on by motion. Then man found means to break the barrier that muscle strain held and was able to apply motion equal to that of the Earth spinning. After that eventful day then came the day man had more motion than what the Earth could provide. This is where nature and man parted their straight line common sharing of the factor k because with the motion that man could produce placed man in a position where man was able for the first time to outperform the earth ability of duplication by motion and thereby can go in disrespect of Earths straight line motion. This is where locomotion generated by steam concentrated gave man more than what man could tap from life and sweat. From then on man produced his very own straight line gravity in such abundance that mans gravity generating ability is no longer in harmony with the Earth's gravity ability and with mans very own straight line man could eventually leave Earth altogether. But let us get back to the straight-line man moved in, as that straight line no longer followed the Earth spin. At first with mans first attempts it showed a diverting form the straight line of the Earth and we even gave that diverting a nice name as we do with all things. We called this diverting flying and flying is proving what Kepler said what gravity is in so many ways again and again. The space grows as we increase the motion with which we travel while complimenting the space through which the space travel in which space we are in locomotion.**

If mass was the major factor in generating gravity, the mass will play a major part in the time it take a body to fall from any given point to the surface of the Earth. It just has to because the mass then does the pulling. Since Galileo proved otherwise the concept of mass being the producing factor in gravity comes across as rather less thought through and more than a bit silly. In flying a certain criteria must be met by involving the motion of space and as that motion is running through space in time the motion is relating to space by the scale of time. This then has to indicate that there is a restriction in all motion running through space. That we can observe by the speed of light slowing down in denser space than what the speed of light is in less dense space. When an object travels through space the density affects the motion. The slowing down by relevancy as a factor comes about by bringing into the equation where a smaller space must negotiate travelling through a denser larger space. By being in a "thicker" density the motion has to manoeuvre through more restriction say in comparing the space-time we find as the Earth atmosphere being denser than the space-time that the Universe grant. In denser space-time there is more reference points being spots of singularity to contemplate as space distances where the singularity concertina in the time given to fill the more or the less space being contemplating. By lesser dense space there is lesser virtual singularity forming less restricting to the moving object. It is very obvious when looking at an object that is coming from outer space into the atmosphere of the Earth. As the factor one has to consider that the space represented by the line **k** increase in the density it represents there is more space to relate to motion. It will be the same as if the space **a**$^3$ is moving faster because of relevancies re-applying to conditions changing. Then space **a**$^3$ would reduce in the relative factor presence within the larger and newly introduced **k** that comes about. Then that means that the faster **a**$^3$ travels the more **a**$^3$ will extend **k** away from **k**$^0$ and by that increase in motion that increase will also apply to the time relevance by reducing the space **a**$^3$ there will be a smaller **a**$^3$ because there will be more space that **k** has coming from the being in the larger **k**. A larger **k** will bring a relevance that reduce the space because the holding space increases in relation to the lesser space occupying a part of the holding space.

There is always a double relation to space, which is the space the travelling object occupies and the space in the circle of time that is in control of the object by motion because the object is in motion. But this relating is affecting on both sides of space and because of that mans quest to travel at the speed of light is totally unrealistic. Those comedians hiding behind science from where they are trying to pretend to be all the wise about the Universe because of they're accomplished scientists has a smaller chance of going even one tenth of the speed of light than does Little Red Riding hood has herself a bigger chance in finding her talking wolf with the ability of eating Grandma wholesale as one unit than they're having any chance flying through space and achieving $3 \times 10^5$ km / sec. In the fairy tale Little Red Riding Hood has a bigger chance in finding her talking wolf than the chance any cosmic traveller will have to go into space flight and achieve $3 \times 10^5$ km/sec.

By establishing motion and such motion is bringing about certain contact with space by increasing motion such an increase can bring about much more space to be in contact with the moving structure and rebalance the dynamics of such space in motion. The space **a**$^3$ in motion **T**$^2$ will establish a larger area **k,** which is then contradicting the gravity of the Earth's motion in descent and this will count for a larger area present in a smaller time relevancy. While all this is happening the result is that a stronger motion line in a 90$^0$ directional change will come about and since the Earths motion remains the same it gives the flying structure an advantage to increase the relevancy in favour of the accumulating the dynamics of the balance in space-time by space contact increasing by the contribution of more motion above and on top of that of the Earth motion.

Gravitational Constant
(Symbol G) It is the constant that appears in Newton's law of gravitation. It is the attraction between two bodies of unit mass at unit distance apart. Its value is 6.672 x 10$^-$ $^{11}$ N m$^2$/kg$^2$ when the distance is expressed in metres and the masses are in kilograms. Although it is described as a constant, in some modes of the Universe G decreases with time as the Universe expands (see Brans-Dicke theory), but there is no evidence for this.

## Gravitational Field
**The region of space around a body in which that body's gravitational force can be felt. Within this region, other bodies will experience a force of attraction that diminishes with distance from the body.**

## Internal Mass

Inertial mass is a measure of a body's resistance to change in its velocity or state of rest.  Inertia is a direct property of the mass of a body:  The greater the mass, the greater the inertia.  Although mass is formally defined in terms of its inertia, it is usually measured by gravitation.

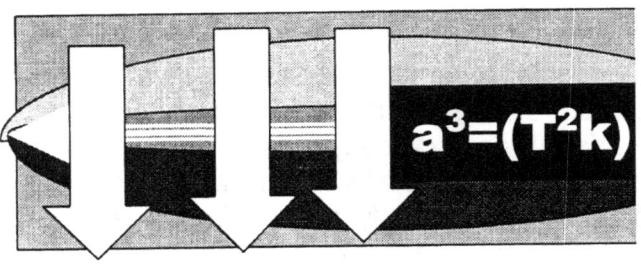

Mass is the refusing of any object to dismiss the form it has and to join the Earth solid structure. Mass cannot and does not contribute to the establishing of gravity except by depleting space through motion and such numbers of the protons in a space forming an exclusive unit.

Kepler answered the question of flying and airflow dynamics before Newton gave us a name for gravity. Kepler said the motion of space must form equality to the motion of the space. When the aircraft is maintaining flight height the motion equals the mass and equilibrium ensures a constant flight height. As long as the speed and the mass are equal the aircraft will be in equal gravity balance.

## Gravitational Mass.

To overcome this breaking effect that the smaller object has on the larger object that we named mass of bodies being on the Earth the smaller object firstly has to transform and transmit singularity from the centre of the Earth to the Centre of the motion wherefrom a new balance sets in. This we call the Coanda effect and it works either on the linear aspect of gravity or it works on the circular

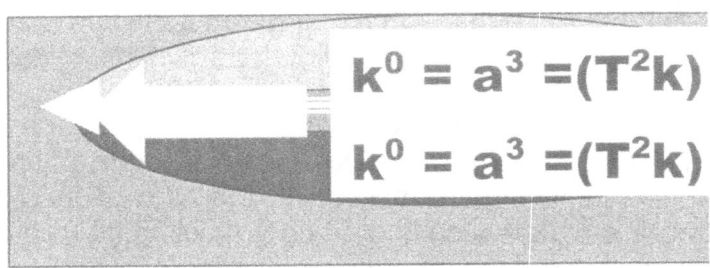

It is when the motion exceeds the mass the aircraft has the ability to break the sound barrier. Galileo proved that no mass is present in falling, which is also matter in the process of flight and because of that can the sound barrier become some form of constant.

When the aircraft increases its motion, the motion changes to accommodate both space $a^3$ in the motion thereof $T^2$ that the applying $k$ factor would allow.  The space $a^3$ is a fact influenced by the Earth but directly dictated by the atoms forming the aircraft.  The time factor $T^2$ is directly derived from the motion the Earth dictates. The Earth rules on the distance that $k$ would produce. That is gravity. That is what Kepler said gravity is when he said $a^3 = T^2 k$.  That is what no one in four hundred years cared to take notice of or refused to recognise.  What no one ever apparently saw what is that gravity and recognise that gravity acts that is precisely in the manner Kepler stated gravity is. If the motion of the aircraft $a^3$ increase the space increases, which the aircraft influences but also the space influencing the aircraft changes bringing in new alterations.

If the motion decreases the aircraft space relevancy increases and if the space decreases the motion changes the contact in space therefore it changes the volume of space which is reducing the relevancy of the space in motion which is gravity applying by motion in space forming space in motion. The more the motion is the more space contract is produced and the more space is affected by the increase or decrease of the motion of space.

The more space that is duplicated the more space is produced but also the more space is reduced through such duplication. Seeing gravity acting in this manner does make nonsense of Newton's $4\pi^2 a^3 / T^2 = G(m + m_p)$ changing of Kepler's formula as it has no part to play in correcting the formula of Kepler and it is only Kepler's $a^3 / T^2 = k$ that comes into the equation since $k$ is $G(m + m_p)$ in any case because $4\pi^2$ indicates an individual structure encircling the Sun centre when one use the cosmic relevancies which I later introduce.

We gave this forming of separate gravity coming about by means of performing individual motion which is disguised as a very well known and commonly occurring phenomena which is burdened by carrying yet another name of being called the Coanda affect. The Coanda effect depends on singularity forming when a solid and a liquid is in relevant motion where such motion of either the liquid or the solid or both factors has to move and such moving contributes in selecting a singularly centre point that will secure the control of the space-time, affected by such motion.

The Coanda principle forms a circle and motion establishing an independent singularity in such a circle centre. Then the singularity cuts the Kepler formula in two parts where space is following motion and motion leads space. Being as human as the next person and showing as much human tendency as anyone else being human I changed the partly to the Coanda effect because that is what humans do best.

Humans decide they have no idea what they discover and hide what they don't know they discovered behind But with a fancy name other meaning behind the discovery gets less important and the naming becomes the accomplishment a name will scare away any one also in mind of finding out what was discovered. Like calling heat plasma when plasma is the same as heat gone liquid. Well in my case I use the name as the Coanda effect because it is a process where motion is having an effecting on space-time.

The prerequisite to flying is creating motion because the motion establishes antigravity. The motion reduces the friction that mass creates and in that reduces the gravity that creates the mass. While the aircraft is motionless on the Earth the Earth motion create gravity and that motion also applies to the craft structure albeit that the aircraft seems motionless to us. The mass of the Earth establishes motion of space spiralling down in time. With the space of the Earth being in motion the object resist surrendering the form its construction has and with that refuse the accepting of the joining of the Earth solid structure.

This resistance we believe to be mass. It is where $k^0 = a^3 /(T^2 k)$ unconditionally. My argument about gravity being motion becomes prominent when we consider the motion versus space occupied by airplanes using wings to fly. Without motion the aircraft and all that it carries have mass or weight. As soon as the motion overcomes the space restriction by defying gravity affecting the aircraft to a stand still which the Earth enforces on the airplane the airplane loses mass and becomes airborne. Notwithstanding the corrupt argument Newtonians bring in about mass remaining a factor. To prove their corruption in this argument, let those that disagree with my stating them being corrupt answer the following. On their admitting we know that mass increase as gravity in stars increase.

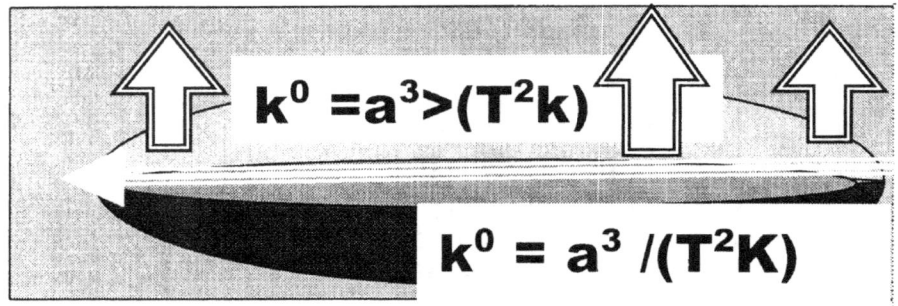

$$k^0 = a^3 > (T^2k)$$

$$k^0 = a^3 /(T^2K)$$

By decreasing motion the mass of the aircraft will tilt the balance towards favouring the gravity the Earth applies and the favouring of the dismissing factor of space-time, which then overcomes the duplicating effort of space-time by motion will contribute to the descending. In order to apply a perfect controlled landing the wing must establish additional space-time dismissing to

The establishing of independent motion of the craft secures an individual gravity and such individuality leads to the breaking of the sound barrier because the one gravity can no longer subdue the smaller motion, which is producing gravity

allow the steady descent and the perfect landing. Even when performing the landing under the most stringent conditions the balance still rely completely on the balance Kepler gave us of $k^0 = a^3/(T^2k)$

The more the gravity is the more the particular mass will be. But then the very opposite is true where in space there is micro gravity. Then there has to be micro mass which means by their own admission, mass disappears. It is only when applying the illogic use of mass and weight differentiating and insist on proving the incorrect correct in using a method which in any case that does not make sense by any standard of arguing that one can argue about in order to prove the nonsense about mass still being a factor. Thinking about this I feel delighted that those being so very incoherent about mass see my argument about space and nothing being incoherent. The protons spinning are supposedly bringing about the mass. The protons form the centre of the atom and by spinning 1836 times more than the electron the protons dismiss 1836 times the space that the electron does. Because there is an increase in contact with space by the body/s in motion the dismissing of space-time does not only become fully substituted by the duplication but also totally overwhelmed by the motion. It is the dismissing effort applied by the combined unit of all the atoms in the motion in relation to the contact made that tips the balance. There is a way to detect gravity and that is in looking how we fight gravity or eliminate the effect of gravity that is securing objects onto the ground. We know that $a^3 = (T^2k)$ locks objects in position on the ground and flying eliminates the flying device including its cargo being locked onto the ground. Let's us look at why we fly.

The Earth applies motion in the atmosphere of rotation but from the human position we hold on Earth we have to follow the Earth in motion. From our stance we are moving in a continuing straight line and the straight line we receive from the Earth giving the impression of a straight line for us to see. Before investigating the principles of flying we first once again must define why the mass is keeping the object secured on the soil. The space the object claims within the space the larger object retains. This then is what should be overcome to fly.

When the aircraft is gaining lift the motion exceeds the mass and with that is adding heat at the bottom of the wing to create more mass $a^3$ added than the speed $(T^2k)$ can create motion above. Below the wing there is more space in contact with material that is improvising to dismiss more space by collecting space compressed with heat by restricting more of the motion. On the top of the wing the motion that accelerate the flow of heat and by doing so dismisses the possibility of having more space-time dismissed as is the case applying at the bottom of the wing. In that way the wing is at the top creating an environment, which favours extensive duplicating. At the bottom of the wing we have $a^3 > T^2k$ and at the top of the wing we have motion outranking space accumulation by restricting motion therefore changing the balance on that side to $T^2 > a^3 / k$. As the speed gains, the wing will strike a balance and at a certain flight height the motion will equal the dismissing going on and equilibrium ensures a constant flight. The motion of the craft establishes individual gravity that is surrounded by the Earth gravity but the independent motion grants the aircraft some individuality and exclusivity.

By decreasing motion the mass of the aircraft will tilt the balance towards favouring the gravity the Earth applies and the favouring of the dismissing factor of space-time, which then overcomes the duplicating effort of space-time by motion will contribute to the descending. In order to apply a perfect controlled landing the wing must establish additional space-time dismissing to allow the steady descent and the perfect landing. Even when performing the landing under the most stringent conditions the balance still rely completely on the balance Kepler gave us of $k^0 = a^3/(T^2k)$

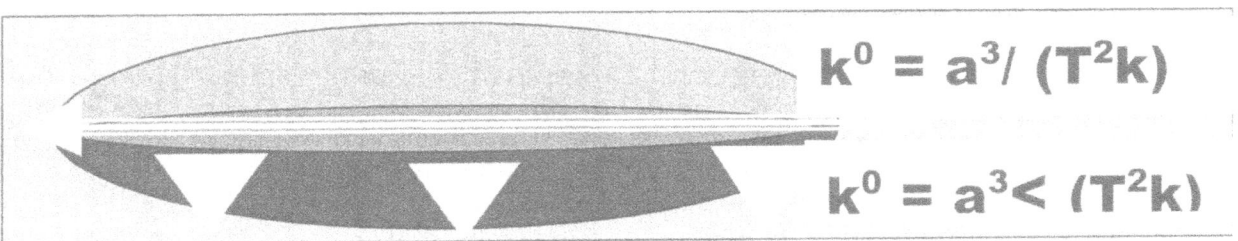

$$k^0 = a^3 / (T^2k)$$
$$k^0 = a^3 < (T^2k)$$

At a height of 31000 km above the Earth the mass of the wing becomes compensated only by a motion of a relevancy that comes about at 2500 km per hour. In that case the craft has to apply motion at a rate of 2500 km / hour just to create the required velocity to keep the aircraft in motion in the sky. Motion creates gravity just as Kepler said when he said gravity is about $a^3 = T^2 k$, which translates to the dismissing of space and the motion, duplication establishes a centre that controls the balance that the newly secured singularity will provide.

When the aircraft stands still the Sun provide such a pivoting centre but when independent motion comes about the point shift from the Sun to the Earth centre where there is a line contact between the singularity that the Earth holds which then forms a new relation in respect to the singularity activated by the independent motion of the moving body which the aircraft takes on motion that the relevant singularity is claiming are released to the minor space. The Earth provides a point from where space depletes completely within the centre of a sphere from where gravity is securing the centre spot in the form and the space surrounding the form that controls the space and time in which the independent object moves (in this case it is the aircraft). When a balance comes about between the departing object and the space reducing only then does an orbit establish a balance of speed serving time duration and space dismissing evenly. That is gravity and that produces gravity only when motion creates a centre to form a sphere.

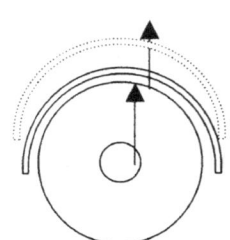

The new **k** that is applying the relevance, must link the space $a^3$ being equal to the motion $T^2$ to singularity $k^0$ in order satisfy $k^0$. The flying of the aircraft is then unequal to the motion in the previous relation that was in place where it was part of the sun and the Earth motion relation and the new motion will bring a correcting in the relevant distance **k** to put the motion in balance with space. Space will always demand a correct establishing of the miss-interpretation of the equilibrium that is needed to sustain the effected singularity because of the space-time factor.

This is very important if one wishes to understand the sound barrier. The two positions in **k** depends on the motion of $T^2$ in relation to the **k** the object is maintaining while the object stands in the space $a^3$ and related to the motion the Earth provides.

When the motion amplifies the status of the object elevates as the motion increases the relation to singularity, which is located in the centre of the Earth $k^0$. The more motion that applies the higher the velocity of the flying object will be and the bigger distance **k** it can sustain. If the sustaining of **k** is not required, as is the case with racing cars the space $a^3$ needs relative improvement and that is done by allowing wings on the car to contact more space improving the dismissing part of space-time where this contact will proportionally reduce **k** by amplifying the relevancy of $T^2$.

In our normal posture we are moving much slower than the Earth is spinning for reasons I shall come to later on. With our bodies not applying the motion we apply the thrust to reduce the motion the Earth has in order to achieve the much-needed equilibrium the cosmos requires.

To overcome this breaking or restricting effect that the smaller object has on the motion which the larger object provides and that we named as being the mass of bodies being on the Earth as being in the role of the smaller object firstly has to transform and transmit singularity from the centre of the Earth to the Centre of the motion wherefrom a new balance sets in.

This we call the Coanda effect and it works either on the linear aspect of gravity or it works on the circular aspect of gravity. By applying motion either to a liquid in the presence of a solid or supply motion to a solid in the presence of a liquid a new point in singularity establish a new centre elected by the motion creating the defining of the space $a^3$ that is in ratio equal to the motion thereof $T^2k$ bringing dominance and the process applying the rules I shall explain a little later on.

All the principles that I make use of to explain my theory is part of nature. I base my theory on heat becoming stabilized through collecting more space using motion to produce cooling. This idea is most basic and that I admit. It may sound basic, but Mainstream science is also most guilty of their departing from this most basic principles through the employing of terminology and terminology has covered many of the crudest, most basic meaning behind the most basic principles in nature. I do not applaud a principle Mainstream science underwrites in the sense that matter in the beginning was coming about and anti matter came to destroy the matter. It is moreover the disappearing from the Universe of the dissolved by product which antimatter somewhere not in the Universe. It is this vanishing being the result of between the two opposing materials that I strongly reject. If anything ever was part of the cosmos, it still must be in the cosmos because there is simply no other place to go to outside the cosmos. The friction that once produced the heat in the time before and during the Big Bang period still today actively participating as mass.

Mass is the result of let us call it "stationary friction" which is the relation between two cosmic objects having motion inequality but still sharing space within space. By creating friction through the bringing about of any form of motion discrepancy between objects in any such a test performed today such friction coming about will produce heat and the heat will result in space forming. In such contact between objects in different speeds that such motion discrepancy produces to cause destruction of matter in space and heat comes about. In that the net result eventually leaves space created when overheating material no longer fills the space after cooling sets in.

In the picture we see a moving aircraft create its own atmosphere within the boundaries of the earth's atmosphere and this atmospheric split is the result of the movement of the aircraft creates.

The cracks showing in the cooled material afterwards is a result from the overheating of the material that created the extra space and then reset the occupation to what it was before. But the material that is reducing from retracting used space with the becoming colder again leaves cracks behind on the surface that proves that there was more space filled when the material was heated than what it had before the material was heated compared to the decrease in space after the material; again cooled down back to what it was before the material was heated. Evidence of this is evident in all supersonic aircraft as the fuselage forms cracks in the body structure of such an aircraft.

The outcome of this heat is that when cooled the material occupies slightly more space than before. The grown space then tries to fit into the area it did fit into before the heating but with the extra space it employed when it was during in the heating process it shows afterwards as cracks. This takes us back to what Kepler said. In Kepler's formula it is the extending of the distance $k$ that influences the time aspect $T^2$ which the supersonic aircraft does by its going supersonic and by shifting $k$ from the previous location the Earth prescribed to the new $k$ the aircraft implicate in according to the $k$ that comes about since the distance in affect becomes longer.

The aircraft produce a new time value $T^2$ in accordance with the Earth time factor $T^2$ because of the fact the Aircraft still shares space in space of the Earth with the Earth. The aircraft now has a bigger time vale $T^2$ in the space $a^3$ of the Earth using the Earth time so therefore the fuselage of the aircraft has to reduce its space $a^3$ it claims compensate fro the extending of $k$ which it does by going faster as the extending of $k$ will introduce a bigger time factor $T^2$ that will reduce the aircrafts occupying space since the Earths atmospheric space will not compromise and the aircraft still remains in the space of the Earth.

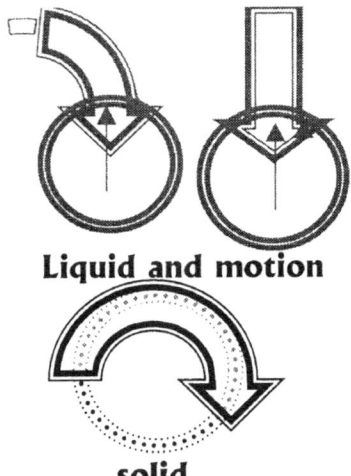

**Liquid and motion**

**solid**
All the results coming
from this we also find

The object fights to retain independence while trying to slow down the Earth either as to accelerate the motion of the rebelling object or to slow the Earth down. This is all an attempt by the Earth to increase the motion that the secondary object can apply or by friction reduce the object space-time to liquid heat. There is a specific border or a definite barrier where the motion differentiation becomes so critical that the incorrect transforming of motion can have the same effect on the incoming object as hitting a solid wall. Later on, I shall show that it is the equivalent that is matching a solid wall that the object will have to break when the object falls into the atmosphere. When the object enters the atmosphere, it is through the small door there is in the area we call the atmosphere.

*By having more heat per volume in ratio the material will claim and introduce new space that formed. Heat establishes space that expands. This truth science does not recognise. The claiming of more space and disposing of the space after cooling shows new space formed in the process of heat multiplying where there were no space before that which the material in the cool state afterwards cannot fill because of the void that came as a result of the material getting cold and contracting space, reducing the space as the space filled when the material was overheated. If material employs this as a basic technique today it was a basic technique back during the Big Bang. That evidence we can see in when material having a heat level amplifying upwards when motion difference brings on friction and such friction brings on heat. Two opposing issues came about, but both opposing issues are still present in the cosmos somewhere in a place where we are missing the presence thereof. Material is energy and energy is indestructible. However energy can change form…yes that we all know and energy may even hide appearances. Therefore we have to search for the new form in disguise. I believe I can show that it is a motion discrepancy that produced matter and anti matter and we do not have to go and look for non-exiting positrons and negi-protons (if I may be excused for using such bizarre terminology but it is fitting a bizarre statement which the first one is not my doing as my brain I did not make it). But a positron must produce a negative proton and such a performing sub atomic structure cannot be possible. By changing legions the proton must then perform gravity by rejecting material or if I am correct, producing space! I am about to prove that antimatter is in fact a process where the heat that became formed heat, which forms space, and therefore space has a valid substance other than being nothing. The motion between particles in a cramped space as the case was during the initial stages of the Big Bang would have brought on friction in space we cannot even calculate.*

*It is the combination that forms gravity by motion. The result is that in the very beginning some matter particles produced gravity in their sustaining of independent singularity by applying motion which in some cases lead to the demise of some forming space-time by converting the where some compromised solidness. This route the one side took resulted in plasma forming on the one side and material on the other side. This was done because there was less control that confirmed the space and the volumetric space grew. By having some softer than the other harder ones the softer one became a liquid. The notion or defining of a liquid is very relative because as solid as the Earth seems the Earth vibrates as a seemingly liquid during an Earthquake. It forms waves many meters high just like it would be a liquid like the see. Afterwards when those in charge of damage control come to assess the damage it is hard to digest the destruction and damage because all liquid-likeness disappeared.*

*This electroplating motion is possible since electricity is gravity to some intense extreme. Electro motion or electro flow is the concentration of gravity to the limit where we will find gravity has the same intensity in the centre of the Earth as electricity has in the open. By removing material from the less dense and electroplating that which is removed from the less dense and then galvanise that softer material onto the harder material (which by the way is a very natural process taking place all around as a corrosion) the density of the liquid will demise in the liquid sector and the material will grow in the solid sector. I believe even to this day and throughout the rest of the Universe wherever there is space such space has to have motion and space cannot be what it is without having motion. With that in mind that is space-time. Space-time is space flowing on.*

*Where there is motion in space, the motion through of all space is carried along by time in space. The plasma is transforming to material through the motion we named gravity. By being electroplated onto material. By duplicating space in the process of establishing gravity the object does not reduce to a standard in occupying space that it had before the motion took place but by placing liquid heat into the form of solid matter the matter use the newly acquired heat through which to cool. By absorbers the liquefied material onto the solid material is thus freezing it into a solid to secure more material in the fight of combating overheating. In other words in the present time in our Universe gravity is freezing space to first become dense and form a liquid after which it then solidify the liquid heat by freezing the liquid into a solid state within the substance that is the atom. However, I do prefer to use heat as the term of choice and not plasma. The process I just explained was the manner used by the cosmos as the cosmos came about and this is the manner that will repeat until such time as will the cosmos conclude its final motion. I believe that the first motion came about as singularity was without space and found irrepressible heat levels rising. By overheating it moved into space that was still non-existing and that had therefore produced motion to rebalance the heat. From this I also believe some material that came about from singularity overheating remained as particles forming atoms where there is this relation between the solid proton, the liquid neutron and the gas electron.*

*The development of space from liquid heat such fluid was becoming gas that is space with the ultimate gravitational relevance that space can carry. This was all contributing to the lack in contracting gravity promoting expanding gravity to those particles that applied lesser motion helping the extending of space to turn into heat that again turned into space. In this, of uncontrolled release of heat performing as softer space-time such release of space-time is the destroying of singularity secured in a unit, which again I believe (within reason) I do prove. I show that on the one side singularity introduce space-time, which confirms singularity and space-time makes contact with space-time not directly controlled by singularity or that, which is directly confirming singularity. This I conclude from studying Kepler's formula. I believe heat is the destructed form of material that overheated and this confirmation the atomic thermo explosions give us. But to realise that we must beforehand find what any and all space is and we have to accept that space is made of something.*

*As the space surrounding the Earth, which we call atmosphere reduce in volume of space the heat content rises as much as that the space holds heat having the heat rising by the same token. By becoming less the space also become hotter. The ratio there is between space and heat increases as space in measure reduces. We have to learn to see heat where the heat in the space has two different identifiable substances. We also must see material holding space to be different from the space holding the material. We must see material to be different from the heat covering the material and compromising the space that produces the format of material in being a solid, a liquid or a gas. This changes in the state of materials holds a direct relation to the heat that also claims a steak in that space.*

**Galileo proved space-time is functioning in the manner that space diminishes as space has to compromise to sustain the flow of time but time slows down in stronger gravity reflecting again on density changes rather than mass influences.** $a^3 = T^2 k$   $\frac{1}{2} a^3 = T^2 \frac{1}{2}$

**As the space surrounding the Earth, which we call atmosphere,**   $\frac{1}{4} a^3 = T^2 \frac{1}{4} k$ **reduces in volume of space, the heat content rises as much as that the space holds heat having the heat rising by the same token. By becoming less, the space also becomes hotter. The ratio there is between space and heat increases as space in measure reduces.**

**Galileo substantiated Kepler's findings that space $a^3$ correlates directly to time $T^2$ when space $a^3$ compacts with the decline of $k$ reducing the swinging arm of the pendulum that maintains time $T^2$. (1) The swinging arm will not relinquish relevancy, but reduces the space it moves through while it moves slower because of density rising when the distance $k$ changes. (2) That proves that space and time $a^3 = T^2 k$ is directly related.**

On the outside of all material the density provides a distance in space vowing between objects. That density also introduces heat as part of the distance of the space that is in place under the specific conditions applying. This is density because in the cold particles will be closer and when hot (3) particles will be further apart. By performing motion through pumping air into a container the pump collects particles and rush the particles into the cylinder, which is just a cylindrical metal container. By (4) removing air from the atmosphere and squeezing that air into a container it leads to the reducing of the space between the particles *(5) in the cylinder when compared to particles outside the cylinder. (6) As the container fills the space (7) that was space meaning it was keeping particles away from each other at a certain distance turns to heat in a ratio to the square (8) as the compressing removes space and the material density within the space increases because the material density rising forms liquid heat. This further aggravates the heat brought in with the material in the pumping process because the compressing of space is adding to the presence of the heat by accumulating even more concentrated liquid heat..

To elements hot and cold as influencing substances are outside influences that do not apply to the core of the atom. The atom constitutes of densely frozen space flowing or liquid space and releasing(9) the liquid space into gaseous space  (10) When insufficient control leads to uncontrolled expansion of such liquid heat. This is all singularity governed from all centres involved. Heat and space are influences outside the proton but we may imagine that with in the proton nucleus it is bitterly cold. The heat or space will surround the atom on the outside, but has clearly no influence on the inside of the atom and therefore of the star.  The star is in every sense the atom forming the star.   Atoms will reluctantly compromise by reducing space but this compromise in solid structures such as the atom is will totally depend on the singularity that rules on the applying conditions. The heat or space is a state that somehow extends beyond the electron, which does not influence the proton or change the proton.  (1) Only gravity coming from massive numbers of protons working simultaneously can remove space from the inner atom.  The atom cannot compress but will withstand the worst pressure there can be.  Remember a star cannot have pressure.  Neither is there any possibility of atoms touching.  Only space in the form of gas surrounding individual atoms can to a certain measure compress.

The heat in concentration or the manifesting as space is neither hot nor cold because the proton presents eternal cold. Heat is an exterior influence bringing about influence between atoms within the star.

A gas will allow influences to charge as loosely connecting bonding the positioning and forming other objects occupying space next to one another. The gas will allow a lot of flexibility and compromise because in the case of gas space never becomes a premium. I should think placing space at a premium would apply when fusion comes into the picture. If a gas surrounds the hydrogen the gas will withstand as much compressing as can be induced by whatever force bringing about such compressing.

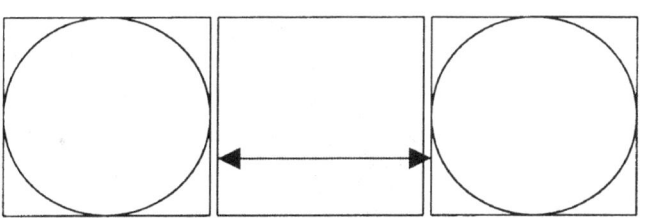

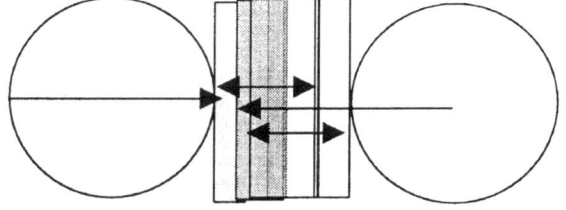

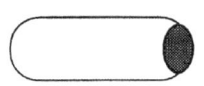
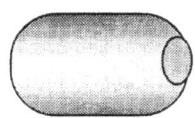
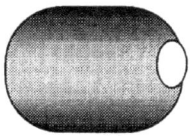
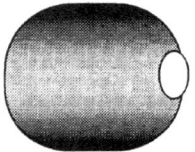

There is no correspondence between the air container and a star. The container in relevance grows while the inside in relevance is shrinking. That is pressure being in the container and the walls shrink as the pressure grows. In the star there is no containing but a time in spin bringing about space-time reducing and not space alone.

Every particle that is of the airborne type is of the airborne type because it favours an association with surrounding itself in an envelope of heat. That makes the airborne particles airborne particles. With that they disassociate with close proximity to others because they are airborne particles and airborne particles are what they are because they are cooking or boiling.

Hydrogen, oxygen and nitrogen are airborne because they are in a worst state of overheating than water is when water is vaporised. The "gasses" are "gasses" because like "gasses" are "boiled vapour" even below 200℃. Repressing the airborne elements as "gasses" is the same as conforming vapour in a cylinder. I witnessed once what happened when water vapour was accidentally contained as a vapour. The house is no longer classified as a built structure. When pumping the "gasses" the "gasses" brings in heat because in the vapour state they are highly associating with heat. When the airborne particles enter the cylinder they are no longer able to dissociate according to will, but forced to cluster while they will bring with their "boiling" "gas" status. This heat enveloping the elements is taking up space with particles that are enveloping them with heat as the compressor fills and that is what is compressing by compromising space where space. Space reduction becomes the norm needed to fill the container because the actual solidness of the airborne elements in own rite cannot compress in such a manner. That it places the burden of compensating to allow the increase in density on the space not occupied. The unoccupied space between the restraint particles deliver a gain in heat levels between all the particles and the next particle. Since all the pumping of air which is going into the cylinder the space soon gets cramped and compressing space forms heat. That is evident in all eternal combustion engines.

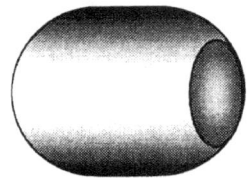 **By pumping in air the molecules into an air-compressing container the molecules bring with every molecule entering a specific amount of heat from the outside that covered the molecule because all molecules entering is airborne molecules.**

**Scientists apparently in the know how of why what works express their opinion as one being that molecules run wild on the inside of the walls of the air container and bump each other while colliding and going mentally berserk in anguish and excitement. They maintain the argument that it is the clapping, fighting and general applauding of hostile, locked in particles that are in general behaving like British soccer hooligans that is going on inside the container. According to those Superior-Wise the general agony and anguish that the particles suffer while highlighting their complaints as they knock each other around ... well that is what is causing the friction, which is causing the heat to fill the container and fill the container to such an extent the heat that filled the container, is flowing out through the container walls. I wonder who thought this lot up? The person obviously was a tremendous thinker with true potent ability to dream an impossible dream with not a lick of logical argumentativeness mixed into any or in all of his thinking ability.**

Since the heat is still under the specific conditions that was in place on the outside which was set by the atmosphere on the outside of the cylinder at the time when the pumping started the conditions in the air did not change by the pumping alone. As the pumping commenced the conditioning went on in the same manner, as it was the outside and taking those conditions inwards. The air takes the same heat with when the pump takes the air on the pumping journey. The pump is taking heat surrounding the elements with the elements and the conditions they were in, in the atmosphere into the cylinder. The pumping changed little about the heat applying under the conditions as it was brought in with the elements. In contrast to such conditions that were applying on the outside there arte then new rules that changed considerably.

New dynamics on the inside brings about the applying of change to conditions. On the inside of the cylinder pumped with air forms new singularity elect that will enforce new controlling standards where all electrets are controlled by another set of laws. In the atmosphere where from the pump removed air the particles was widely spaced and the particles brought those conditions associated with a widely spaced atmosphere into where all are clustered in the inner confinement. In the atmosphere there was different elements all bonded by a heat holding the space between the separated particles apart. These conditions was set by the Earth governing singularity ruling space-time at that point in the atmosphere. The ratio the particles were apart was precise and specific and preset by the Sun burning heat into the air or not burning heat into the air at that specific point. But the heat was the fabric that connected the material because it connected to the material by a specific ratio that set distances we call air pressure. The heat between the particles were contracted to a certain degree by gravity and was not as utterly expanded as it would be in outer space nor as compressed as it will be in the container. Therefore some heat was in the space parting the particles in the conditions in the atmosphere at that very specific location. The Earth singularity brings about the gravity that determines the density of the "gluing heat" holding space between the particles in space.

The solidness is the material is actually surrounded by heat that brings with the atom seclusion from other atoms by covering the atom and the surrounding space of the atom with the heat surrounding the material. It is the density of the heat in the density of the space that specifies the conditions defining every position every particle holds as space occupied relating to space unoccupied that is separating particles and filling space between the particles. As the solid "stuff" is pumped in the solid "stuff' which is particles is covered by liquid "stuff" which is heat and together the solid and liquid "stuff" is pumped into the cylinder containing. This combination forms the composition we think of as pumped air. There is coming in as much heat with the pumped element material in the form of flexible space. That flexible space is that which is containing a specific degree of heat. As there is that much material coming in as a result of what the pumping is bringing in the ratio on the inside change somewhat. A lot of space turns to forming heat on the inside. The heat is in fact the outer part of the material coming along and is accompanying the particles as separate space but forms one unit. The particles in essence is solid when being without the heat containing the solid atoms and the atomic unit cannot compress as much as the particles cannot touch because all particles hold negative charged electrons to the outside. The electrons are all negatively charged. Excited negative electrons will not touch each other notwithstanding whatever any argument the wise in words wish to create arguments about confirming this idea. Negative particle subatomic or not will repel one another and the repelling will become more fears as the particles come closer. Unless scientists of high standings and other fiction writers can bring along antimatter and prove that the antimatter is hosting a positron, or what they seem to see as a positron there is no chance of electrons meeting. All electrons we no can only be what we then must call position because it is the only way that such a meeting can take place. Without the likes of a negatron and a positron meeting each other the two electrons sharing charges will not form a company. Without the likeness of such proof, (finding positrons next to negatron or what ever they will be named) the argument of touching atoms fall apart and will impress not even one clear minded thinking individual.

We have to recognise that with the space in the form spaces come, the space that is holding heat is part of the electron status. The heat in space is a part of the un-compromised space that is not yet fully extended to the maximum as space we find in outer space. Such space carrying pure heat within the space concentration also enters the cylinder chamber. The only flexing is the space containing heat as the space accompanies the molecules entering in a cover of heat. The space entering can enter as heat or as space since space and heat is the same thing. However in all circumstances there is a limit as there always is some limit to everything we find in nature.

 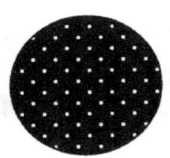  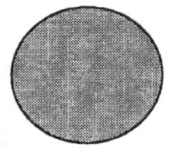

**The gathering of particles is also accumulating space, which then turns to heat by reducing the available space**

When this comprising of space between particles should carry on within the compressed chamber there will come one point where the heat turns from space to liquid and then the liquid exceeds the limit that atomic bonding of the container can contain. What happens next in such a container that formed liquid heat puts us all in the centre of the star. There is only one exception and that is that where we are there is no control of a governing singularity that we will find in the centre of the star. The heat turning to liquid will exceed the atom bonding of the container walls. By exceeding the atomic gravity the heat between the particles overwhelm the bonding of the elements forming the cylinder wall, which takes the space as the atomic particles that forms the wall. It will supersede any limit the atomic gravity can bare in that bonding between all and every atom as they form a union that we see as the cylinder wall.

The union that forms as the cylinder wall also become the union that forms the cylinder wall that holds the resisting imploding as well as exploding. The wall is as strong as the particles forming the wall would permit and the atomic gravity will permit the union of the wall that formed to be. The collapse of such a wall under pressure that will come after the heat in space liquefied has to come about on the side of the material bonding together to form the wall. The wall then is the weakest spot because the liquid heat will never give way once the space turned to liquid. On the inside there is no governing singularity elect to contain the heat within the unit since the whole idea of a pressurised cylinder is artificial to the cosmos. The liquid in the space will cut any and all cylinder walls to shreds. It liquefies the space in the cylinder wall. The space forms heat and the heat forms liquid and the liquid cut the cylinder walls bringing about the explosion. If any reader is of the opinion that he or she has heard this before I advise such a person strongly to go back to Mainstream physics and find the official explanation about the material providing the pressure to bring about the explosion. Pressure has little or nothing to do with the whole procedure. It comes down to density and this process is not the same or equal too the process that applies within the star. The star is filled with heat being in a liquid form and it is not filled with gas that is true but in the star there is a singularity elect that is strong enough to contain the liquid heat. With the likes of a strong enough governing singularity, the particles as such can become small enough to fit the whole Universe into the container because some time back the lot did fit into a space that was at the time the size of the container. While saying this we also must add that this may only be possible on the condition that we can find a container able o hold such contained heat. It is the release of the particles that concern cosmology since we accepted the Big Bang theory as the correct interpretation of facts, which of course it is.

What happens inside this cylinder is most important to cosmology because through the pumping of the air into the cylinder science consider conditions inside such a cylinder as the same as a star. Science holds the opinion that that cylinder is on its way too becoming a star. Such an idea is totally flawed. There is a big difference in the way the material inside the containers is contained. In the cylinder a big metal wall that is sometimes inches thick contains whatever is building up by means of pumping air from the outside to the containing inside. In contrast to this in the case of cosmic structures there is a massive pumping coming from a space less centre which does the pumping by contracting the space that prevent the escaping of whatever is contained.

While the contained is contained there is a centre able to contain what the star wishes to contain. That is not the case with the air container. In the case of the star there is no requirement for the walls to contain with the any walls taxed by keeping it all on the inside. The inside of a star does not need a large metal jacket in order to contain what we find on the inside of the star. With that in mind it must be true that the Earth as all cosmic structures do, does not have pressure. For instance that which the Earth holds is contained from way within and there is very little chance of that which is contained has the opportunity to be escaping. This we appreciate but also we know that whatever is inside the metal jacket cylinder container will eventually escape notwithstanding the best human efforts of containing. The escaping is just a matter of how long the escaping process will last. Clearly the two processes are so far apart in managed control; the two processes oppose each other totally. There is no comparison between the two.

Even as far as the use of the term "pressure" is concerned. Therefore if we use the term pressure to indicate what the air container holds we are stuck with an idea how the system functions. This is totally contradicting what happens inside the cosmic structure. With the container not forming a visible border in the star there is containing coming from the very centre and this has the opposite of pressure within the limiting walls of the container in question. The air that came in to the cylinder as a gas will turn to liquid if serious pumping continuous and whenever the pumping exceeds safety limits the gas then turns to heat forming a liquid state. In the air container the density increases changing the cylinder on the inside from a gas to a fluid that becomes much more intense than the heat we think of as liquid heat produced by say an industrial cutting torch working with oxygen that is mixed with acetylene. The oxygen acetylene also forms a liquid flaming heat when the mixture is ignited but in the case of the cylinder the igniting will be spontaneous and burn through all the atomic particles. The point of noteworthy is that at one very specific point in time the space becomes liquid and the heat returns back to space as it cuts through the cylinder wall bringing about "an explosion".

The exploding is actually the space that heat returns to space and takes space from times when time was much closer to the Big Bang to what space is at the present. It is a Big Bang that is coming from then to that which is presently applying. The main issue to realise is that the pumping produces a density time factor that increases and that the density increases which turns the inside from gas to liquid. The cylinder moves back in time when the pumping starts and the moving aback in time realises another not containable Big Bang in the micro. It is not the oxygen or the hydrogen or whatever that fills the container that is a gas or a liquid that explodes, but it is the amount of space that turns to liquid heat that turns the container from a "planet" into a "star".

Even the Earth has already some flimsy liquid atmospheres in comparing to outer space. This is the only difference between planets and stars if you insist on having planets and having stars. The stage that the cylindrical container reach when space goes liquid is re-enacting the star with the difference that with the continuous pumping of air into the cylinder there is no sustaining or governing singularity and when uncontrolled the gas will turn to heat and the heat then all goes bang. By this measure mentioned the star is different because in the star such a liquefying is a long process that the star was awaiting for one eternity. That is the difference between compressor having a metal jacketed wall containing what is on the inside and having a pump on the outside that becomes most apparent in comparison with the star, In the case of the star the gravity within the centre of the star is where the star is having an inside pump. By the drawing or the gravity of the star, the star really shows its worth. This is a reaction coming about from the centre the star. Where space and motion ends is truly where the star starts to contain the inside without a possibility that the wall will burst as it goes bang.

The star does not allow escaping of material whereas the heat escapes from the container as it does from the star. When saying this we have to add that heart only escapes from young and incompetent stars such as the Sun is. The star shows no relaxing in the process whereas in the event of no further increasing of pumping heat into the containing cylinder in any way possible and in the event that the pumping does not restart or continues again the process will rectify itself. Condition inside the container will once again return back to room temperature if pumping of air into the container stops. A balance will come about leaving the heat to escape through the container walls and then release as space outside the container.

However, there is no loss of materials such as atoms that rushes out through the cylinder wall. This reversing of the flow of heat will escape through the wall relieving pressure without taking any of the confined particles along. This is apparent because one can feel the heat coming through the walls as conditions inside the container walls strive to become equal once again to conditions as they are on the outside of the container wall. As the cooling goes on, the conditions on the inside of the container will again in time reach room temperatures inside the cylinder too to match the temperatures outside the cylinder. Once the pumping stops, the particles colliding and causing the friction then does not play a further role. They either become very calm unexplainably and without any apparent reason or the principle taught by Mainstream physics lacks truth.

Never once could I ever find out what brought such calmness back to the material in the container without the particles having the benefit of a serious drag dos song that will be of a paralysing tranquil medication will bring on or having a real potent Sangoma (an African male witch or a concept thought as being a witchdoctor) present to enforce a calm. After all the science that teaches us that this heat increase comes from particles colliding and that is what is producing heat inside the air pressure container, the effort in getting them still and calm again after all the excitement must be a tricky operation. What is truly amazing about the becoming calm again is that the already over filled container does not continue the friction between the particles rubbing and brushing and through so many collisions stirring up more heat which was produced within the cylinder container as a result of what science teachings lead us to believe.

When the compressor is left by itself with no further filling or any releasing of the filling substance carrying on the temperatures inside the walls of the container stabilize as they go back to the same levels as what the temperature was on the outside. When all stabilizing comes about from time moving along and allowing the continuing of the stabilizing of the heat is under broken because there then is suddenly a motion of the air being released through the opening of a valve, such a sudden controlled releasing of the air under controlled conditions produces motion of the air as the air is relieved from the container through a release valve. The particles are unable to escape in the same process as the heat does. The heat flows through the cylinder walls and that means all the particles are still present and accounted for in the cylinder.

When the particles are released through a releasing valve the release will create a flow of air, which starts the motion of air, which then will bring cooling. The pipes where through the air flows when released will cool to such an effect that pipes can and does freeze and block all airflow. Two American Submarines were lost in this manner. It is not particles that have to take the blame but heat being released or admitted. If the cylinder is left undisturbed for a few days the stabilizing in the air within the cylinder walls will lead to re-establishing a total equilibrium and such establishing once again will take temperatures back to a freezing state as it cannot retreat heat that escaped space of the cylinder before the valve release came about. After the pumping subsides the space that is compromised then is allowing equilibrium to set in where the equilibrium equalises the particle density within the cylinder to that which is in the room temperature.

The problem that results from this is that there are far more particles in the cylinder. That causes a much lower heat level or space in the cylinder than was the case when the heat was at room temperature. By stopping the flow and allowing an escape of heat the heat leaves space at levels where the levels in the cylinder is forced to be at a lower density then than what was applying on the outside. This flow of air in the releasing pipes show motion and such motion cools down space occupied. Motion is gravity and gravity is cooling of material into the oblivious. Through the motion coming from the releasing of air there is a much lower heat level in the cylinder. The motion of released air will force heat levels down when comparing the heat levels inside the container to what the comparing heat levels are outside the cylinder. The conditions will be directly in reverse as the space inside the container then is much colder by air motion than the conditions of the heat is on the outside. The natural flow of air from inside to the outside will change the conditions on the inside so much in spite of the air being overfilled on the inside and the comparing lack of density in the air on the outside.

If the container and pump is left without further human intervention or human influencing any circumstances developing such absence of interfering in the conditions as far as heat distributing goes will become and remain equal on either side of the cylinder wall. The compacting of air molecules is then still much higher on the inside than what the denseness is on the outside but this difference will not produce the same heat levels inside the cylinder than that what was achieved during and immediately after the pumping operations was performed. The temperature will only become affected when more motion contributes to changes in the balance. The releasing of air will extract heat from the process to a point where it will lead to freezing coming about in the narrow pipes where air is released and such airflow is the fastest.

The rebalancing will go the other way as was the case when the pumping was in place but when the released air is in natural motion the flow contracting heat flows from the outside to the inside of the pipes. But since the motion of airflow will be much faster as the air is released such motion contributes to cooling whereas the motion of pumping gathered heat. Although if science is correct then the heat could have started only when particles made contact and went bumping into one another with causing friction by their bumping and dancing. It is paramount for cosmologist not only to gather mathematical proof but also to apply such mathematical proof in amongst stringent natural laws and find not only what but also why certain events came about.

The spin correlates time by supporting the flow of heat unoccupied to feed heat occupied. There is a definite correlation to establish balancing in space-time and one thing we know for sure is that if it applies today it is because it applied way back when it all started. There is a correlation between heat and space and heat in space, which is very far from being the same thing. Space contains heat because the atom is secluded including heat captured by individual singularity but space can also accommodate heat by producing expanding of space not captured directly by the control of singularity in the absorbing of heat.

Heat will always flow from the hottest to the lowest region because the density of heat will bring about a flow such as water does in gravity. The mere fact that such flow of heat from hotter to a cooler region can take place makes it clear that heat will flow just like liquid does and find stabilizing as liquid does. Realising this must take our thinking back to a time when space was preciously little and heat was souring almost beyond control.

If we tackle an issue as unimportant as a container being full with air in the correct line of thought then it might enable us to draw direct parallels from this. Such parallels might enable us to see how fusion comes into the picture. If a gas surrounds the hydrogen the gas will withstand as much compressing as can be induced by whatever force bringing about such compressing. Gas can compress like nothing else can.

By heating an element the density will deteriorate releasing heat to form extra space and the decreasing density will bring about that more space becomes available within the confinement of the compressed area that is heating. But the more space does not come in the form of more material therefore the extra heat is extra space claimed by the same material occupying the space. If the confinement is such that it will not allow the growth of more space then the heat level will rise. By heating without allowing the liquid to form space then we find that liquid is even more uncompromising than material can be. The heat will grow by increasing heat levels in the liquid heat until a state will arrive where the liquid air will cut through any container that is man made.

On the other hand it is true that gas will always compromise by giving away space to produce heat. With such knowledge we have to look at the functioning of stars once more but this time with a much more critical view. In fact having this realising it prompted me to look at pictures of the Sun with much more critical gaze. What I saw was, the Sun being a sloshing bowl of liquid flowing as fluid once more and then I realised from what I saw that the Sun is liquid. If the Sun is a liquid bowl then all other stars has a liquid inside!

We also know the density levels rises extensively as the Sun's space decline towards the centre of the Sun. That means the liquid will become more and more denser as the Sun reduces space towards its most inner centre and the Sun would have heat in a sub solid state down towards the centre.

This changes every conception there ever was about the inner workings of stars. One can compress a gas to a state of liquid but then the gas is no longer a gas.

The gas transformed from a gaseous state to that of being liquid. Gas can never compromise enough space to allow fusion to take place because there is far too much flexibility to bring about compromise, but liquids are quite another story. Before we come to that however, we must reclassify that which we think about when we think about a star as in this case the Sun.

In the process there are relevancies. If outer space is $- 276^0$ C or $0^0$ K then it can only be because other conditions apply elsewhere where the elsewhere will bring about that under the elsewhere conditions there is something else at $0^0$ C and that can only be valid when another object is in an environment that is allowing that something in the other environment to reach $100^0$ C. In other words there has to be a scale fitting all possibilities that goes from densely deep frozen hydrogen at $- 269^0$ C to a boiling iron at $3000^0$ C.

That is a spectrum but such a spectrum only has validity under very specific space-time applications. But if space stood alone and we gave outer space a temperature value of $- 276^0$ C with no other references to compare such a statement has no validity because the number could be anything just as much as it means nothing. If there is no correlation between two factors that produces a range from zero to a maximum such a scale is as meaningful as much as it is senseless. There has to be a relevancy to validate any figure in temperature used. If we pronounce anything being whatever it can only be valid if it is because being whatever requires not to be another whatever and therein then lies the value of being whatever.

To bring about the relevancy is only valid when allowing us some comparing between the two of different and non-equals. With the one number standing alone such a number by itself has no meaning, not withstanding whatever connection there are between the two.

The objective of a star or any sphere for that matter is compromising by compacting the inner space as one comes closer the centre. As the star is reducing the space it claims for individual use outside the wider Universe it is also condensing such space to the inside where such space claimed is providing a reference but what reference and to what will such a reference point refer? If the sun was a gas on the inside of the sun as science declare, then there was a lot of space going around within the confinement of the sun but that is also true that there then is not much filling the in between the solid substance also going around within the sun. I have seen where comparing was made between the sun and a tire. In such comparing I wondered what was in the mind of such a Master cosmologist when thinking of the star as a tire and a tire as the sun. If the sun is an air inflated tire what then is burning? The burning is the purist liquid of all, it is raw heat flowing like the liquid it is. That is the case where a liquid fills the space because the in between space is filled with a nicely flowing liquid.

With the Sun being under another singularity in charge of different rules setting different standards to the markers we use on Earth the Sun will have a totally different heat standard of freezing or boiling coming from the controlling singularity than we find applying in the governing singularity of the Earth. The Sun we have to admit is very different from the Earth and Earth standards just will not apply where the Sun is concerned. Making a statement that outer space was $10^{34}$ degrees in the shade during the Big Bang festival is rather meaningless because what was the other boundary then?

There is a question arriving from this issue which is: Am I allowed to draw the comparing that the Sun was $6500^0$ on the rim and $18 \times 10^9$ on the very inside at the time the Big Bang was $10^{34}$ in outer space? If that is the case and it has to be the case if science wishes to apply any meaning to $10^{34}$ in the shade at the time of the Big Bang, then the Sun on the inside was an ice box at the time and during the Universe presenting the Big Bang presentation and on the inside the Sun was one of the most potent freezers at the time. We then may regard the Sun at the time of the Big Bang, as being a freezer storage facility, which was colder than the human imagination, will allow our perspective to go. And more compelling is the fact that this then makes more sense than anything anyone said about the Big Bang or that was ever previously said about the start of Creation.

Liquids in the Sun prove to be much more practical because liquid is more dense in certain ways that will solids can ever be. A solid will rapture when asked to compromise because after all that is what fusion is about. On the other hand will any a liquid just become more and become more liquid without giving way. Liquids did the compromising which lead to the Big Bang and after that there is no further compromising under pressure possible (if I am allowed this once to call the inside of a star pressure). That is why the strongest power engineers can use is hydraulic cylinders. By using a liquid within the star too the comparable gas outside the star and the solid being within the very inner star core, we can begin too see comparisons emerging. Remember that outer space gas is hotter than the inner space fluid of the star not withstanding our perception on culture formed by schooling and with memories that our culture is shouting to us in defying our wits because the truth is quite the opposite to what our minds are telling.

When something gets hot space is added and inside the Sun with all the filling of liquid and material space truly is an issue of scarcity. If it was not true that space is little in going around in the star then fusion could not have taken place. Fusion is the product of space in the demise thereof and the demise of space brings about a need for space. That is the prelude in conditions applying upon which the Big Bang followed.

However, with such little space going around then the absence of space makes that the space which should be available, then has to be a priority and with space that little to go around in the Sun, the Sun must be bitterly cold. Again I stress that heat in abundance form space by the volume and that is one thing that is very absent in the Sun. If space was in abundance fusion was not possible. Heat brings about space and looking into the Sun we find that whatever the space there is, there is not plenty of that in the Sun. With no space it means there is a cosmic cold raging in the Sun. Do not for one minute think of size of the Sun that we think is taking up space because that type of space is not anywhere in the cosmos.

There is no big as much as there is no small and the space which we think of in terms of size and what size depends on is the precisely the incorrect human way of thinking because that is the human instinct to measure by size in terms of big or small. If we do measure by giving size in terms of big or small, hot or cold, near or far, and we then include our importance that we shower ourselves with as we put our being the centre of the Universe in the centre of the Universe. Then we set a standard judging from what we control but by putting us in a position in the Universe that makes our point irrelevant by which we use and whereby we judge the cosmos. If we make ourselves important in our eyes we become part of the cosmos and that makes us incapable, as we then are irrelevant about our judgemental concepts.

Having a position in the cosmos will translate to our incompetence as humans, which goes directly into our irrelevancy what we can use by which measure we judge. By placing me the human in the position that I, the human think I am in, then we give our position we think we have as forming the centre of the Universe such a relevancy by which we measure. By seeing us as life being the centre of the Universe we loose perception of what is valued in the Universe. That tendency we then have we must destroy to find meaning to our thinking about cosmology. We think of hydrogen as a gas in the atmosphere of the Earth and we transform that standard by which we think to the Sun.

If hydrogen is a gas here on Earth where we are then we are convinced that hydrogen is also a gas where the Sun is and then the Sun is a gas structure. How much more Biedermeier can we be when we are thinking in such terms...well atheism gives such thinking of the Biedermeier manner quite a go. If we continue to repeat that line of thinking we might as well move back to the cave because such backward thinking belong in the minds of cave dwellers. The Sun is as liquid as the sea is liquid with the difference that the sea uses water and the Sun use pure liquid heat. That makes the Sun a giant hydraulic pump and that removes the pretty weather system that we grant the Sun from the Sun. There is no winds blowing but there are rivers flowing at an astronomical pace. There then are no winds in the Sun but there are rivers of flowing heat running and raging in the Sun.

By using hydraulic power within the Sun instead of the presumed pneumatic gas as suggested by Mainstream science the rules on physics that is applying within the Sun changes as much as day is different from night. In a hydraulic system the hydraulic power will only fail once the weakest spot in the solid breaks down. With a tough enough cylinder that will withstand all pressure pushing at it, something will give way and we know one hundred percent it will and cannot be the hydraulic fluid. The hydraulic oil will produce more fluid when overheating by acquiring more space from the heat asking for more space. However in such a case where we think of oil as a liquid the hydraulic oil is the solid substance, which we humans consider as being liquid and it is only performing as a liquid. It is not the liquid we should think of as being the substance in the Sun.

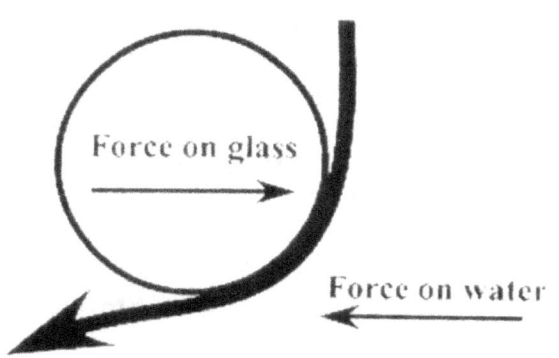

We also can see that the liquid that the oil is re-enacting or standing in on behalf of a true liquid. Being the liquid can only be valid when conditions apply to form and not to the state of the material. If one considers the qualities of liquid then we can ask what will happen when the liquid is pure and uncompromising heat in the purist form the cosmos can provide? What happens when that which is flowing is heat unable to break down? In that case heat can only compromise by becoming more of what is containing the space and not less of the containing of space. By becoming more it also takes up more space and that is less uncompromising than what solids are. In a star such as Jupiter that has gas within the atmosphere, (let us call the atmosphere a gas this time because gas and liquid is a very grey issue) the pressure within the structure cannot apply a solid base to secure solids compromising space that will bring about fusion.

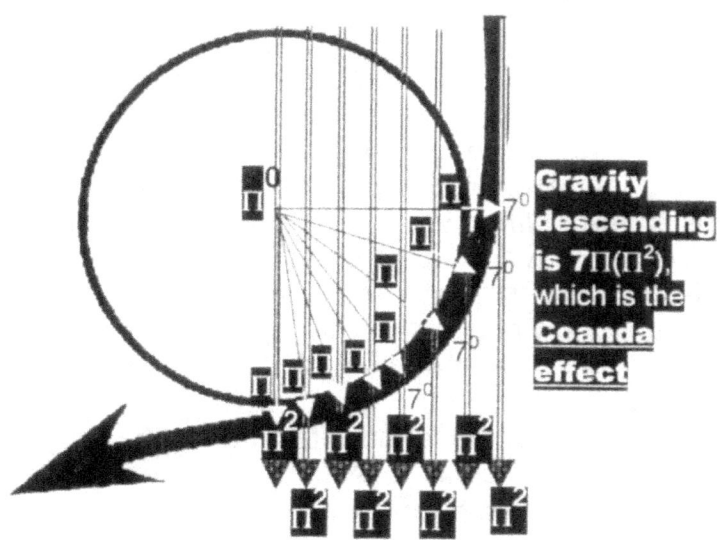

The solidity that eventually brings about the collapse of occupied material holding space is unavoidable when confronted by an uncompromising liquid and by increasing the volumetric space the liquid holds it must come to a situation where something must break. In the case where fusion is in place everything is pushed to the extent where nothing can break. In this environment we know from the characteristics of hydraulics that the final collapse will not come from the hydraulic heat.

The final collapse is that what happens when the solid atoms collapse from being two to being one structure and the solid singularity has to compromise independence. There are two in the space the one had previously and by fusion, the one gives way by denouncing its independence. This action can only present fusion when cold is at the limit there can be cold and the Universe at that point is as cold as it shall ever be and heat is at the very other end where heat can only peak as heat touches eternity at that very spot.

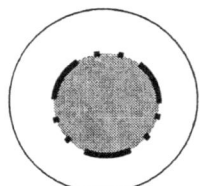 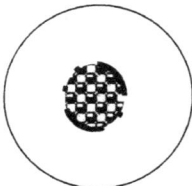 

When outer space is $0^0$ K then the sun is estimated to be $6500^0$ K. That is the relevancy we bring in on a scale between the sun and outer space. On the one side the gas in outer space is $-276^0$ C or 0 K and then the sun changes the environment to what we know the sun is. The sun in our minds is the other side of any extreme where the sun fills space. But the material within is neither -$276^0$C nor is a frozen liquid on the atmospheric rim where outer space meets the suns atmosphere holding a freezing 6500 K. Deep within the sun the temperature seems to be $18 \times 10^6$ K. It is only a relevancy with substance when humans relate the information to circumstances seen from the Earth. If standards applying to the sun were transformed to standards on Earth then no substance known to man including man in person would have any chance to exist because the sun harbour conditions that is beyond frightening as the conditions in the sun is hostile to life of any sort. Actually when thinking about it in a scientific manner this statement includes all conditions applying anywhere except on Earth and then not even all of Earth. By placing life anywhere and at a penny a dozen that notion put the manner science consider life and their life's ability to judge into question. When any person and in particular a scientist is assuming a role as being a judge of conditions in the cosmos such judging can only be introduced when the person in judging divorced him or her completely from they're associated with life. We have to disregard whatever we feel about the importance of life and only pursuit with what importance the cosmos regard life. We have to eagerly pursuit the sun as our nearest star after we dislodged all mind concepts with life attributing to our way of thinking. This is not just necessary but it is a necessity.

Where Mainstream science so desperately wishes to find life all over the Universe they should realise that even on Earth conditions differentiate where certain life can be and where other forms of life cannot be. Even the Earth is at some places hostile to all forms we think of as life. Life cannot be either in outer space or on the rim of the Sun or in the very inside of the Sun. When investigating that which forms the cosmic spectacular one thought that must never leave our minds is that of whatever is out there, we have this tiny spec and that is all. With that realisation we have to come to terms by excluding life when judging the Universe as well as all conditions we may be familiar with because such conditions will fit in and will swing our judgement when such thinking is suiting life. With standards applying in the Sun however there must be one point that is zero to give the Sun an accepted range. Say the Sun would be $0^0$C at the point we give it the norm of $6500^0$ as that will be the lowest temperature that can be found within the confined space in the boundaries of Sun. This will be because the Sun is a secluded, isolated Universe detached from all other Universes except the immeasurable Universes we think of as atoms that group together and form the single Universe we call the Sun. The Sun holds all that is inside the Universe that is the Sun.

From the singularity in charge of the Sun the Sun and the content the Sun holds is the entire Universe and there is only that one Universe. That Universe is the Sun and whatever else is there, that which is there, is there to feed the Universe of the Sun. Under such conditions water that is so absolutely crucial in the mind of human's would be an unknown substance holding an unknown quantity as a reference. Outer space must then be minus 6776 K if the Sun whish to use the scale we apply. But it does not work in that manner either because most of the substance that is providing life the support life depends on and life therefore have no claim to space-time within the Sun. All readers might be agitated by my comment about life not being able to survive in the Sun, but go back and investigate what rules scientist use when we investigate the Sun. All rite I do admit that I am exaggerating about life being part of the Sun but they (them in science) put seas and winds and even fair weather in the Sun. We find winds blowing, gas clouds and such things.

All conditions outside the Earth atmosphere are completely hostile and dangerously destructive to life. If we wish to obtain a clear mindset in our discovery of the cosmos, we should not try at first to discover other life wherever we wish to think we can find life in the cosmos before we discover the cosmos. We have to set standards in regard of very Universe by setting a standard by which other following our thinking afterwards can truly apply our thoughts.

The standards set can only match when all conditions are met with the criteria that that any specific Universe dictates to allowing life. We must then look for life by exactly the very same standards applying on Earth where we find life. We have to find a dot that holds water in all three conditions and we must be able to see the structure with our eyes. If we cannot see the structure those scientists then cannot boast that they are only using truths as science. They must then admit they are in fairyland and introduce their standings as such.

We know water boils at a much lower temperature 100 km up in the atmosphere than it boils on the surface of the Earth and down below sea level it boils even at a higher temperature. This again supports yet again my argument about outer space being much hotter than anywhere else in the Universe because it needs less heat to apply to get water to vapour. We take water as the absolute element of prominence to our survival therefore in all the scales we apply we use water as the centre value and then in one single step further we think about it as being pivotal requirement for cosmic scaling not deliberately but just because we forgot to switch off out human connection. We focus on water as the standard applying whereas water in not found in any other structure.

We even go as far as taking water to boil at $0^0$ C because our thinking that way suits us in our human requirements but even that is a total abolishing of the truth. Water boils where water is located and according to the density applying at such a point where that density in space-time will have a heat to space ratio that determines the boiling point of water. True as it may be that we have little use for water boiling at $5^0$ C it does nevertheless boil somewhere at an altitude of $5^0$ C and some places higher up at far less even less than five degrees C.

Putting water and our totally artificial manufactured perception of the boiling point thereof to where it is in a centre stage and we project that centre stage we think about manufacturing it even further to fit the entire cosmos underlines the desperate inadequacy of our reasoning about cosmic matters. We should not see how life might find the use for the boiling water but we must focus on the rules bringing about the boiling of the water. If water takes less time or heat to boil then water must be closer to boiling point in the highest atmosphere and that can only be if water up there is hotter from the beginning before the boiling process started.

The reason why water boils at so low a temperature at any high altitude must be from the more heat present in the space where the boiling is taking place. With more heat in less space less heat is required to heat more space to get the water boiling. I am referring to the actual space, which heat and material occupy, and is apart from the occupant heat, which is in a fight with material for space to occupy separately. I am referring to heat expanding to the point where the heat occupies all of the space and the space has more room to expand. Therefore the space takes in all the heat it can use to bring about expansion.

If ever life will find a way to go down into the inner Earth say at a distance of 1000 km we will find that at that point water only boils at many hundreds of degrees Celsius notwithstanding the fact that the average temperature will also be many hundreds of degrees Celsius everywhere around. Conditions on Earth have such a variation, yet science has this tendency to standardise everything in the cosmos by applying constants that will fit the conditions we find applying in downtown New York on a pleasant Sunny day in mid spring. If water boils that quickly the closer we get to outer space it should be a big indicator that water is naturally much hotter from the start in outer space. When the heat surrounding the outside of the molecule increases the heat inside the element has to reduce because there is a relevance attaching to the two opposing limits without the one opponent having any precise limit to show.

As the outside fluid heats up, there is no breaking down of the substance because the fluid is the purist fluid there can possibly be. It is pure liquid heat. The heat will be more solid than the solid elements can be because the heat cannot give way any more than it already did at the event of forming conditions that realised the Big Bang. Just before the Big bang arrived it took all the heat occupied or not to form heat that would be able to bring about the forming of space as a compromise to the heat that was bound to destroy the Universe. By sustaining the conditions that was applying even before the Big Bang when the conditions changed there was not enough gravity left to demand the liquid heat to produce more compromising flexibility of supplementing space by reducing density in the liquid state from heat in a liquid form.

By heating the fluid the space the fluid holds becomes more and by heating the fluid the element becomes colder reducing the space the element holds. Space reducing is synonymous with becoming colder and becoming colder is about compromising space occupied. On the Earth material will reduce space occupied that much and no more because the relevancy establishing the edges can only push the reducing that far and no more. But in the Sun the conditions applying is a lot different and the relevancies can push that much further.

As the elements enters conditions suitable to allow fusion one of the two factors surrounding the fusion will insist on all compromising of all space that the elements may have in separating between them or in holding the structure of the atom and in that act it supplies the biggest compromise in the relevancy between the shrinking of material as cold or the space as liquid that removes space by heating more and that is to give space over to the fluid side that removes space from the material side. In receiving the space the heat sacrificed heat for space by acquiring a sudden space. By acquiring the space it receives a cooling that will translate to dark spots on the surface of the Sun once the heat again surfaces to the top. The dark is only contrasting the light because the dark allows less density therefore has more space. There is no actual hot or cold to be found anywhere in the cosmos. In sacrificing heat the liquid obtained space and by sacrificing space the elements in fusion reached the ultimate freezing temperature it could ever achieve. The elements never became hot or never frozen but it only became relevancies between solids and liquids. There is no hot or cold, it is all in relevancies. This is because mass has no role to play in the fusion process. Changing brought about condition allowing the heating and cooling to take place without ever taking place.

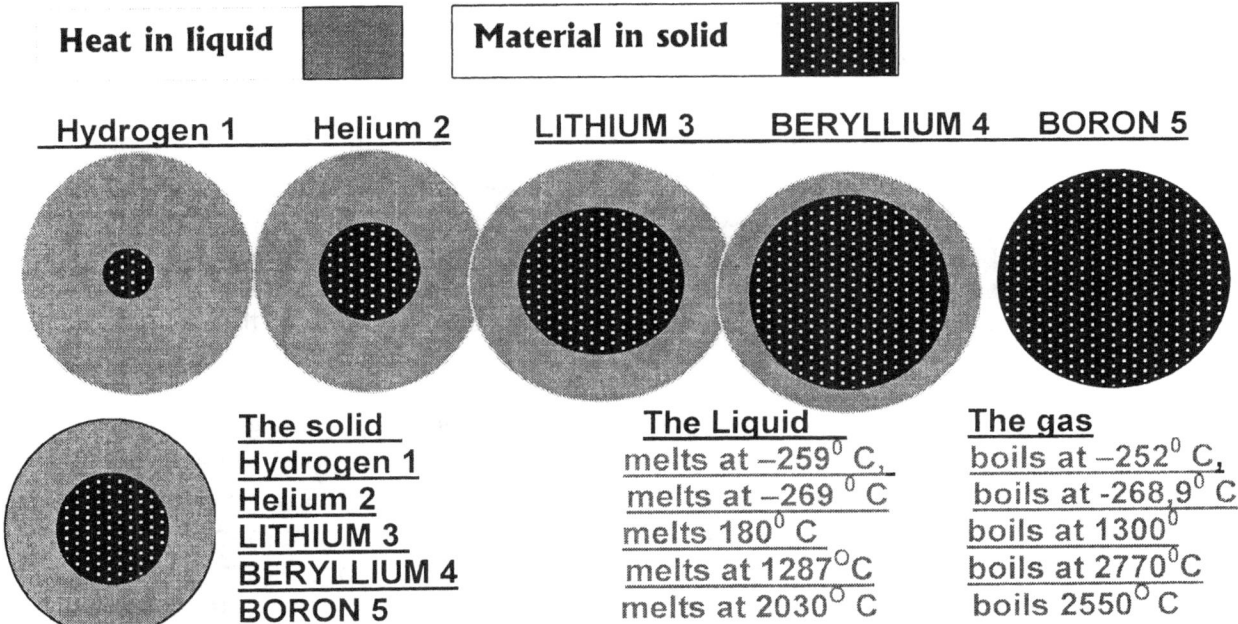

| Heat in liquid | | Material in solid | |
| --- | --- | --- | --- |

**Hydrogen 1     Helium 2     LITHIUM 3     BERYLLIUM 4     BORON 5**

| The solid | The Liquid | The gas |
| --- | --- | --- |
| Hydrogen 1 | melts at −259° C, | boils at −252° C, |
| Helium 2 | melts at −269 ° C | boils at -268,9° C |
| LITHIUM 3 | melts 180° C | boils at 1300° |
| BERYLLIUM 4 | melts at 1287°C | boils at 2770°C |
| BORON 5 | melts at 2030° C | boils 2550° C |

It clearly shows that there are groups of elements with much different relations to heat than other groups have. This is such a dominant part of Creation I truly am surprised our distinguished Academics never investigated the evidence in hand of elements forming different conditions that obviously totally devastated Newtonian claims on mass and weight producing gravity.

It is believed that mass produces gravity but according to my reckoning it is gravity that produces mass and mass is only the result of gravity. The less space a mass holds within the stronger the mass is subjected to the more mass there is in less space it to holds. Mass increases when the mass is surrounded by atmospheric heat where the heat will increase the mass as the additional heat will add onto the mass and as the heat will influence mass. By adding heat the object becomes more massive but also more spacious. In such an event the adding of material albeit heat increases the mass. But where gravity increases the mass increase as the space reduce. That shows a complete different tendency. Therefore mass does not produce gravity as science indicate by their formulas used in calculating gravity because gravity does not increase when heat influences mass.

Heat stored in motion produces gravity. Any one not in agreement convince yourself by comparing the neutron star with the massive red giant and by you're acquiring a logic conclusion that is not tainted by your opinion about big and small you will be convinced of my correctness. Science go about in order to measure by calculation a Black Hole as follows: they go and throw $C^2$ next to the dividing radius and throw the square onto the C by removing the square from the position the r normally have and they're doing that should then present the speed of light acting as the retaining diameter. What the hell the speed of light, which is pure motion has to do with a diameter, which is merely a measure of a space distance and to top that their reconciling of the two is beyond my ability to substitute fact for imagination.

With that they can enjoy much of an evening's fun and games manipulating understanding because then they can cheat enough to make nonsense of the whole lot of mathematical science they ultimately play with as a toy to enjoy. Then they sit back and feel smart in the way they manage to cheat once more to prove their incorrect views correct because after all who will ever fly down a Black Hole and return to support or deny their calculations. The gravity applied by the Black Hole is a speed, measure comparing space in ratio to time taken because all gravity is speed. Then the speed that light has is gravity.

The gravity of the light can be gravity as much as it at that very same time can be antigravity. What the hell has $C^2$ got to do with a Black Hole because you can pop what ever nuclear device far away from a Black hole and it would be at the most and at the worst very much insignificant. The light will not even escape form the gravity of the Black Hole but that has nothing to do with the diameter of the Black Hole except on the condition that they agree that gravity is the reduction of space and that brings about a link.

They never even suggest what the speed of light and the diameter of a star has in common. The one function is for measuring a distance and the other is calculating a speed. Obviously the one has no tidings with the other. When this became apparent that the radius of a star reduces as the stars develop through progress, someone was supposed to say: hey there is a dead rat I smell. For my saying so I am the clown in the courtyard, and the Academics see me as the one with the two dead brains cells and have no more to use as spare.

What one can clearly see from the element table is that every element carries a different surrounding coming about because of different speeds and in that the proton number has no coherent reference value. Every element carries a different value of heat in relation to the atom solidness. Much more important than mass is the density factor which means the volume of space an element takes up in an overall distribution in a specific quantified space volume. It is the number of atoms that can occupy a certain volume of space that is much more pertinent than the mass erected.

Airborne elements just cannot stick together and share space, as solids can. It is their qualities that carry the reason for a star performing and not the magical pulling of quantifiable man made mass that is so easy to calculate. Every element holds characteristics inspired by form and prefers placement and these characteristics make up a performance of space-time motion balance within the star.

The way that an element interacts with heat in space and space is having heat as the containing substance where heat has as much a function to perform by assessing with duplication as does the dismissing of space have a function to perform. In that manner every element performs differently and to specific characteristics. We are able to see some features of element characteristics versus the misconception science have about contained air when looking at the way a compressor cylinder are working.

In this process one will find the very principles that may apply as matter and anti-matter. The spin synchronises from singularity form the centre, but also equal in demanding space and time. By commanding space and synchronising time one can see that all singularity does not demand equal spin therefore it does not command equal space under equal conditions. From this situation heat will arrive, as friction will devastate all material. In particles sharing space there is a motion ratio putting particles in a rotation sequence that seems to be a gear-locked action.

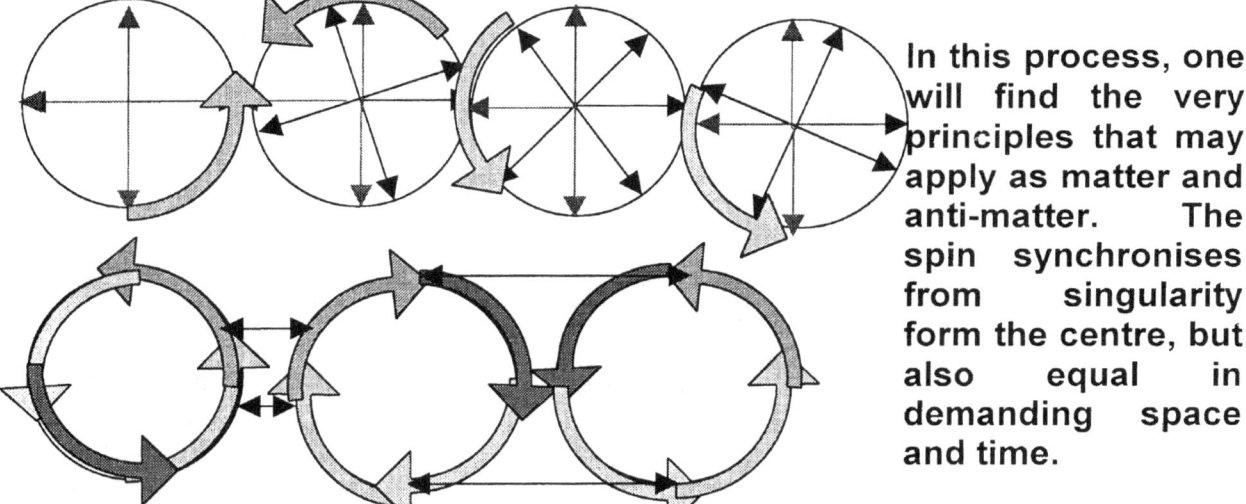

In this process, one will find the very principles that may apply as matter and anti-matter. The spin synchronises from singularity form the centre, but also equal in demanding space and time.

At first in the beginning far back the heat was high but to have heat that high something had to be cold otherwise the concept of what is hot or cold and high or low and what is little and what is a lot in heat margins is very much meaningless. If one takes what we have now then that being what is in place at present when compared to what was present way back only can have any meaning when comparing delivers standards and when that brings across a meaning. But what we now have was not in place back then and what was then is not with us now. If what was in the cosmos back then was being that hot then one has to focus on what is applicable now, but what is applicable now did not apply back then because what is now was looming in the background, waiting for now to arrive in the future.

There has to be a ratio or a relevance that places the utmost in correlating fashion to one another. If the temperatures was $10^{34}$ then that which contained the heat at $10^{34}$ had to be at another level to produce any other field marker in reference at $10^{34}$. With no zero to be found there was no $10^{34}$ around. If there is a zero to this day there was a zero back when things came about since nothing alone can be added or removed by science from the Universe at will and at their pleasure. The rest of whatever is and was has to remain to this day on condition that it was already in the Universe as part of what is in the Universe.

A large part of my proposed theory is that in singularity heat and space was joined as one thing in one unit. We are not aware of how or what that consists of because that with the knowledge about the margins of heat and cold joining is outside this Universe. We cannot conceive what lies there because we sit within the result of what came about when space and heat parted their unified ways. That was when Creation came about and space held heat in a part being apart of space. How much ever the all-controlling Science community think them equal to God, they must swallow and accept they are not God and this knowledge will forever be beyond their inquisitive reach. Heat and space divorced the eternal marriage and space produced heat apart from space and was after that separation no longer being a part of space. The release came about when space gave birth to heat, which formed space-time in the form of structured material that remained solid or where the other second dimension of Creation lost density and became liquid heat.

At the first release of singularity creating space-time which established space-time it went about by putting all on an equal footing. The equality did not favour all and some became marginalized. Then gravity and antigravity formed and this forming brought about that some particles conformed into less dense heat, as some of the material was unable to produce gravity and secure their form. One should realise that the period, which I am referring to, took place when the electron was still inconceivable. There was no cubic Universe with square edges. That came much later. Those less firm were the ones committed to motion producing antigravity. But antigravity is as much part of gravity as gravity is a part.

Other particles formed as singularity retaining structure by their applying of gravity and constructed a valid way of retaining form as it then developed to become material by retaining a specific form that locked in heat from the other liquid sources. In the process material applied gravity as a means whereby it removed heat from surrounding space created by all singularity that was overheating. Some malformed through their motion while others contributed to contain form also by motion but the motion this time came about in the spin they accomplished around their singularity. This spin of particles through space formed gravity functioning as gravity within the space that the new particle was in as well as the surrounding space the particle withdrew heat from. As the particles removed heat from another sector it found material by which it could manage to bring about duplication in motion because the duplicating also meant the securing of material onto the form it upheld as it maintained singularity. By this application of securing form and abandoning form a trend was set where some particles was placed in relevancy and through the action of the relevancies in motion some particles could freeze heat onto already frozen matter.

Matter grew as it galvanised heat (I have to use the term galvanized because that is precisely what it is.) It is used in this sense because the process we use to bring about galvanising is the same process gravity uses to apply heat onto matter. It is for the lack of a better term or my reluctance to create one more meaningless term that forces me to use the term in hand and an attempt to prevent yet more useless naming to come into disuse. Such galvanised of heat is the precise same means used by the electromagnetic plating process but only at the rate of very concentrated gravity. The fact that we can use electromagnetic plating is proving the fact that it is a principle life may be able to manipulate but it is what nature created.

Gravity, electricity, electromagnetic fields and electronics all amount to the flow or the displacing of heat. It is the very same thing with only having the concentration levels in each case and distribution of charges applied differently. This plating of heat onto space filled the already frozen solid material with more substance, which is heat that it by means of plating freezes onto and into material that then formed a sealed or frozen structure. We call this structure the protons. This process is continuing to this day as it did when creation started. Some singularity remained protected and formed material where as others was unable to bring about motion through spin or gravity but had to resolve to motion coming from expanding by overheating and relinquished form by liquefying. This statement I shall prove as the writing progresses and as I introduce more substantiating facts. To us in life this first event that ever took place is now known as the Roche limit.

One should see why did the heat come about when singularity produced space-time for the very first time. Saying this we must understand that there was two phases at least where heat came about. The first was the moving from the spot to the dot, which was a spontaneous growth but then there was the motion bringing about space. This was when the dot became gravitational or anti gravitational and each dot responded according to personal as well as group relevancies. In this second or motion heat was mainly a result of friction between created particles and exposing came due to the lack of space. If there is no space to contain the material the end result is friction between the particles. Once again the result of friction is further heating. This fact is so extra ordinary because we know that billions are spent in industry to combat friction heat. That was precisely the conditions that brought about the forming or deforming of different particles in the pre or post Big Bang second. Heat will always flow from the highest value to the lesser value.

This is the concept I use for the basis of my entire theory. I base my theory on gravity producing cooling and contraction while heating produces motion by expanding and creating more space, which on the other hand also leads to cooling. By being without rotating motion cosmic objects retain heat and expand. The motion will be whether it is the motion of the unit as a whole or whether it is within the unit forcing the unit into more space. When overheating no amount of force can retain the container from becoming too little and with the heat coming about forming the expanding the space produced. With this action it will destroy the space any container may have not withstanding whatever type of material we have at our disposal and is therefore used to manufacture the container we are able to use or the container size that we can use or even the force the container in use is able to contain.

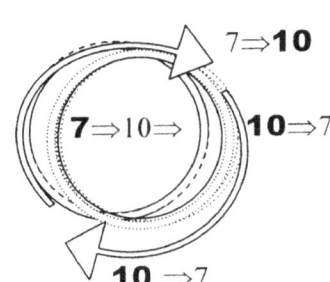

In this writing I will show how the Titius Bode principle in conjunction with the Roche limit form gravity and I show with the evidence of that how nature uses the law as a jig where it becomes a template forming space between the different planets in the manner in which they are evolving. The Titius Bode law should be seen as the process where gravity left its mark. Titius and later Bode found marks that gravity left for us in space growing but Newtonians failed to see how.

Since not even Newton could claim to know what gravity is and therefore nobody was able to read the markings of gravity this tendency was dismissed as coincidence, by those who should have investigated the process.

But since not even Newton could claim to know what gravity is and therefore nobody was able to read the markings of gravity this tendency was dismissed as coincidence, by those who should have investigated the process.

During Creation before the Big Bang the compactness of the particles produced motion discrepancies between different particles bringing about friction where some particles overheated and formed heat. Heat then formed space in a process that became the Big Bang where heat produced space and formed the motion we see today in the Universe as the Hubble constant. In space there is different dimensions that we call sides of which there are six opposing one another. Those sides that form is forming from the heat that became space in motion but the sides play a much stronger role that is anticipated presently. The sides are dimensions and the dimensions come about since motion created the discrepancies. Let us think what the verdict at present is using current Mainstream Physics on the question why a water drop would float by forming a sphere in an outer space capsule housing people in outer space in the presence of micro gravity. Why would a sphere always form when left on its own will to capture free form?

The most important issue about gravity is that gravity is the strongest where space is the least. But when space is the least heat within that space where the strongest gravity is, is at its highest possible point. This we put to human advantage in the working procedure of the internal combustion engine. One thing that is true is that if a process is there for us humans with intelligent life to manipulate then the process is a cosmic fact used throughout the Universe in some process within the Universe. Man can create nothing and that is all man can create.

Man may construct what was not constructed before, but in that effort man can only re-arrange what is already created. I suspect the process we use are the same process used by very young stars use to get heat levels rising within those young stars however the fuel that is processed is not necessarily the same. I suspect in very young stars the atomic ignition process, is similar to the process which Oppenheimer "invented" to ignite the two Japanese bombs used after the Americans got the nuclear fuel from the Germans but did no know what it was or how it worked. This nuclear manner of igniting extremely heavy elements that is found in all stars' core is mainly used in extremely young stars with a desperate need for heat and motion growth. It is reducing space to increase time coupled with non-nuclear fuel that ignites the uranium and possible natural plutonium in the core of the young star.

We all accept that the true cosmic form in Universe as a whole would have and most probably does use is the sphere...but why would the sphere form as the original form. Newton acknowledged the fact but was unable to explain it except for some vague nuance about it. If liquid material is left to a natural outcome material in a liquid state will take on the form a sphere has. While saying this we must acknowledge that all stars are only massive liquefying machines hanging in space with nothing more to do than to liquefy everything they can get hold of. With that in mind, the question about the sphere becomes extremely valid. What is that special in the sphere to secure the favourite form or form of choice of gravity in the Universe as a cosmic structure?

By merely blaming gravity pulling from the centre is rather avoiding the question with simplicity because the question arising from this answer is that since we see the Universe taking on a form as a sphere then where then can we locate the centre of the Universe? When applying Newton's standard cosmic formula whereby it is used to calculate the gravitational field the formula in use is $F = G (Mxm)/r^2$ the issue in the formula becomes the square radius and the radius is forming the centre whereto the mass directs its full force compliment. The more growth by which mass increases and the more mass there may come about the more the pulling motion applies by the centre towards the centre that will secure the motion of mass moving towards heat within the centre. The more mass there is bringing about the pulling force of gravity pulling that secure all gravity force within the centre by the radius square the stronger the domination of such a centre will be on the surrounding space-time. By the securing of a square radius in the centre the reducing of **k** will produce less space occupied $a^3$ but much denser space occupied bringing about a much stronger rotation of time $T^2$.

## The piston in the exhaust/ intake position stroke

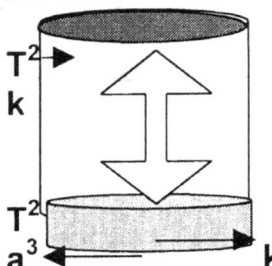

**Compression / ignition**

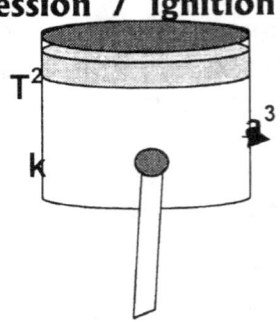

One must see the piston engine in the same manner as stars in operation or the Big Bang. By reducing the stroke **k** we find the space $a^3$ decreases as the temperature $T^2$ rises in ratio. Take this scenario and compare that to conditions applying during the period we think of as the Big Bang.

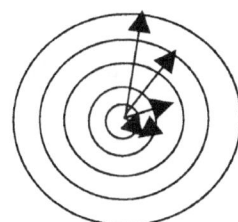

Compare the cylinder conditions to the conditions of the Earth in comparison to outer space or any star. As **k** reduces from the outside towards the centre the space represented by the declining **k** also reduces in ratio and by ratio equal in measure. However, a smaller distance **k** brings about more heat concentrated in the reduced space. If $a^3$ reduces then **k** must also reduce but then $T^2$ holding the gravity or heat concentration must increase. That is just what is happening inside the engine and inside the sphere and with the Big Bang where with the Big Bang the situation is reversed. This is what Kepler said with his formula $T^2 = a^3 / k$. If **k** gets smaller it will increase $T^2$ as much as it decreases $a^3$

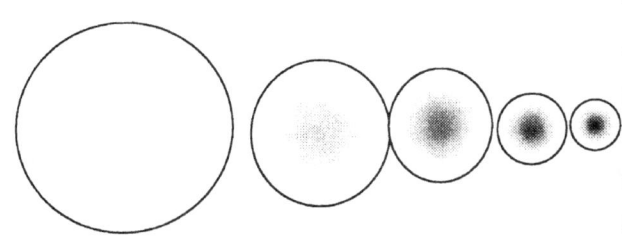

The closer **k** reduces space towards the centre of any star the hotter the star is. It is the same with the engine. The internal combustion engine teaches us the implication of gravity.

By reducing the **k** factor the $a^3$ factor also reduces but the heat in $T^2$ surges. The power computed by the engine is covered by the principle that compressing space is reducing space and that generates the heat required to ignite or contribute to the igniting of the fuel. The space has to turn liquid to command the maximum power the machine produce.

When investigating the process in all cosmic objects the gravity collects heat and accumulates the collected heat in the centre of material forming a sphere. All structures with a superior gravity collects heat and more so progressively towards the centre because in all objects with a strong gravitational field there is a collective quantity of heat towards the centre of the object. It does so by concentrating space and as such produce heat. By producing more concentrated heat it secures independence to the unit forming the individual container unit and the independence will produce motion to move away from the singularity in control. Gravity has two factors influencing space, which are a straight-line **k** and a circle going around the centre $T^2$. It is a balance $a^3 = k\ T^2$ that forms $k^0 = a^3/k\ T^2$

Gravity $T^2$ increases as space $a^3$ reduces by the reducing of **k**. Much more to the point that implicates gravity within the star would be the heat the star concentrate within the centre of the star. Such heat shows much more relevancy than the mass it holds in the sphere of the star. The more space there is between the particles forming the mass within the sphere of the star the less the gravity output is because the less motion such a star can achieve and the less space the star will dismiss. We should find more collaborating evidence when we investigate the working principle behind the eternal combustion engine.

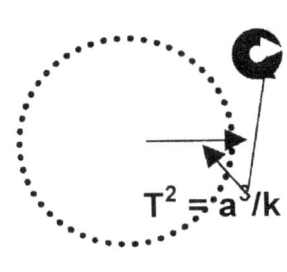

A smaller factor **k** will result in much more compact area. The smaller **a³** becomes the larger **T²**. This will be from the reducing by a larger k. The compactness is the result of the density of the atom distribution increasing and by such increasing the mass or occupied space becomes denser and more intense. With that kept in mind it is quite silly to go about measuring the diameter of any star and from that detail, try to determine the "mass" because the diameter could have reduced a billion times with the motion that the inner star achieved.

$$T^2 = a^3/k$$

The space reduced within the star centre is existing space that was the result of the left over of material, which was unable to establish gravity and reduce therefore established heat becoming space. The space that the star then moves into is that space which the star then contracts. It is existing space made redundant by motion applying the gravity the motion command. When Mainstream Science investigates stars and finds the true giants being very small as well as very massive every one acts surprised. I wonder why...? Has it got anything to do with the ignoring of Kepler as the cosmologist? I wonder why...? Has it got anything to do with the ignoring of Kepler as the cosmologist by favouring Newton as the mathematician? This question is supported by another question.

Are stars really about mass producing the fundamentals or are stars about reducing space by gravity to fit into space more particles taking up less space and thereby producing the more mass which then is the result of much denser particles holding more mass by more compact particles because of higher motion achieved? Is the increase of the density of the atoms reducing the space but at the same time increasing the mass claimed in space occupied by material? Can it be the result coming from stars with much higher motion potential producing much higher gravity?

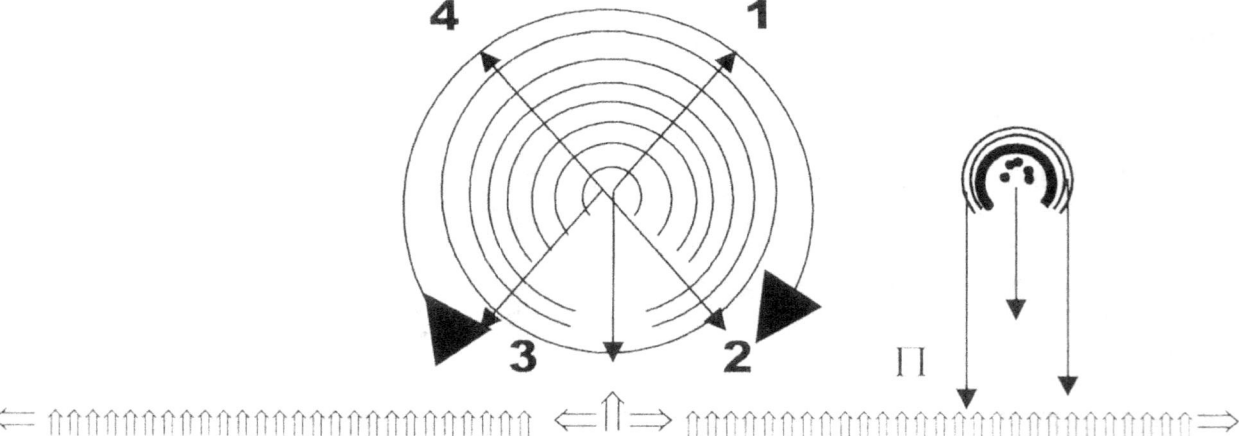

We know that the Black Hole is spinning the oncoming space faster than the speed of light... and that means the gravity is contracting faster than the speed of light... As gravity intensifies in true large stars the space vanishes from the equation where the star becomes forever smaller. That means gravity destroys space but does not pull matter. We all are very aware which star in nature is the mighty gravity producer. It is the smallest one as far as it holds space. So then has mass the least say when gravity is generated? But it also is the one producing the most motion.

That means gravity is not about space and all about motion. It seems most likely to be true. In such a star there is no more mass as all the mass turned into momentum. Later findings proved Kepler as being astonishingly correct. The diminishing of **k** will reduce space **a³** but it also will advance gravity **T²**. It says so when reading mathematics correctly. By redirecting Kepler's formula mathematically: $T^2 = a^3/ k$ shows that gravity grows where space demise. Every time the diameter diminishes the gravity produced by the star becomes immensely more. The space claimed decreases as we increase the motion with which the star holds relevance in the space it occupies and which it claims.

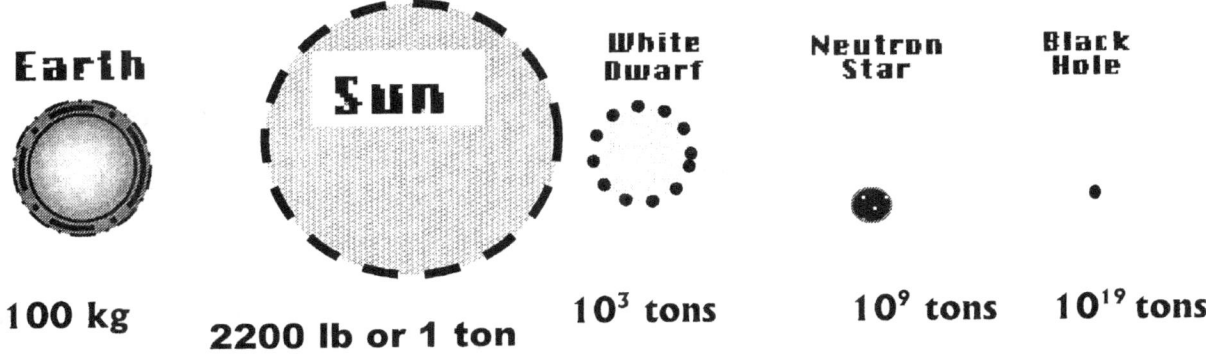

| Earth | Sun | White Dwarf | Neutron Star | Black Hole |
|---|---|---|---|---|
| 100 kg | 2200 lb or 1 ton | 10³ tons | 10⁹ tons | 10¹⁹ tons |

The Oxford dictionary of Astronomy defines Gravitation Collapse as follows
The collapse of a body that is unable to support itself against its own gravity. Gaseous bodies undergo such collapse if they are not hot enough for their gas pressure to balance gravity. This can happen in the early stages of star formation, or when nuclear burning ceases in a star's core. The time taken for such collapse decreases rapidly with increasing density, varying from about 100 000 years for the birth of a new star to less than a second for the formation of a neutron star. Star clusters may undergo a similar collapse if the random motion of their constituent stars is insufficient to offset gravitational effects, either during their formation (see violent relaxation) or at an advanced stage of their evolution.

If mass was an issue in the applying of gravity then the mass will play a major part in the time it takes a body to release from the motion when releasing from the Earth. Mass increase when accelerated and in that then the body must actually grow bigger to have more mass. By increasing motion the actual virtual mass increases therefore motion can increase mass by gravity increasing. One has to accelerate an object to navigate release of the Earth. But in such acceleration the mass increases many times over but notwithstanding that mass increase the accelerating motion secures the release in any way.

The difference in velocity that the star produce and the matching velocity the object in outer space must produce in motion to match the motion requirement of the planet's motion relating to the outer space factor such motion creates a velocity differentiation but why that takes gravity is one of the biggest mysteries I can find in the Universe. Again I must press the issue that since Galileo proved otherwise where otherwise is being that what Newton stated everyone agrees with Galileo and then completely ignore Galileo. Galileo said mass fall equal...that means mass draw equal notwithstanding mass differentiation in size or mass and while every one is admiring what Galileo said then by the same margin every one is completely ignoring what Galileo said. Newton said Mass one and Mass two draws space according to mass, which means mass has the ultimate influence in the process.

Galileo said all fall equal notwithstanding mass...Newton said mass does all the pulling and Galileo said mass has no influence on the drawing of the object...yet everyone is totally ignoring the fact that mass, be it big or small, proves not to influence the drawing of material that is falling! If Galileo is correct something else than mass is doing the pulling...and for the life of me I can get no Scientist to see what I say in the contexts that I say what I see. The concept of mass being the producing factor in gravity comes across as rather less thought through and more than a bit silly when considering the above. All evidence points directly to the idea that mass generates gravity and such presumption is most inaccurate. It is completely incorrect to think that it is mass that produces gravity since it is a notable fact that as space shrinks or reduces the stars ability to generate gravity excels. But space has little to do with mass.

**Red giant**

**Betelgeuse**

Dia. 1400000000 km

Diameter relevancy is

**35.2 km.**

Before we can ever dream of understanding stars we first have to define what the purpose of stars are. What is the role that stars play in the evolution of the Universe? Even before doing that we have to define the role that the Big Bang played and what brought about the Big Bang The Big Bang was about producing space-time and by space-time developing stars release from galactica. The galactica produce the space – time from where stars arrive into the Universe.

Gravity is strongest when and where space is least. Where space is least motion is producing most relevancy and change. More gravity leads to smaller space and in such a smaller space the bigger gravity can hold more shrunken mass. This is even truer, since motion produces gravity. With the stringent reduction in space more particles will fit into less space and by having a smaller space it can manage a higher quantity of particles. But in the end the reducing of the space held by the material in the star is the result of motion contributing to space decline By producing a larger area to fill by using a specific space occupied the space occupied will have to spread thinner to compensate for the larger area commandeered to fill $a^2 = T^2k$ should $T^2 k$ increase $a^3$ must reduce.

By this it is clear that even referring to a force applying is directly suggesting motion occurring or a tendency of a serious effort to bring about motion restrained that then is a blocking of motion occurring and that makes that mass is the restricting of motion trying to continue in a specific direction to come about. If one takes away the motion or the tendency to bring about motion then it is clear that gravity disappears. The Black Hole contributes the strongest gravity since the Black Hole places all motion in space and no motion in the star. Nevertheless the motion we see comes as a response to that whole ending of space-time is in essence the product of the motionlessness of the star. Since all stars apply motion and the Black Hole reveals the ultimate form of motion the Black Hole shows that gravity is space in motion by reducing of space towards the centre where there is space less ness and motionlessness. In the invisible centre is a point even beyond where space and motion is at its least. Without space there is no motion and without motion there is no space. Kepler tells us this as $k^0 = a^3/T^2k$.

**Yellow dwarf**

## Sun
### Dia. 1400000 km.
### Diameter relevancy is 38 meters

**White dwarf**

### Dia. 16000 km
### Diameter relevancy is 300 mm meters

● **Neutron star**
Dia. 19.2 km. Diameter relevancy 3 mm

. **Black hole**
Dia. 9.8 km. Diameter relevancy 1.5 mm

It is in this vain that we have look in order to find the essence that stars have as a product in the Universe. The star is a contracting space-time devouring machine that combines and unites atoms. It is about reducing. The more atoms the star unites in a process called fusion, the less space will be available to occupy by the local inhabiting atoms. The bigger the atom numbers are in proton clusters the more gravity the star may produce because of having less space to deal with. More gravity is equal to less space holding material. The purpose of the star is reducing what the Big Bang was increasing, because we are still in the Big Bang and we are still the Big Bang that is moving away from a center towards another center. The less space a star has the more gravity it has and the more gravity it has the more it will reduce space-time.

However a point arrives where the star reduces more space than what the photon can produce. At such a point the contracting motion within the star, which is gravity exceeds the motion that light can achieve. Bang goes Einstein's prediction that nothing can go faster than light and therefore the speed of light is equal to time. In the process we see stars develop way beyond what the Big Bang restricted the Universe to have. These stars are at first in a desperate struggle with light and we see that eventually they develop past the barrier that light provide. In the end the star is about motion becoming motion less by compromising space. From the evidence stars leave us we can again see gravity is motion by movement through time as much as it is motion by the reducing of time.

If one is infinitely small, all is infinitely small. There can be no space when motion is at its slowest possible speed. That forms the ultimate relevancy available and space-time or space in motion is all about relevancy. But the enormous gravity falls outside the star. In the Black Hole singularity controls matter and space applies all motion that is in fact the time factor to space occupied where the motion aspect is more commonly known as gravity. But the space less ness of the Black Hole shows that space less ness is the location of strongest gravity. It is in the place that the heat is the most, which is in that centre area of any sphere. If any one does not believe me then test nature. It means that mass has the least say when gravity is generated. According to Kepler mass in motion within space in motion and gravity is the same thing. $a^3 = T^2 k$.

The motion that the Earth resists which fills the space that applies the effort to move is the result of what ever is in effort of trying to commit motion. It is precisely what Kepler said when he said space-time is $a^3 = T^2 k$ , with the only difference being motion that can also be a tendency and not only an established move. Looking at evidence we find in the cosmos it seems most likely to be true but the radius **k** distinguishes the space $a^3$ required and the gravity $T^2$ produced. $k = a^3/T^2$. Later findings proved Kepler as being astonishingly correct. The diminishing of **k** will reduce space $a^3$ but it will also advance gravity $T^2$. $T^2 = a^3/ k$ shows that gravity grows as space diminishes. Every time the diameter **k** diminishes and the space $a^3$ acquired by the star reduces the gravity $T^2$

produced by the star becomes immensely more. In Kepler's formula $k = a^3/T^2$ the smaller **k** becomes the smaller $a^3$ becomes and the bigger $T^2$ then gets. $T^2$ represents the gravity that positions the space $a^3$ at distance **k** from the centre capturing the structure through gravity applying $T^2$. Looking at the illustration of the comparison between the mass and the available space in stars we have to come to a conclusion that we all are very aware which star is the mighty gravity producer. Now after some rethinking we can again ask the same question we asked a while ago.

Looking at such clear evidence one is struck by the question that we arrive at next: Has mass then the least say when gravity is generated? I would say that mass is the result of motion breaking down and therefore is a factor of and not a factor presenting gravity. The more gravity reduces space is the result from the accumulative effort of every individual atom according to proton mass (number) as a unifying effort of all the atoms in the star in accordance with mass applied. The idea preached by those that should know better is that mass is the same everywhere and is never changing. Why would there be such huge mass increases in the bigger or should I say smaller stars. Mass is supposedly another constant because a body has the same mass even in outer space where it absolutely has no mass. What would entice the material inside let's presume for instance the Black Hole stars to grow more massive if mass comes about from the pulling of one particle closer to the next particle. If it was about pulling on each other the mass of the particles could not increase through applying such a method because the honeycomb would grow bigger in size and not smaller. Even by combining the mass of two individual atoms the increase is already in the equation.

By reducing space to the ultimate in the centre of the star it happens where space is little and where gravity peaks and where the heat sour to the ultimate. As heat rises and space reduces it is surely true that the particles in such an area must take up less space so that more particles (atoms by numbers taking up less space by volume) will fit into less space by reducing the inside volume of atoms and thus in occupation totalling they then at that point in space-time are taking up less space. After all such evidence proves that more material fit into smaller space. This establishes the cosmic situation such as the conditions we are able to calculate showing what was present during the Big Bang. What was then part of the Big Bang is now prevailing in the centre of the massive star.

A star is the cosmic device that is placed in space to reduce space back to what was the case before the Big Bang event and so the Big Bang is once more repeated but this time the star is taking the Big Bang in reverse by compacting the space within the star. Within the space within the centre of the star being as little as possible the heat is as dense as possible but in that scenario more protons pack a smaller space and more protons in a smaller space accumulate more motion per space unit. An electron accompanies every proton and where the proton represents the flow of space towards the centre, there is the electron responsible for the duplication of space by motion. This brings about more space diminished as more space transforms to more heat in less space. While the star is reducing space the Universe is expanding space to comply with cosmic equilibrium. This is in place, since the heat that the star gathered through gravity will return back to the centre of space less ness and motionlessness where it was when Creation started. If the heat accumulation by motion control is not well controlled by the governing singularity we then find Super Nova events occurring.

EXPANDING Universe
Any model Universe in which the space between widely separated objects is expanding. In the real Universe, neighbouring objects such as close pairs of galaxies do not move apart because their mutual gravitational attraction exceeds the effect of the cosmological

expansion. However, the distance between two widely separated galaxies, or clusters of galaxies, will increase as the Universe expands.

KANT, IMMANUEL (1724 – 1804) was the German philosopher, which proposed a cosmogony, published in 1755, in which the Solar System forms, via a disk, which condensed out of primordial material. The Solar System was part of a larger system (what we would call a galaxy), and many of the nebulae seen by astronomers were in fact other galaxies, which he termed island Universes. Kant was influenced by I. Newton's theories, which he termed island Universes. Kant was, as everyone ells up to now, influenced by I. Newton's theories and by the English philosopher Thomas Wright of Durham (1711-86). This did deliver the ideas but still we have to be practical.

Where all motion is ending in circles, which is the result of gravity, it puts everything in circles that are evidently growing by expanding. We have to venture back into our past. As we venture back, we must take a circle along to find that the circle is representing the cosmos and reduce the circle to see what comes from there. Of all the models prepared by all the wise during the past this model was never made in reducing the straight line that runs between intervals in the Universe. That is what I did and by following the suggestions arising from Kepler's formula I uncovered a hornet's nest. The hornet sting depends on the individual person's view of what a sting is. The sphere as any circle and all circles does have two parts. One part is a line that indicates the distance between the two opposing sides being $180^0$ apart. The second factor is the pi indicating the result of the square value of the other factor being the straight line which by the square produces the measure of the shape it indicates. But where as in the normal square the line matches and have angles it connects touching lines that connects one another and in the square one will find the surface of the cube in the lines. With the circle in comparison, the square falls inside the circle but the lines cross at an angle and at such a crossing of the lines in the very centre, the circle takes form from the securing of all possible corners and contains them in one point. The absence of edges hold the form value of the circle being $a^3$ coming about as pi cuts corners literally. But the overall prominence comes from the fact that the circle holds a cross inside the very centre as the diameter then doubles by the square and a square never crosses on the outside as in the case of the square where the lines touch at angles. With the square we refer to the lines indicating the borders of the square forming the sides that is the borders and indicate the final measure of the square or the cube.

The sides we face as in example length and breadth that becomes multiplied and form the square or in the case of the cube, the cube. In the case of the circle we use the lines crossing to the inside on the length they represent being a full line or half a line. In the case of the full line the name used for such a line is a diameter and in the case where only half the inside line of the circle applies we gave it the name of a radius. Being fully aware of the various names I prefer to use r to indicate any and all lines that forms a combination to indicate either surface or volume for either the square or the circle.

Whenever I refer to any straight line I would use r as a symbol indictor. To go back into the past where we may find the start of the Universe one have to reduce the radius because the radius of the circle indicates the size while pi indicates form. The Big Bang is all about the cosmic radius that expanded bringing about the filling of more space. By reducing the circle through the radius it becomes another matter of eliminating form because in such reducing of the radius it then becomes a matter of reducing the influence of the straight line. When reducing the circle in size one have to reduce the radius or the diameter because the pi factor is the indicator of the form as being a circle. It then becomes a process where it is just dividing the radius defined by the symbol r by halving the answer of the previous value of r every time until there can be no dividing any further and such a reducing cannot end by becoming zero because in this process of reducing nothing disappeared.  The value of r reduces, although by reducing r did not vanish from the scene.

The lines just went smaller and smaller but if it was part of the Universe then or now it is still part of the Universe as it has nowhere to go but remain in and part of the Universe. One may divide by two, halving the result every time, which is a normal mathematical expressions and by reducing by half the process can never reach zero. The process leads directly in infinity.  Allowing zero to be accomplished through any legal mathematical equation is not a mathematical fact and I challenge any person to prove such a feat mathematically. I challenge any one of those that answered me blaming me of incoherency to that any diversion of any number can end and prove that the line is then mathematically ending in zero.  The numerical procedure may become tiresome or too small for any human to make sense of the outcome or the mathematician may lose all senses of his awareness, but never can such a dividing bring about zero in the ultimate answer when the straight line is reduced eternally to end in infinity. Zero cannot divide nor can zero multiply therefore this statement eliminated the use of zero anywhere and exclude a start at zero should I wish to retrace my steps to where the cosmos started. In using the method of the dividing by reducing the answer to half the size it will forever allow such a process to continue without ending in zero because no matter how small, the next value will be dividable by two in that continuous reducing by half of the straight line will forever allow to be a value in the next place.

This stands as a mathematical fact and I do not have to prove my statement but those persons in academic positions, which portrait me as a person trying to sway facts to fit my convenience, must explain how one can multiply distances filled with zero as a concept and get a valid mathematical answer other than zero.  I know that since they are the Official Academics and I am not, they can be incoherent and be blameless all in one go but they also are in a position to be incoherent and put the blame of incoherency back on me.   By placing zero in as a multiple factor anywhere or either at the start or the end will remove the value of the calculation. Space cannot grow or conclude by the use of the value of nothing. If you put zero anywhere, that act disqualifies the use of mathematics. This alone is the biggest obstacle why there is so little exploring going about the space expansion present overall. In my dealings with Mainstream Science in the past Mainstream Science becomes truly all out aggressive with my trying to indicate that I am criticizing Newton and one such a fact is his way of producing zero as a result of a circle forming an end and at the same time starting with another beginning where after completing the rotating circle the final product is zero. I do not try to criticize Newton on any facts except on what he saw in Kepler's work, which he tried to retranslate but through an attitude failed to realise what the meaning was which Kepler uncovered.   I also criticize the need Newton found to change Kepler's work. Do you as the reader realise that Johannes Kepler introduced an astonishing presentation about space-time, singularity, gravity and the Big Bang in the years of his life between (1571-1630).

By applying his formula we find some answers about questions yet unanswered. I raise the questions and with much study received new answers through dissecting what Kepler introduced to science. It was only possible when taken from the mathematics, which Kepler introduced from his cosmic findings. Kepler's answers were lost so many centuries ago when some Englishman saw him wise enough change Kepler's work and bring in the alterations as he saw fit. All Newtonians followed the lead and continued his act.   I am going to show that the changes brought about in Kepler's work were unnecessary because the changes were unasked for. Such forced changes about the work of Kepler by Newton came without the modern demand on proof because such proof was back then and still is lacking the proof in considering to what extent the accepted norm is of proof in relation to what was required and accepted back then and compare that in the light of what the demand is on establishing absolute proof at present. Back then a man's status was guarantee of proof while today there is no comparing to what is necessary in the terms that is accepted norms today.

I have grown accustomed to outright rejection of members in mainstream physics because of my criticizing of science. Sure they can dismiss me in the light of my criticizing them on science but that they do without explaining why Mainstream Science makes such a performance about Einstein and his views about relativity, when Kepler said it far better and with more to the point explaining and with much less performing. Kepler did the explaining about four hundred years ago. By accepting Kepler one has to accept space-time, Kepler said space-time $a^3 = T^2 k$. Kepler said there can be no space if there is no time. Kepler said space is the equivalent of time. Kepler said when space came about so did time. Kepler said space is time and time is space and the one cannot be without the other. Kepler said space is motion in space in motion, which is space-time. Kepler said $a^3$ could only be present when it is equal (the same as) $T^2 k$ performing in the time aspect. Kepler reported on what he was told by the cosmos. Newton had no need in changing anything because by changing he brought about much misunderstanding as well. Kepler showed what time is but all Newtonians admit that not one Newtonian knows what time is. Time is the motion of heat in space where material is heat and space also is heat in some other composition. That means heat is space and space is motion filled with heat. Time is the spin or motion of heat in space. That means there is no greater all preserving Universe out there. Every point $k^0$ will establish space $a^3$ by applying motion $T^2 k$. There is space filled with heat and the heat applying motion at the point where space is. There is the space $a^3 = T^2 k$ in motion and the motion is time. There is only space-time. There is no liberated space without the restriction of time. There is no space restricting a Universe without the liberating of time through motion. That is what Edwin Hubble saw through his lens. But where there is motion there too is gravity or antigravity whatever relevancy one wish to apply. Motion is gravity, which is motion that is committing space to form in the presence of singularity and that too is antigravity.

The man that (to my humble opinion) took cosmology into a new dimension was HUBBLE, EDWIN POWELL (1899-1953), the American astronomer. He first studied nebulae, concluding in 1917 that the spiral-shaped ones (which we know as galaxies) were different in nature from diffuse nebulae, which he found to be gas clouds illuminated by stars. From 1923, using the 100-inch (2.5-m) telescope at Mount Wilson Observatory, he resolved the outer regions of the spiral nebulae M31 and M33 into star, identifying over 30 Cepheid variables in them. This proved that such 'nebulae' were truly independent star systems like our own – other galaxies. In 1925, he devised the so-called tuning-fork diagram of galaxies, dividing them into ellipticals, spirals, and barred spirals, which he believed to indicate an evolutionary sequence. By 1929, Hubble had good distance measurements for over twenty galaxies, including members of the Virgo Cluster. By comparing distances with their velocities, as revealed by the redshifts in their spectra, he concluded that galaxies were receding with speeds that increased with their distance, a relationship known as the Hubble law. This was powerful evidence that the Universe is expanding. The dynamics of his work was so far reaching everybody (including Einstein had to revise their theories to accommodate his findings. His findings are the most disputed, undisputed observations in all of history. The HUBBLE CLASSIFICATION is a widely used system for classifying galaxies according to their visual appearance, illustrated on the tuning-fork diagram. The sequence is based on three criteria: the relative sizes of the central bulge of stars and the flattened disk; the existence and character of spiral arms; and the resolution of the spiral arms and / or disk into stars and H II regions. The system was originated by E.P. Hubble.

The sequence starts with round elliptical galaxies (EO) showing no disks. Increasing flattening of a galaxy is indicated by a number which is calculated from 10 $(a - b)/a$, where, a, and b, are the major a minor axes as measured on the sky. No elliptical is known that is flatter than E7. Beyond this, a clear disk is apparent in the ventricular or SO galaxies. The classification then splits into two parallel sequences of disk galaxies showing spiral structure: ordinary spirals, S, and barred spirals, SB. The spiral types are subdivided into Sa, Sb, Sc, Sd (Sba, SBb, SBc, SBd for barred spirals). With each successive subdivision, the arms become less tightly wound (but more easily resolvable into stars and H II regions), and the central bulge becomes less dominant. Two types of irregular galaxy are defined. Irr I galaxies show rather amorphous, irregular structure with perhaps a hint of a spiral arm or bar, and can be placed at the far end of the spiral sequence. Irr II galaxies are sufficiently unusual to defy assignment to any of the other types, although this category encompasses only about 2% of bright or moderately bright galaxies in the nearby Universe. The original, erroneous idea that the sequence might be an evolutionary one led to the ellipticals refers to, as early-type galaxies, and the spirals and Irr I irregulars as late-type galaxies. Colour and amount of interstellar material vary systematically along the Hubble sequence: ellipticals are red and contain little interstellar gas or dust, whereas late spirals and Irr I galaxies are blue, with significant amounts of interstellar material. The relatively faint dwarf spheroidal galaxies were not recognized as a separate type in the Hubble classification. Some variants of the Hubble classification use plus and minus signs to subdivide classes, so that $Sa^+$ is later than Sa, but earlier than $Sb^-$. The importance of the HUBBLE CONSTANT is still to this day, underestimated. This "constant" is well explained, for the first time, I might add, in this book. The Symbol $H_0$ is the figure that relates the speed of an object's recession in the expanding Universe to its distance in the Hubble law. It represents the current rate of expansion of the Universe. This important cosmological parameter is usually measured in units of kilometres per second per mega parsec. In the Big Bang theory, $H_0$ varies with time and it is therefore more properly known as the Hubble parameter. Its value is not accurately known but is thought to lie between 50 and 100 km/s/Mpc, recent research tending to favour values towards the lower end of this range. In the HUBBLE DIAGRAM, a graph plots either the redshift, or velocity of recession of galaxies against their apparent magnitude or distance from us. The Hubble law appears in the form of a straight line on such a plot. The original diagram, presented by E.P. Hubble in 1929, was the first indication that the Universe is expanding. The Hubble diagram is mainly used to test the geometry of the Universe, since at large distances any departures from the simple linear form of the Hubble law should show up as a curve. The HUBBLE FLOW is the general outward motion of galaxies resulting from the uniform expansion of the Universe. All motions lie in a radial direction from the observer, and the velocities are proportional to the distance of the galaxies. The real pattern of galaxy motions is not exactly of this form, particularly close to us, because of the mutual gravitational interaction between galaxies; some nearby galaxies are even moving towards the Milky Way. At large distances, however, the discrepancies are small compared with the Hubble flow. All these findings are incorporated in the HUBBLE LAW, which is the mathematical equation of the principle law that governs the expansion of the Universe. According to the law, the apparent recession velocity of galaxies is proportional to their distance from the observer. In mathematical terms, $v = H_0 r$, where v is the velocity, r the distance, and $H_0$ the Hubble constant. The law was put forward in 1929 by E. P. Hubble.

The HUBBLE RADIUS is a distance defined as the ratio of the velocity of light, c, to the value of the Hubble constant, $H_0$, This gives the distance from the observer at which the recession velocity of a galaxy would equal the speed of light. Roughly speaking, the Hubble radius is the radius of the observable Universe. Depending on the precise value of the Hubble constant, the Hubble radius lies between 9 and 18 billion l.y. This data is the basis on which the age of the Universe depends and is the HUBBLE TIME. The time required for the Universe to expand to its present size, assuming that the Hubble constant has remained unchanged since the Big Bang. It is defined as the reciprocal of the Hubble constant, $1/H_o$. Depending on the precise value of the Hubble constant, the Hubble time is between 9 and 18 billion years. In the standard Big Bang theory, the actual age of the Universe is always less than the Hubble time, because the expansion was faster in the past. The Universe is a combination of many material formations holding positions in space. Some of such material was covered in the blanket of heat, distributing into more spacious surroundings as the material expanded from the centre flowing outwards. Hubble's constant is proof that the space between cosmic structures are departing from many centre positions between such objects and this is a trend being located between all the objects through out space but also indicating a definite growth in the radius and such radius growth follow a patter where the growth seems to flow from any such a centre point away from the centre. Without the absolute and undeniable proof coming from the Hubble constant bringing proof beyond any possible doubt in any one's mind that expanding is very much and a very big part of all Cosmic activity the accepting of the Big Bang would not be in place. $F = G (M_1 \times m_2)/r^2$ is in essence a big issue about contraction while Hubble showed the space was not dividing. The space was multiplying. The stars are growing apart and so is the galactica. This then brings in the question of space available. The discovery placed mainstream Science in a spin that still spins the daylight out of all accepted cosmic conclusions. It now even initiates a search for dark matter and all theories are in place to cover the egg on the faces of Newtonians while Newtonians only have to look at Kepler to find the correct answers.

With your accepting of modern cosmology your deciding of accepting the Big Bang principles and therefore rejecting all the other cosmic possibilities being proposed in the past as such factors coming from say the steady state Universe must be reviewed. Mainstream Science still cling to views, which still form part of other theories that now must be rejected in view of the accepted Big Bang theory. The accepted mainstream views therefore should be rejecting all the other possibilities that may lead to double viewpoint or incoherence. Then that will also bring along a new perception about cosmology that could be endorsed by the minds of the common folk of which I am a member. The changes include altering the widest picture to match the Big Bang concept and only the Big Bang concept. Such a new perception must include a new perception about Kepler. In your reading of this book and seeing what I try to show, I shall lead you along as you will investigate Kepler and such investigations will introduce facts. The facts and views I introduce may seem as if no one before came to realise these facts belonging to Kepler. But then one feels surprised that, in the past the views were bluntly ignored, however obvious they seem to be. It becomes clear when we go about analysing Kepler's formula somewhat differently that the Big Bang theory might be exceptionally correct in the way it is presented. But at the same time the Mainstream view is incredibly flawed when an overall view establishes a universal picture. For instance Science accepts that space expands but science do not permit any view about material expanding. Science will not commit their view to the accepting that the two are linked. Try to fit the Earth as it is and the moon as it is in the present or if you wish to split hairs the material used that produces the size of the structures in their current form. The atoms forming the star must be holding space-time and when the size the atoms hold at present is projected to the past and placed in the past size Universe we had back then one will quickly realise that it was not that long ago that the two structures are holding the volumetric space that they currently do could fit an entire Universe into the space they have now. What we find in the solar system at present was representing all the available space in the entire Universe. Understanding the cosmos helps one to realise another part of Kepler for the first time. The realising involves issues that seem beyond answering given present facts because at present science is without such analysing of Kepler and therefore beyond the realising of such evidence. By scrutinizing Kepler again we find that when re-aligning such an attempt to reinvestigate Kepler and what Kepler really said. Without investigating Kepler, that which Kepler said now forms the unknown part of science. From the investigation the insight gained becomes a new vision and all the unknown becomes surprisingly clear, thanks to Kepler. But it insists on a divorce separating Kepler's ideas from Newton's ideas about Kepler's ideas. It is necessary to give Kepler the recognition as a mathematician without Newton belittling Kepler's skills as a cosmologist and a mathematician. The difference is about finding what Kepler really brought to science in relation to what Newton saw what Kepler brought to science, while Newton had no vision about Kepler's work at all. It also puts a new appreciation on what Kepler said and diminishes Newton's effort to change what Kepler introduced.

The expanding of space $a^3$ is the establishing of space $a^3$ from a specific point at the length of $k$ by the applying of $T^2$. The motion Hubble saw was space being established by time. It is how that every spot within motion around an invisible group of dots redefines a new time $T^2$ releasing space-time in the presence of all aspects of the visible Universe in that region. That region is the only Universe there is in relation to that specific Universe centres. It was space-time because without space there is no time as much as there is no time releasing space from singularity by means of forming space-time $a^3 = T^2 k$. Science was all impressed with Einstein's effort and some are still sceptical of Einstein committing space to time. Their scepticism runs so wide that there is still, at this time of our liberated age a group that in wisdom rejects Einstein's space-time. They refuse to connect the Black Hole as being the most poorly understood star structure to some situation that was apparently part of the Universe in the beginning. It captured the space-time before the Universe went exploding with space-time expanding. They propagate a limited space Universe wherein that era there was no space-time and refuse to connect the limit ness of space being in the beginning to the space less ness within the space that

singularity holds in a Black Hole. The problem about this poorly understood or misunderstanding about space-time brings on a concept that space-time is something we must rather use as some negotiating method to a situation that either was or will become part of the cosmos because no one knows where to place it in the current standings. We are surrounded by space-time.

I have heard cosmologists declare that we must "somehow" not think of space but rather think of space-time, but while saying that they mildly refer to something to say to please Einstein and rather nothing else. There is no corroboration between connecting space-time and reality in their explanation of what is behind the suggestion of space-time. They go about presenting space-time in such a manner as if they wish to promote an idea that will install a fashion or a trend to secure the use there of. The impression I got when I listened to those in the profession of cosmology when they talked about the using of space-time was as if an idea was promoted that was loosely connecting with something bleakly understood. It was as if they were promoting some trendy fashion amongst students without the students or the academics teaching the students seriously understanding why. It was as if they were trying to convince themselves before convincing others about their belief in and understanding space-time as a cosmic reality. They were in method more about self-convincing in their attempt in promoting space-time. It was as if to please Einstein when they were committing themselves to the promotion of space-time and their effort in the use of space-time as a concept. The way they put it was equal to promoting a fashion or the start of a new trend. During the three hours of the debate the three professors failed to give one motivating reason why anyone should start using space-time as a term instead of just space. They miserably failed to convince that they knew what space-time really was about. It is as if they were comparing the use of an expression such as "groovy" to the use of this new concept of space-time and to talk about a statement called space-time as being very similar. What was very apparent during the debate was that these people lacked understanding of the concept of space-time themselves, which is why they were unable to define it. Being Newtonians, they have not yet realised that space is the motion thereof and time is what it takes to bring such motion into a cycle where the motion of the space starts and ends at another point. Since motion from singularity is bridging one Universe to the next Universe the time it takes is in the square. That which such motion crosses irrespective of size is the bridging of space using time to go from singularity to singularity and that entails the crossing of an entire Universe. Motion establishes space-time by cyclic periods bringing about the motion of space in motion that contributes to the forming of time. They're referring to space being space-time in a fashion statement has very little convincing about their sincerity of understanding all the reasons why. This they say while the man that started cosmology some four hundred years ago introduced modern cosmology as space coming about through time. Kepler said in his formula that if there is space $a^3$ then that space is in motion $T^2k$. He declared space could only be if space was liberated and restricted at the same time by motion and the motion establishes time as a factor being as much part of space as space is part of the concept forming motion. But in the very same event the motion is the restricting of space by singularity forming time $T^2$. The return of space to singularity and the returning becomes a rotation forming the second part of the time factor. While space is the liberation of singularity through motion $k$ that is part of time, the restricting of the liberation of space from singularity is time $T^2$ in which singularity achieves the return of space to singularity. In this comes about four cosmic laws, which I named the four cosmic pillars. 1. Titius Bode Law; 2. Coanda Effect; 3. Roche Limit; 4. Atomic relevancy.

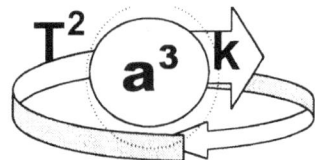 Space is the forming of motion $k$ because space is the liberating of time from singularity where the hottest part of space will find a way to move away from the rest of the cosmos and that motion forms the time component $k$.

When accepting Kepler's work as the very basis of cosmology one has to accept that space is time and time is motion and motion is heat and heat is space. The one cannot be without the other because time and space is the very same thing, it is space-time and it is $a^3 = T^2 k$. Singularity is a point where there is no space and that point can never have motion simply because there is no space to rotate about. Thinking of the cosmos must exclusively be in the form of space-time and in accepting there can be no space if there is no motion causing time. Science must review the past where concepts still hold questionable dogma by scrutinising previous concepts. If the proof that was presented then is accepted as unquestionable proven fact today to the same degree as demanded of scholars today it can again be given to scholars and only then should it be passed on as accepted evidence.

But in cases in modern science where there is such evidence but the unquestionable proof was never produced in the past researchers should go back and investigate how the old dogma fits into modern evidence. It has to be investigated notwithstanding what other implications may be involved. Science still clings to previous cosmic concepts while trying hard to incorporate the Big Bang concept. Science still thinks about a mixture of theories and mix some facts about conditions applying in different theories about cosmology.

If and when material is being part of the conditions presented in the steady state Universe theory to what will be present in the Big Bang all sorts of mismatching concepts arise. By trying to consolidate different theories from the past with the Big Bang rival theory such consolidation spreads considerable confusion. Trying to promote Einstein's vision and trying to bring that thought across to fit the Big Bang theory is what constitutes too much confusion. The acknowledging of the growing of space while thinking that only the space in-between structures are expanding and not the structures in their own space are expanding is a very typical of what I am referring to. Take the growth of space as one example where Science does not agree to the fact that the Earth and all in it are growing by becoming sizably bigger. The fact is that every part of the cosmos is growing rapidly. If you do not agree with my statement...well test the following: The Sun at present is 1392 530 km in diameter.

Let us only concentrate on the Earth in relation to the Sun, as that is what concerns us humans intimately. The Earth is roughly 150 million km from the Sun. The Big Bang theory wants us to believe that everything once was so small all fitted on top of a needlepoint. The Sun is at present in diameter 1392 530 km. There then is a distance of 1.39 million km separating the one lot of atoms within the Sun from the other atoms on the other end of the Sun. Just as there is at the present 150 million kilometres between the particles that form the Earth and those forming the Sun. If we are to believe the Big Bang and agree that the lot were less than one millimetre apart at one stage, we can appreciate that the distance there was between all the atoms that now forms the Sun and all the atoms that now forms the Earth were one meter apart. The Sun was spinning one metre from the Earth.

It seems rather trivial except that it also means the Sun in as much as being the space occupied by all the matter that now forms the Sun was some time back reduced to a speck of what now is present. The Earth too has to be a lot bigger at present than what it was some time back. It is rather unrealistic to believe that the structures remained the same size when the space placing them apart grew from millimetres to what is now applying. Kepler brought us the insight that the distance of **k** is at present $150 \times 10^6$ but believing the Big Bang then proves that **k** once was one meter and much less than one meter. With **k** at one meter it forces the size of the Sun and the size of the Earth to respond to that and shrink in sympathy. This also brings a realising that if they grow, then why do they grow because surely the idea manifesting as nothing can't grow. But even moreover is the importance of realising that material is also growing. But if so then what is material using to support and sustain such growth. It is inconceivable to place the Sun as it is in the present day into a Universe that was in place just after the Big Bang. Our minds function in a rather dubious way of reasoning because if space is nothing then space can't grow because nothing can be duplicating nothing and in the process nothing adds to nothing then its nothing that is becoming more.

We then had no reason to explain why nothing can grow and accumulate because nothing is nothing, so nothing can add to nothing without any reason to explain where the additional nothing is coming from. Nothing is a concept, by which the using of nothing as a term is what makes the accepting of the idea so easy. By giving nothing a character and not a value is soothing the idea that what ever is part of the cosmos came from nowhere and that answer then must be so meaningful it better satisfy everyone regardless of reason. Silly as it seems it is also very true. The idea of using nothing to build space with must be seen by all as ridiculous as it truly is. The Sun couldn't fit into a

Universe of the past because the Sun out grew such a Universe. The growth we find in the Universe comes from the growth we find in atoms accumulating by using gravity.

We have to bring all space being **k** in $a^3 = T^2 k$ in realising there is growth of particles as much as particles moving apart. The growth is everywhere and not selected to the nothing part. If the Sun and the Earth were that much closer back then, the Sun and the Earth was that much smaller. If a Sun with a diameter of 1392 530 km in diameter and an Earth with a diameter of 12756 km was one meter apart, then the distance separating the two was enough to allow the two objects to join. As we can see they did not join. That brings across that by accepting the growth of space it then brings along the growth of all space. That includes space that is holding material as much as space not holding material. It includes the space holding subatomic particles as much as the space not holding subatomic particles. It includes all space occupied or otherwise. That leaves us with one fact being the principle issue of this very book: If space is not nothing and yet space can grow, then how can space grow and what is space using to grow?

If the space we think of is constituting of as something because it not only holds material but has to be formed by something for the sheer fact that it is there in the first place and by being formed by something it then can grow as much and in relation to space we think of as nothing then the nothing has a lot of the something within the nothing because both is growing. It is an argument of complexity and it can only become apparent when Kepler becomes apparent and when everyone can see that, Kepler separates and stands apart from what Newton saw what Kepler was saying. Kepler must come into his own as a mathematician, a cosmologist and a scientist because of his tremendous achievement. If we accept that Kepler's formula is mathematically $a^3 = T^2 k$, then it must also be $T^2 = a^3/k$, $k = a^3/T^2$ and $k^0 = a^3/(T^2 k)$. What this then states is that Creation according to Kepler started at a point coming about from singularity $k^0$ and grew out of singularity into space standing directly and undividable related to time where the time aspect is the motion created by and continuous creating of more space.

Reading this book will introduce facts that no one before came to realise during the past almost four centuries ago. This becomes clear when we go about analysing Kepler's formula somewhat differently. In a way for the first time it helps one to come to realise another part of Kepler. But I must press the fact once more that this realising demands a divorce separating Kepler's ideas from Newton's ideas about Kepler's ideas. We must recognise the Master Kepler as an equal to Newton and not just someone with vague ideas and the ideas he had, later needed revising by a better Master.

The difference is about finding what Kepler really brought to science in relation to what Newton saw what Kepler brought to science. It also puts a new appreciation on what Kepler said and diminishes Newton's effort to change what Kepler introduced. I hope by now, you as the reader of this book, have come to see why I say that Johannes Kepler introduced an astonishing presentation about space-time, singularity and the Big Bang in the years of his life between (1571-1630).

By my reinvestigating Kepler in the state Kepler's work was seen in, isolated from Newton, I received new answers taken from the mathematics by which Kepler introduced his cosmic findings. Kepler's answers were lost so many centuries ago when some Englishman saw him wise enough change Kepler's work and bring in the alterations as he (the Englishman) saw fit. I repeat this because it is most critical to realise that Kepler saw $a^3 = T^2 k$ as space **a** by the sphere $^3$ in the third dimension in motion **T** by the square $^2$ shifting along a line **k** from a distinct centre $k^0$. This was no mathematical referring of a sphere in a drawing on a paper but a cosmic expression space-time, which is space by motion thereof.

The changes were unnecessary because the changes were unasked for, yet all the intellectuals that came afterwards and held any position of prominence in academic circles, was and remained impressed by the part that Newton changed. All academics coming afterwards were in total agreement about the changes, notwithstanding being the unnecessary changes that they prove to be, which Newton also unnecessarily changed. When reading the rest of the book, please keep in mind that it was the cosmos that spoke to Kepler directly by using mathematics and it is that what the cosmos said about itself to Kepler that Newton did not understand and then changed to what he (Newton) then understood.

Newton did not just change Kepler's work but he tried to re-invent the cosmic code in mathematics which the cosmos used directly to explain to Kepler matters concerning the Universe. The Universe used the cosmic principles used in the Universe by the Universe. Then Newton changed what Newton could not interpret because the cosmos did not reveal the code to Newton but the cosmos did reward Kepler by giving the code to Kepler. But now some three hundred and fifty years later it comes to our attention that Newton completely misunderstood what he thought he understood about what was revealed to Kepler.

The cosmos spoke to Kepler about space-time coming from singularity. Kepler gave us his findings. Any discomfort that may come when we read what is revealed must be set aside, because we must remember it is not me, or Kepler, but the Cosmos that is doing the revealing and lending us the tools we can use to decipher what the cosmos is trying to make us understand. Kepler translated what the cosmos told him (Kepler) as $a^3 = T^2k$.

Translating Kepler's mathematical expression $a^3 = T^2k$ correctly to the verbal statement in English Kepler said that there is a space $a^3$ which is equal **=** to the motion in the time duration $T^2$ thereof between two specific points which holds a relation to a centre where from there forms a straight line **k**. What is there mathematically not correct in Kepler's expression and why is any changing thereof necessary in any way? It says where there is space such space has to move. Test the following symbolic values in the mathematical expression and test the principal behind the expression in which Kepler stated them. Convince yourself about the evidence that Newton saw what Kepler saw where the translation thereof that was done by Kepler is mathematically incorrectly translated by Kepler's interpretation from mathematics to English:

$a^3$ The fact that any symbol uses a value to the third power indicates space or a volumetric established and separate unit which is serving an under dividable dynamically separate space being within a space. Although being apart the two in space sharing a unit can never be apart but serves as a unit by division of motion. It is space because it is volume using the third dimension. But since the space is smaller than the Universe it must be space being within space, which is within space. There a relevancy is forever present.

$T^2$ Is an indication of space apart from the surrounding space by granting the independent space by establishing borders through motion, an ability of moving from one point to another point or following a flat distance between two points. It is motion that is taking time in the second dimension.

**k**[1] Is the symbol used to indicate a straight line between two points with a definite beginning and a specific end position. The two points is valid only by re-aligning an eternal straight line to the figuration of a circle through alternating as well as recognising the control coming from such a centre. It is Pythagoras by the triangle, half the square and the straight line sharing value in the $180^0$ they represent. Kepler introduced this absolute basic mathematical principle.

This leads to the question: "What formed the grounds for any need by Newton to change Kepler's translations from the cosmic given to mathematics and then from mathematics to English? The space-time that the cosmos introduced was so brilliant it took the likes of a genius such as Einstein and another few centuries of mind development to realise the presence thereof that is of space-time many hundreds of years later. What did I not translate correctly from the mathematical expressed to English? When I used Kepler's mathematics by my translating Kepler's work correctly I came upon answers not yet uncovered by Mainstream Science.

Kepler gave the World a means to use mathematics in order to translated cosmic mysteries and reform such mysteries to answers that Kepler uncovered long before Newton, Einstein and others got wise about cosmology... Such is the advantage of recollecting Kepler's facts that it does answer many questions, which went unnoticed and therefore not spoken about up to now and some questions that is answered were previously never even thought about. Mainstream Science never previously thought that through any examination of Kepler's work the scrutinizing would uncover these facts that I present. Subsequently by ignoring Kepler's uncovering of decoded cosmic massages Mainstream Science elected not to ask the correct questions and in the process Mainstream Science never found the correct answers.

By not asking the questions Mainstream Science could not decipher any of the decoded mathematical messages, which Kepler received directly as a mathematical message spoken by the Universe and coming from the cosmos. We all know and appreciate that mathematics is just another language and the professional mathematicians have the responsibly of translating mathematics to a verbally competent language that is understood by others that are not fluent in the language of mathematics.

They did not read nor recognise Kepler's mathematical translation and thereby was unable to translate Kepler's mathematics to the other communication forms being all verbally spoken dialects. In other cases human natural study methods brought along a cultural of Academics forcing students to comply whereby the students will accept the knowledge through our inherited past which when tested by modern standards is not that highly proven. In Newton's time the accepting of accurate representation came a lot easier than what is the case today. When we use information coming from the past we accept the answers as questions that is without doubt already fully answered because culture demands the accepting thereof. Can any person in a sane mind for one second think of the fate awaiting the first year student that will question Newton's findings or the accuracy those findings represent? Let us see where the motion takes space.

The first of the three laws of Kepler concerning the orbital motion of planets is focussing on the orbiting route of planets always form an ellipse with the Sun at one focus of the ellipse. This is in the total of the totality pointing to a relevancy acclimating a position that alternates the prominence of both parties in the relevancy and more specific allowing the dominant role of either on applying at any one time.

The second law states that the radius vector that connects the Sun to that of any and all the planets sweeps out equal areas at equal times. This is another total relevancy establishing two factors that contribute in forming space being specifically related to the time component, which is space-time as they combine in securing a definite ratio the time period in relation to the space in motion during the time applying such relating to the space involved in the bigger picture. There is a space moving through a space in a specific time period.

That is space-time in any one's book. It points to the time it takes for a space to move through a space in relation to the centre of the Sun. The third law of Kepler is also indicating relevancy as it states that the square of the orbital period which is nothing less than the time factor because if it does nothing else then it indicates the years in orbit proportional to the centre of the semi major axis of the planets in orbit. But it does more than that because it positions a precise location of another space also holding $a^3$ in relation to each other, if the larger space indicates the space factor it shows motion in relation to space where the space holds the motion every time.

By positioning the lesser space the motion applies to the lesser space running through the larger space. This forms another relation where if we then were focussing on the larger space while the larger space is gravitationally centralising a position in relation to the smaller space. In this the principle is very obvious adhering to the Coanda affect where motion indicate a valid centre for gravity to focus. The Larger space shows a motion coming from a different position every time in relation to the changing position of the larger space. But motion indicates space in motion and the motion of one space is the other space being in motion or proving the existing of another space that also holds motion. The mere fact of proof of both spaces sharing one unit is the contradicting both has to each other by motion.

In fact we know that there is the motion of the Earth in relation to the motion of the moon in relation to the motion of the clouds that in turn is in relevance to the motion of the moon. The relevancy of motion in relation to space filled in space not filled is an endless ratio of different objects in space in unfilled space that is at different rates. One may focus on the moon and find the clouds to be motionless but have the moon travelling at a dazzling speed.

Even when we think we are motionless say when we are sleeping the fact is that we then are travelling as fast as ever before because the Earth is taking us on its journey around the Sun which is going around the Milky Way. Every aspect of the cosmos has to move to be in the Universe. Every time there is motion, there is a relevancy to such motion. The Universe is in motion because the Universe is motion in motion about motion. In every individual case of the three laws of Kepler there are a larger space relating to a smaller space within the larger space. The smaller space serves as much as an indicator to the being of the larger space as the larger space serves as an indicator to the position of the smaller space. It shows space being in relation to motion of the space, which is in the space holding the space relevant through motion. It is $a^3 = T^2k$ where relevancies point to either in space in the relation of the motion. What may be linear motion in relevancy to one is of another part of the same space unit. As a whole the two factors represents space, which is the circular factor of the other indicated by linear motion that produces the border of the circle.

Nothing in the Universe is without motion because "nothing" as a factor is "not any" factor in the Universe. Other than nothing everything else is in motion because motion brings separation and points boundaries that establish individual space. Therefore what is in space is in motion. However, as Einstein pointed out motion is a relevancy between factors of different motion in space through space.

Space consolidating by claiming independence by as space 1

Space contracting to centre operating as a holding space

space. When looking at the moon on a semi cloudy night we can place the relevancy on the clouds being without motion and the moon travelling at a dazzling speed or we can put the motionless relevancy on the moon and find the clouds travelling at a large speed.

Any line coming straight from the planet running directly towards the Sun will fall victim to the rotating action of the Sun but also the combination of a group of such lines dotting the motion of the planet as the smaller object travels will prove to become the circular rotation in the end The factors representing $a^3 = T^2k$ can turn around and become interchangeable. When one candidate takes the position of the space, the other space becomes performing the motion part. The issue is that neither the holding space nor the space being held has a validity of existing without the other being a factor part of the equation. All this information above Newton blatantly missed while considering his opinion of such a standard as to bring changes to the work of Kepler.

Newton is institutionalised academic culture force fed by the generations of Academics passing by as they go from the present over to the past whereby the Academics in generation after generation introduce Newtonian ideas to the following generation of Academics. At the time the following generation fills the role of students on their way to becoming Master by promoting Newton's ideas about forces. As they progress from students to the Masters of the day they insist on the unquestionable accepting of Newton as they did as students taught by their masters of the previous generation. They were taught to learn Newton off by heart if they wished to be found sufficiently knowledgeable and intellectually competent to become the next generation of masters.

All they required was to prove they understood Newton's ideas about active forces as being correct above all else. The passing on of such accepting of Newton's forces lies above and beyond and passed the accepting onto the following generation of future Academics. Providing that the student "understand" Newton by repeating Newton, is the biggest and ultimate achievement of all Academics that is presently in charge of Mainstream science. The next generation of Academics, which is now the present day generation of students, is conditioned mentally to accept Newton or die an academic death. The academic death will be in the form of cutting any further mind contamination by terminating all further study possibilities. If Newton comprehension about and the proving of the understanding of accepted forces is incorrectly accounted for by students repeating statements in examinations of precisely duplicating that was made in the past and carried to the future generation of Academics. This is not a precondition set to students.

It is the ultimate condition of acceptance by the present day masters to all students when the students are performing the answering of their examination questionnaire. If and when the student will not accept such culture without reservations that his peers teach him that student will fail all further acceptance and this then is disallowing the student the right to attend further classes in any of the classes given by the institution. I personally have experienced such bias from Academics in charge of institutions. By reading what Kepler said correctly so many centuries ago the effort brings all the answers to the questions academics at the present are incapable of answering. But it does not involve looking at what Newton said about what Kepler said... it is all about looking at what Kepler said. To understand Kepler one has to include the opinion of Kepler and remove the opinion of Newton about Kepler's findings. Answering the main question about gravity is locked in answering why would a water droplet form a sphere when floating in a space capsule in outer space? Kepler gave us the answer in much simpler mathematical answers than Einstein did.

Kepler said $k^0 = a^3 / T^2k$, which means that $k^{-1} = T^2 / a^3$ which means that by reducing gravity $T^2$ space will disappear. Kepler gave us the answer about the water drop forming a sphere even before Newton thought about the question. He gave us the answer three centuries before Einstein got the question wrong about his Universe going flat. Why would gravity always result in forming a sphere when gravity is left free and unhindered to capture form? Let us recollect Kepler's statements. He said that $a^3 = k / T^2$ and from that the mathematical relevancy guide one to the answer that $a^3 / k T^2 = 1$ and $k^0 = 1$ bringing about that singularity is $k^0 = a^3/(T^2k)$ which is the smallest space being in singularity produces gravity in forming space $a^3$ relating to time in motion $T^2 k$.

Indirectly Kepler said that mathematics prove that $k^0 = a^3/(T^2k)$. $a^3 = k / T^2$: That is what Kepler brought into civilization for all time but mathematically Newton saw Kepler's $a^3 = T^2k$ as incomplete and therefore he had the urge to correct what he saw to be incorrect changed to be $4 \pi^2 a^3 / T^2 = G (m + m_p)$. In this, the value of T and "a" are the period of revolution and semi major axis of the orbit of a planet of mass $m_p$ about the Sun of mass m, and G is the gravitational constant. Newton added nothing but duplicated everything in his suggesting completing what he saw as incomplete. Can the symbol "a" be reckoned as a period of revolution as Newton suggested and therefore Mainstream science still suggests? Is the work of Kepler therefore incomplete? Let's dissect it once again because it is very important! Newton diluted Kepler's formula by insisting that all factors only contribute as motion. By re-examining one can clearly see the finding of Newton is most incorrect.

$a^3$ As I said before any symbol using a value to the third power indicates space or a volumetric established and secluded unit. It suggests the using the six dimensions allocated to space. It is space because it is volume using the third dimension. There is no other valid interpretation or translation allowing for another translating from mathematics to English than by categorising that space as a volumetric separate identity. If it is cubic it is space. One measure a fridge or a stove or a room by the cubic measure without including a square because of the volumetric content there is in the third dimension. An aeroplane flying is a cube flying and the square part falls to the repositioning of the motion of the cubic space. The fact that there is a line connecting the space to a specific allocated centre and enforcing a rotating motion around that specific allocated centre connects whatever volumetric measure the space has in the cube to motion. The cube as a separate issue is independent from the space in which the independent cubic space rotates, which brings about the circle and yet there is no need to implicate pi because the rotation brings along the circle after the cyclic completion of the rotation by the independent cubic space.

$T^2$ Is an indication of motion, the moving of an independent space that is holding $a^3$, where the space in motion will be measured as $a^3$, as the space $a^3$ that is occupying a separate part of the including unit forming a space by opposing motion within the unit of the other space that is travelling by using the second dimension of motion in $T^2$. Both sharing the unit is moving as independent space from one point to another. The opposing motion combines to form a relevancy that produces gravity. Both sharing the one unit is following a flat distance between two points. But which form one point. Because the motion is coming from both ends sharing a point for that reason time can never stand still because although there may be a single point the point refers to the square of both aspects in motion.

The distance and only the distance of movement is $T^2$ where the distance is established as $T^2$ by both aspects of the space $a^3$ but is not the space $a^3$ and as the independent space $a^3$ in each case of the space that is going from point $T$ to point $T$. By both matching a point there is a reserved position applying on instant by both and from both the value is $T^2$. However when one aspect alone is considered there remains the $T^2$ square of motion of one single party contributing to the unit as motion making the distance $a^3$ travelled $T^2$. It is space filled with material consisting to allow independent space within the surrounding of an enveloping space of bigger proportions $a^3$ being in motion $T^2$ and that motion $T^2$ is the filled space $a^3$ that is taking time in the second dimension moving the space $a^3$ in question from one point of choice to another point of choice by which time will be established.

$k^1$ Is the symbol used to indicate a straight line between two points with a definite beginning and a specific end position. It is also the mark by which the straight-line motion is altered to accept the rotating motion. The change of the motion is directly linked to the change in direction and the change in direction amounts to the gravity strength. This indication of a distance is an indication of a bigger space that is big enough to include a smaller and separate space $a^3$ within the bigger space running all the way from the start of $k$ to the end of the line of $k$. One has to see $k$ for what it represents because $k$ indicate the presence of a larger space that is large enough to allow the smaller space to rotate all the way as the circle that runs from $T$ to $T$ in the full diameter of $k$. It also indicates where the smaller aspect in the unit in motion crosses with the larger aspect of the unit and in that way form $T \times T = T^2$.

Kepler introduced this absolute basic mathematical principle that all others failed to notice. It is positioning the independent space in motion in a specific relation to a controlling dynamic situated in a domineering and controlling centre. It is indicating that the space in question is in motion acknowledging the centre in control of the motion and therefore in control of the space location. It proves that a larger space is holding the space in question as part of the larger space where the larger space is in ratio to the other part so much larger that the space in question can effortlessly go in motion and therefore full rotating motion. By the rotation it then is concluding a rotation within the larger space that $k$ indicate. The presence of $k$ is not merely a line but points to the start and the end of the space containing the space $a^3$ within the space $a^3$. But $k$ also must therefore indicate its coming from a very small start that is centralising the motion of the space in question. It is using the space that $k$ produce from where that specific space is controlled in the realms of the larger space. Nobody before saw it in as simplistic manner as I just showed it to be and yet in the simplicity is the sensibility of it all...this argumentatively is indisputable reasoning.

What in this is there to dispute...yet when I say these facts Academics find grounds to dispute my saying this about Kepler's saying that! Kepler gave us the answer but no one ever took notice!

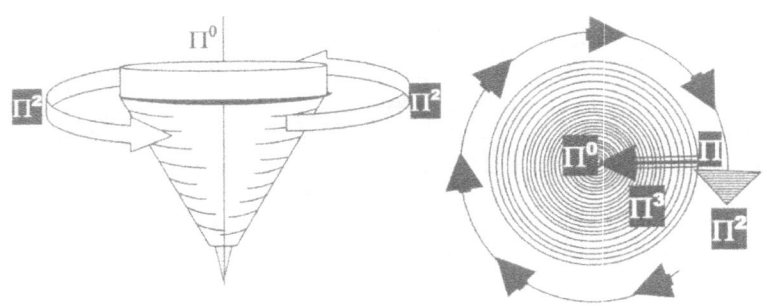

Kepler was the one that discovered **space / time** and Kepler announced it as $a^3 = T^2k$ which can translate to as $k=a^3/T^2$. Kepler was the one that discovered singularity as $a^3 = T^2k$ that also translate $a^3/T^2k = 1$ which then is $k^0 = a^3/T^2 k$. Kepler was the one that discovered gravity holding space-time relative $k = a^3/T^2$ the contracting part that Newton claimed gravity as a force **is $a^3 = T^2k$** that translates to $k^{-1} = T^2/a^3$, but that is as vague as saying humans are life. If gravity is a force, then what is a force? If gravity is the product of the elusive graviton what is the graviton and where is the graviton? Mathematicians get stuck by using mathematic rules and laws.

| Planet | Mass per Earth unit A | $k^{-1} = T^2 \div a^3$ Movement B | $a^3$ of space volume | $T^2$ During time units |
|---|---|---|---|---|
| Mercury | 0.06 | $T^2 \div a^3 =$ 0.983 | $(a^3)=$ 0.059 | $(T^2)=$ 0.058 |
| Venus | 0.82 | $T^2 \div a^3 =$ 0.992 | $(a^3)=$ 0.381 | $(T^2)=$ 0.378 |
| Earth | 1.000 | $T^2 \div a^3 =$ 1.000 | $(a^3)=$ 1.000 | $(T^2)=$ 1.000 |
| Mars | 0.11 | $T^2 \div a^3 =$ 1.000 | $(a^3)=$ 3.54 | $(T^2)=$ 3.54 |
| Jupiter | 317.89 | $T^2 \div a^3 =$ 1.000 | $(a^3)=$ 140.6 | $(T^2)=$ 140.66 |
| Saturn | 95.17 | $T^2 \div a^3 =$ 0.999 | $(a^3)=$ 868.25 | $(T^2)=$ 867.9 |
| Uranus | 14.53 | $T^2 \div a^3 =$ 1.000 | $(a^3)=$ 7067 | $(T^2)=$ 7069 |
| Neptune | 17.14 | $T^2 \div a^3 =$ 0.999 | $(a^3)=$ 27189 | $(T^2)=$ 27159 |
| Pluto | 0.0025 | $T^2 \div a^3 =$ 1.004 | $(a^3)=$ 61443 | $(T^2)=$ 61703 |

**Therefore Newton's** $$F = G \frac{M_1 M_2}{r^2}$$ **becomes invalid.**

Kepler goes further by correctly using cosmic mathematics and investigating the formula that Kepler introduced intensely but without Newton interfering and telling Kepler what he (Kepler) should have found. Instead we come to a part, which takes the Universe one step further back than the Big Bang reaches, to a time before the Big Bang to an era where no one in modern science previously dared to go before. We can reach that point by tracing what Kepler said to the time where gravity started. Newton had all the information we now use to his disposal. Instead of using it correctly Newton chose to change what Kepler said because he (Newton) did not understand what Kepler said. Newton should instead have been looking at what he (Kepler) found then he (Newton) would have seen what gravity is.

He (Kepler) said that the cosmos said that space is time being space-time. $a^3 = T^2 k$. The space is held in check by motion from a centre and that is gravity. It becomes more than clear that space $a^3$ is time by dimension $T^2$ and time is space $a^3$ without dimension $k$ Gravity is $a^3 / k$ but $k$ is an addition of motion $T^2$. Motion $T^2$ of space $a^3$ being apart thereby forms $k^1$, which produces gravity. It is gravity that keeps the Sun and the planets at a specific distance and apart while the planet remains in motion around the Sun. It is **space-time $a^3 = T^2 k$** that keeps the space in motion and at a distance whereby the space of the Sun is parted from the space of the Planet.

That is gravity…what else can it be? After all it is space-time keeping the structures apart and space-time is a result of gravity. Once singularity was found the rest was simple but was finding singularity really that difficult. Not if one was guided by Kepler's formula. It is merely retracing **k** until **k** becomes **k⁰**. It is so simple to reach singularity that it is almost ridiculous.

**r /2** By dividing the radius r by the half of the value that then reduces r to a point where the left edge of the line reducing will be at the very same place the right hand edge of the line that is reducing will be. At one point the spots that formed the two ends of the line will be at the same spot where the original centre between the two points were. The two points would have moved evenly towards and in the direction of the centre by reducing all the space on both sides of the centre. By moving towards the centre they will at some point have to reach such a centre point notwithstanding cultural concepts favouring nothing to be filling that spot. Reaching that centre point will land all the sides on the same side and because of the presence of all possible sides such presence of all possible sides removes nothing out of any further possibility.

Any further dividing will land the left hand spot past the right hand spot in the opposing half where it then will grow once again but in the opposing direction that the specific spot previously represented. All possible dividing then ends on one spot where such a one spot that represents the perfect centre point and that divide the left side from the right side and the top from the bottom and the front from the back will land on one spot. At that spot all the sides just mentioned shares a location with all other possible sides. The centre that then is holding all the previous points in one spot then physically is in the single dimension applying as one spot to share a location for all sides. At such a point there is no further dividing possible. That point cannot be zero because that point represents an eternity of possible growth in an infinite number of directions available to grow. The line starts in infinity and not in zero or nothing as is taught to scholars by teachers' worldwide. Trace that centre while the top is spinning and one will find a centre that favours no side since the centre divide equally all sides while spinning. That centre proves to be no specific side because such a centre proves to be all possible sides. On several occasions in the past I have been accused of manipulating the argument to produce none-existing or overrate facts. That is not the case. I am not manipulating facts to create an argument as some intellectuals in the past accused me of. What I am talking about is a mathematical fact that any one can prove by calculating. By following a very simple procedure it is within any person's reach to detect the centre of the spinning top which I am referring to, although there is no such a centre to detect, the centre is there for all toe detect. A child is capable of using the two times table and the dividing by two every time that is the most simple form in which mathematics may be used. It is a mathematical fact that a line will reach a point where all sides are at one spot and as such the line cannot divide any more. I have been accused as being dubious about my arguments while it is Mainstream science that dubiously found a way to get to zero as a mathematical starting point of any and all lines. Then they put the double standard blame on my arguments where it is I that has to prove my arguments as not being the dubious one of the two arguments. At such a centre starting point all sides share one specific spot but that spot holds all further future possible growth in any direction of all sides and since everything is in there, there is no room left for zero to be there. That point is filled with all possibilities which prevent zero becoming factor since the sides share one spot and in that sharing they are present and their presence prevent zero from becoming a conclusion. While the different sides are in one place the factor and value is one to all without allowing zero any part to play.

**Tracing the centre of the Universe is still possible by any one wishing to find such a centre. The centre falls outside the accepted Universe since it cannot be mathematically accounted for but that centre is in control of every thing in its influence.**

The centre changes motion to gravity by diverting the straight line to an immediate circle. By tracing the line back to where the circle is no more a straight line will uncover singularity plus one dimension. However, the entire centre forming singularity is still locatable within the Universe we have.

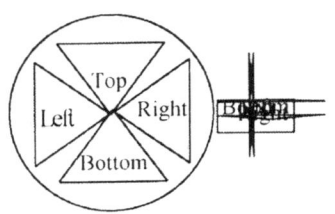

Reducing the radius r from all angles possible throughout the circle will bring about that all possible direction will eventually land on the very same spot with no more dividing possible. Yet zero cannot be a factor since the sides still hold value. In as much as holding all the value there can rise from such a spot. This is arriving at a point where more reducing will land the one side on the opposite side of the line but it will not bring about zero in the equation

What this argument further proves is that the circle reducing must then come from all points because the radius might be a line but that line represents a circle through $360^0$ coming from and accounting for all possible directions. Taking that into account it is important to recognise that notwithstanding the size of a line, which any radius of any size is there is another line (or dot) eternally bigger as well as eternally smaller than the line in question. While we are in the third dimension being part of the third dimension such being in the third dimension then allows that all parts of the third dimension forever can be divided once more until the line in the third dimension is no longer part of the third dimension. When such a line leaves the third dimension it is still dividable because it might not be part of our dimension any more but it can still reduce further as part of the second dimension. By that time it has left our scope by miles but that does not mean that it end there because from our perspective that is where it ends. But our perspective does not represent reality. Yet, even then it can still reduce infinitely more until it has left the second dimension and then at last forms part of the first dimension. Only then when the line reaches the first dimension no further dividing of that line is longer possible. We can never grasp the size of a line that the first line in size that came about when the first motion broke the eternal stranglehold on space. According to our big and small conceptions of what we perceive as large, ultra large, small and microscopic small is just mere words describing thoughts totally unrealistic in the context of what the cosmos sprang from as the cosmos moved out of the spot and formed a dot. Even by the standards of forming the dot, which was eternally bigger that the spot T, as the dot and all the many dots that came from the spot. The size differentiation only between those two exceeds all limits and divides we wish to create forming borders that we can appreciate.

When looking at the circle in the conventional manner, we persist with errors brought about in culture and not by applying some significant modern logic. Take a circle and reduce such a circle constantly to where it no longer can reduce. Reduce it to a point where only form remains part of the circle because the radius has gone beyond human measure and becomes so small it is not noticeable with what ever measuring tools man may use, then what remains is pi since pi does not indicate size but indicate form, and form is all that then will remain. In any circle or sphere the size only depend on the fluctuation of r, as a component to the circle or sphere but that does not affect the form by indication of $\Pi$ in any way there may be. The conclusion I drew from following this process is that from this line can start at zero because that will be a mathematical impossibility since no line can ever reduce to zero. A line will forever be able to reduce further becoming smaller but it can never reach zero because zero is not part of the scale on which we measure lines. If a line cannot reduce to zero it then cannot start at zero.

A line or spot starting at zero would therefore be shorter than the shortest line possible. For obvious reasons can no line, or any line grow or extend from zero because such a line must then quit zero and become something, thus abandon its original value. That would mean the start of the line has a different value to the end and a line holds conformity through out. When any line is starting from point zero and it uses the factor zero, then it can never leave zero because of the influence of being zero disqualifies any possibility of growth. But when coming from singularity $\pi^0$ and the line then had to grow in all directions at the same pace the line must then become a circle $\pi$ or being three-dimensional, then form a multi circle $\pi^3$ we named a sphere. Since the Universe is about circles and lines connecting circles I came to conclude that flowing from this fact is that in the Universe there can be no zero improvising as a filling ingredient for the space of a point or be unfilled space. Zero is no valid factor in the Universe. In the case of the growing sphere the value of the circle is $\Pi$, and that is where creation must have started. That gave me the clue where to start looking for singularity. One would find singularity in the value $\Pi$ and the value $\Pi$ will be in all things rotating in a circle but by measure one dimension smaller. As usual I am again shooting the gun before the hunt started. Lines in mathematics do not start from zero and that is no discovery on my part that was a realisation I came too. The Universe is all about lines and the manner that Kepler pointed to the increasing of the lines by $k = a^3/T^2$ proves growth in the composition of all lines.

## UNIVERSE
Everything that exists, including space, time, and matter. The study of the Universe is known as cosmology. Cosmologists distinguish between the Universe with a capital 'U', meaning the cosmos and all its contents, and Universe with a small 'u' which is usually a mathematical model derived from some physical theory. The real Universe consists mostly of apparently empty space, with matter concentrated into galaxies consisting of stars and gas. The Universe is expanding, so the space between galaxies is gradually stretching, causing a cosmological redshift in the light from distant objects. There is growing evidence that space may be filled with unseen dark matter that may have many times the total mass of the visible galaxies. The most favoured concept of the origin of the Universe is the Big Bang theory, according to which the Universe came into being in a hot, dense fireball about 10-20 billion years ago.

## UNIVERSAL TIME (UT)
A worldwide standard time-scale, the same as Greenwich Mean Time. Universal Time is the mean solar time on the meridian of Greenwich. It is defined as the Greenwich hour angle of the mean Sun plus 12 hours, so that the day begins at midnight rather than noon. It is closely linked to Greenwich Mean Sidereal Time (GMST), since the mean sidereal day is a precisely known fraction of the mean solar day. In practice, UT is determined by a formula from GMST, which in turn is derived directly from such observations of the meridian transits of stars. The version of UT derived directly form such observations is designated UTO, which is slightly dependent on the observing site. When UTO is corrected for the variation in longitude due to the Chandler wobble, a version of Universal Time, UT1, is derived which has genuine worldwide application.

When UT1 is compared with International Atomic Time (TAI), it is found to be losing approximately a second a year against TAI. Broadcast time signals use the time-scale known as Coordinated Universal time (UTC). This is TAI with an offset of a whole number of seconds. The offset is adjusted when necessary by the introduction of a leap second, and UTC is always kept within 0.9 s of UT1. On this issue there is much more to explore than the meagrely mentioned. Time stands related to the position an object holds to a centre such an object refers too while in rotation. Kepler found for instance that $T^2$, which holds the orbit to a rotation specific, is directly dependent on **k** to value the space $a^3$.

By contracting the Universe is expanding and everything is based on gravity providing both actions. The Universe rides on a balance and we have to locate such a balance. To prove my theory I firstly had to locate the centre of the Universe. Even admitting to such a notion sounds like madness or in the least a tasteless joke, but please give me a chance to explain in more detail. I realised that my effort to locate the point holding singularity only stood any chance of success if the reducing of the line enabled me to backtrack the exploding Universe to its origins. By applying some basic effort I have located the position from where all movement came and the direction it took moving forward in time...and yes, while all of that took place, I was also finding the centre of the Universe which I might add I even located at the same time. There are two standard mathematical formulas used to calculate a circle. The one use an r to indicate the radius and the other use a D to indicate the diameter, which is double the radius and therefore needs to be divided by a four to eliminate the Newtonian inverse square law amounting to the difference there will be between the two. This has the significance that it implicates time. The one using the radius is $\Pi r^2$ and the other formula using the diameter is $\Pi D^2 / 4$. At the very start of my interest about matters concerning cosmology lead to investigate the travel of light through time and in particular what Einstein said about time and light. In my involvement as I progressed in cosmology I arrived at the point where I had to understand what Einstein said about light travelling one year in opposing directions and being one year apart instead of two years apart. From this I made conclusion, which resulted in my forming my personal theory about the cosmos. In cosmology normal mathematical principles do not apply that straight forward as we try to envisage. Please allow me to try and explain myself as follows. My understanding of my future theory started when I was trying to understand Einstein's view on light in motion. When light depart in opposing directions from one point jointly shared by all and the light departed will travel in a straight line $180^0$ in direction to each other they are all still relevant as a continuing line.

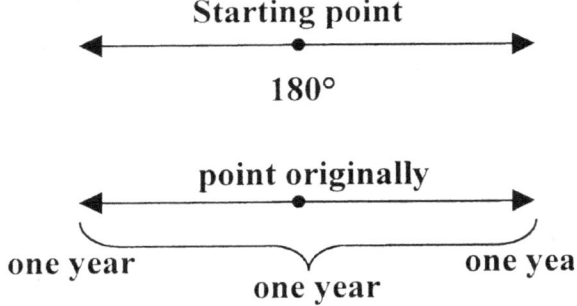

The question that nagged me for many years and on which I spent almost half a year just trying to solve the puzzle is how can light travel for one year in opposing directions and after travelling for one year in opposite directions be in two different points but was one year apart.

After one year of travel both points still are the same distance apart from each other as what both points are from the centre. Both points are from the centre just as far as each point is from where the light came from and all the points are at an even distance. The point of origin has to be evenly matching on both sides of the divide where the divide serve centre point from where the two points came. Under the normal circumstances when applying normal mathematics the light will be two light years apart. If one could stop the light travelling to the right and have that light that is stopped, standing still, while the light flowing to the left can make a complete turnabout, it will take the light coming back one year to reach the point of origin once more. It will take the light one more year to reach the other point that is having the light that was standing still then at the time for two years running. Einstein proved that the normal way humans use mathematical thinking power is not the way rules apply with the speed of light as it is in the case we find with light. Light travelling in opposing directions for one year will be one year from the source it came and the light will be one year apart.

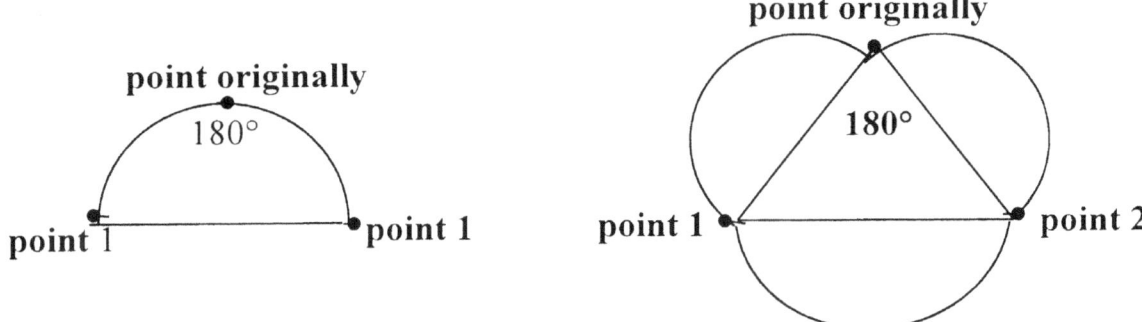

In order to give this argument mathematical logic is to put the light equal to time but this tests logic even further. Einstein's claim that this comes about because the light is equal to time did not make sense to me. Then I came to the conclusion that light became a cosmic factor before mathematics, as we understand mathematics in its fully developed state. It is where mathematics started off with Pythagoras that comes into affect. If the light was time, as Einstein interpreted light in motion to be, then time was no factor to light. In that case light will travel through space in a ratio of one meaning the moment it releases from the source it is on the other side of space notwithstanding the distance the space has. It holds a factor of one with no distance to measure any restraining. That is nonsense because light is restrained by $300 \times 10^3$ kilometres of space in one second of time. Light is restrained by space-time. Light cannot be time, because light is just a simple speed ratio like any car driving or aircraft flying or spaceship launching. Light was forming distance during time duration and that comes down to being pretty fast, but it still remains speed and speed has just the same relation with time than placing time in relation with space forming distance. One should think that Mainstream science would see from this that normal mathematics used on Earth does not apply in astrophysics but apparently that slipped past their noticing. Let us again gauge what is happening when light is travelling.

After travelling for one year the light had a distance of C multiplied by the seconds in one year to each side of the source. The points of light travelled on either end of the dividing line, which was parting the points, and securing their independence. Light had that same distance apart taken from the point the light started from. The two markers are just as far apart from each other as they are apart from the original starting point. With this in mind the use of $C^2$ by science might prove convenient, but it also proves with this mockery how big farce such innovative calculation is. There is no chance of anything going at $C^2$ because there is no exceeding of C by light or any other particle. If that were the case light would not be present in the explosion or antigravity. The speed of light is not a force but it is a speed. It is a ratio putting space in relation to the time in the space density the speed will establish. This whole argument pointed to Kepler holding the straight line in relevance to space and time. From that point I concluded that the link must be the value of a straight line sharing a dimensional value with a half circle and the triangle. If we look at the line supposedly travelling straight we find that the straight flowing is equal to the square relating the triangle. This is completely Kepler indicating gravity being $\mathbf{a^3 = T^2 k}$.

Look at the dimension and not the number implicated. It is $^{1+2+3}$ and transfers that to the line being the $^1$ and the $^2$ being the square being equal to the triangle as $^3$. But it diverts very much from normal mathematics and that is precisely what Kepler's formula also does. With Kepler $a^3 = T^2 k$ and with mathematics the volumetric size of space must either be according to the measure of normal mathematics if it is a cube then three sides form $a^3 = L \times B \times H$ and in the case of a sphere the measure will be $a^3 = 4/3\Pi r^3$. This was a triangle in relation to the square we find in the half circle standing again related to the half circle. It is not standard mathematics and anyone drawing links between mathematics and the speed of light has no idea about what is involves. With that I have again antagonised millions of the most important people with which I have to share a view. I do share their view on the Cosmos but not with their view holding mathematics as a standard fit all and apply anywhere in the cosmos. It is about lines carrying dimensional properties and with that we have to consider the line once again. Before coming to the mathematics I would first like to bring your attention to the practical side. I am promoting a theory in which I am able to prove there is as much contraction (moving in the direction of the Big Crunch) taking the cosmic Universe back to the size it had during the Big Bang as there is expansion (moving apart by Hubble's Constant) and the contraction is as much part of the expansion.

All the difference we find is seated in the human mind. We humans set differences because we look at the cosmos by placing humans and the life we find on Earth in a pivotal centre in the cosmos instead of placing singularity in the centre and life where it belong; only found on Earth. Einstein proved mathematically that in the presence of a strong gravity such a strong gravity slows time down. Surprisingly with that evidence being around this long nobody in science since Einstein's discovery took those statements and made any further progress from there. It seems to have been left in some drawer to dry. Science still sticks to the opinion that time did not change, not even slightly, since the beginning of the time and holds the same pace ever since the start of the Big Bang notwithstanding the implications this concept carries. Before the Earth took one year to circle around the Sun and even before the Sun was there a year was still the same duration of one year.

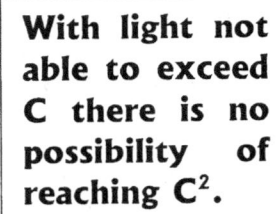

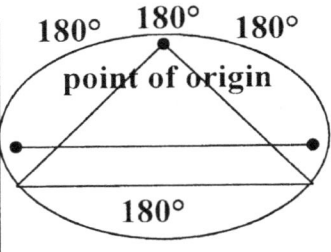

**If light cannot be two years apart when travelling in two directions opposing one another but can only maintain a distance of being one year apart, light is unable to exceed C. How can light then reach the square of C?**

How odd... don't you think ... that the only aspect in the entire Universe that is beyond change is the aspect of time? With the entire Universe including all the gravity now present and not excluding one Black Hole or dust speck pressed in such an area that was possibly the size of a lepton even then the gravity extending from that circumstances must have been beyond what words can ever describe. When everything was that small when the Big Bang took charge, the gravity at the time was beyond light, because even today in the Black Hole the gravity is beyond the speed of light. If the gravity was that high and Einstein already proved that strong gravity slows time down, then there is one logical conclusion and that is that time was in fact at the time of the Big Bang standing still. Mathematically it is incorrect to allow gravity to compress the Universe into a spot smaller that an atom and exclude any other factors and relevancies to change.

There is no chance of anything going at $C^2$ because there is no exceeding of C by light or any other particle. The electron forms the limit of C but after C the space-time brakes ranks with the third dimension and accelerate to $\pi$ and $\pi^2$. Light might equal gravity $\pi^2$ in space 3 becoming $3\pi^2$ but it cannot reach any value above C in the third dimension.

This is the fundamental fact in cosmology and breaking this concept is reducing cosmology to rubbish. If that were the case light would not be present in the explosion or antigravity. The speed of light is not a force but it is a speed meaning it is a ratio of space over time **k = a$^3$ / T$^2$** where space is a distance **a$^3$** = km, **k** is a value 300 and time is **T$^2$** = seconds distance of space in relation to time. It is a ratio putting space in relation to the time in the space density that the speed will establish. This whole argument pointed to Kepler holding the straight line 180° in relevance to space and time (**a$^3$** = 180° **T$^2$** = 180°).

From that point I concluded that the link must be the value of a straight line sharing a dimensional value with a half circle and the triangle. If we look at the line supposedly travelling straight we find that the straight flowing is equal to the square relating to the triangle in ratio. The normal calculating diverts completely what Kepler indicating as gravity being a sphere in motion **a$^3$ = T$^2$ k.**

Look at the dimension and not the number implicated. It is [1+2+3] and transfers that to the line being the [1] and the [2] being the square, which is being equal to the triangle as [3]. But it diverts very much from normal mathematics and that is precisely what Kepler's formula also does. The fact that **a$^3$ = T$^2$k** diverts from the accepted norm of 4/3πx r$^3$ It is a clear indication that what Kepler saw does not in any way translate to normal mathematics. What Kepler saw as $\pi^3 = \pi^2\pi$ is not normal applying mathematics.

That what Kepler saw, predates normal mathematics and it is our duty to investigate why that is instead of changing it to our thinking and our liking. With Kepler's **a$^3$ = T$^2$ k** and with mathematics the volumetric size of space must either be according to the measure of normal mathematics if it is a cube then three sides form a$^3$ = L x B X H and in the case of a sphere the measure will be a$^3$ = 4/3Πr$^3$. This was like comparing a triangle in relation to the square. It predates mathematics to a time when we find in the half circle standing 180° related to the triangle (180°).

It is not standard mathematics and anyone drawing links between mathematics and the speed of light has no idea about what it involves. If I take what I unleashed in the past with this statement, the past is telling me again that with that comment I have again antagonised millions of the most important people with which I have to share a view. I need to get acceptance about my view but if my statement is not well understood I get rejected. It is most important that what I say is understood. Cosmos mathematics is a standard to fit all and apply anywhere all over the cosmos. Cosmology is about lines carrying dimensional properties and with that we have to consider the line once again because what I try to introduce is not what the general perception that one will find in the view of science...

If any further dividing took place such dividing would have brought growth because there then would form space between the sides going in the opposite direction. The dividing brought all there is having all sides literally on the precise same spot, and I have located singularity in just such a spot.

**Let us find the smallest possible line first.** In any and in all sphere on would find a line running from as far as possible towards the centre of the sphere and such a line will always be reducing s it runs inwards. We already have reached the conclusion that by reducing the line , the reduced line will eventually leave all sides on the same spot. Such a spot must be round in form. With the line being the smallest line, such a line will start off as a dot that moved away from a spot. With all possible sides being in precisely the same spot we

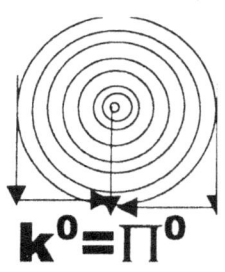

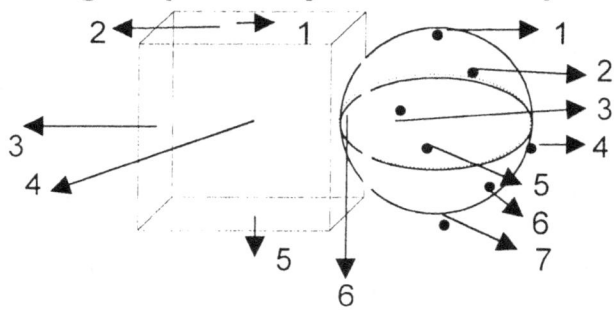

have all possible sides onto one spot. Mathematically the spot is in the single dimension where the space is one and exponentially zeros. There the space moved over to form the dot. We now are reaching into areas only the human mind can venture by understanding and nothing more. The understanding of this concept demands our reaching the point where the mind of the animal cannot reach. If it starts with a line that line only represents two sides being one and as such that is rather a flat Universe. The spot is not yet round because being round are requiring a shape or form and this lies beyond or before a time when any form of shape came into the cosmos scenario. It was in a period where shape and form was a part of the distant future hidden in and beyond eternity. In that time the line must have been so small it had reached a point not yet dividable in any way.

I came to the conclusion that the spot I found had to be singularity purely on the grounds that that spot holds only one side to serve as a start to the starting point of all directions possible. In that side is only one spot where there is only one side applicable and one dimension present. With all the factors given one can only come to one conclusion and that is that there can be only singularity. In such a case more dividing by two will land further positions on the other side of the divide. That point is serving as a position for all possible points and cannot allow further dividing as it is in the smallest line or spot there may ever be. This spot is the result of a most basic process of reduction as the Hubble constant is a most basic process of expanding during a matter of time. By reducing the line constantly the only value that will eventually remain without dispute from any party arguing about the facts is exponential zero. By only having exponential zero instead of a numerical zero and a radius as one in the square (the radius effectively becomes one holding any and all sides on one point) such a point might become any value of any significant measure implicating anything but zero as the radius. By expanding the line, it will be an evenly spaced structure growing into the most perfect round dot ever possible anywhere at the point when it starts to grow.

The reducing of the line is one dimension in six and all though such reducing is representative of two indicators all the other indicators must still be accounted for too.

Therefore the ring or circle is the only way to include all six sides in one aspect. In mathematics there is the formula used in calculating the volumetric inside of the sphere is $a^3 = 4/3\Pi r^3$ which holds two major components that will establish final value where as the rest is indicating ratios. In mathematics there is a line being one quantity and the circle indicator $\Pi$ being the next circle indicator. Reducing the line will erode the value of $\Pi$ by ratio. That will eventually lead to having a circle ratio of $\Pi r^2$ and eventually lead to $\Pi r^0$ but that is not the point where the circle ends. That is where the ration applying factor ends but it cannot exclude the circle. The circle as a concept can still reduce when it abolishes form to the single dimension. It is not the radius that is responsible for the circle but the figure value of pi and by abandoning $\pi$ only then does all the aspects fall back into the single dimension. The circle can reduce one step more when the circle eliminated r completely but the elimination of r as the factor reduced the major factor to the single dimension in $\Pi^0$. That will not reduce the cosmos to zero it will only eliminate all potential lines $r^0$ to potential circles $\Pi^0 r^0$ and from there the circle $\Pi r^0$ will come about as manifesting as a line but that manifesting can firstly only establish a circle $\Pi r^2$.

The only value that singularity can have although the single dimension may host the entire Universe is $\Pi^0$. Pick a number and elevate it to the power of zero and in the process one may have established another point holding all points in singularity but that is not the value of singularity. Only $\Pi^0$ can ever be the accurate value to singularity while singularity will then host the rest of all the possibilities in the Universe. The first value there ever was came in the form of $\pi$. Where mathematics was still an idea in development the Universe granted values of the triangle being 3 circles as $\pi^3$, which was 180° and $\pi^2$ which was half a circle also with the value of 180° and finally the straight line also being 180°. Mathematics was not yet established, but the most basic came about. Science is not taking the cosmos back as far as possible, the are taking mathematics back as far as they can but mathematics does not go all the way. Mathematics presented as numbers and symbols only became valid (as did all other aspects) later on in development. But the most basic of mathematics was in place when the spot moved on to form the dot by going from $\pi^0$ **to** $\pi$.

The reactions of those in charge of producing official policies which are responding to my argument is of the opinion that my argument is silly, but should that be your personal opinion too then test where the silly part applies. Bring the zero into the calculation, the zero that science so eagerly place in outer space and see the mathematical result. The forming of densities is once again establishing certain relevancies and when one remove one factor with a zero the density relevancy goes incoherent. By applying the distance one accepts automatically that the figure become calculated with a one. Since one is a representative of a factor that is having a value and not being without any value because as a factor it represents at least one in being part of the calculating process of the cosmos. The calculation as all calculations normally are is in order to calculate something and the something will at least stand in as one in relation to the rest being part of the calculation. When replacing the one with the nothing that science do when they say they are calculating that which is contributing to space then you can see that nothing is not what you may find in a Universe filled to the point of overflowing. But saying that the factor of one in fact represents the nothing which becomes a name and not a number since nothing is then a factor of one as it is that much the part in the calculation being calculated, then the one has to replace the zero as the fact of the factor of being calculated. You may also think that Nothing can connect a half circle, a straight line and a triangle except their sharing of a value but I try to prove that your granting of nothing is in this case a calculatedly value being something.

The claim becomes obvious when observing the connection between the half circle, the straight line and the triangle, which could also promote all the qualities lurking behind the pyramid. Consider the connection between $180^0$ sharing three different forms all part of mathematics where each is different in form, but equal in value and then one may realise in considering the very basic in mathematics being the Law of Pythagoras on which all mathematics are focused. The triangle stands in for one factor represented by one at a value of $180^0$. So does the straight line become a factor of one and the half circle also becomes one where the factor of one equals all $180^0$. All three are most seriously part of shapes in the cosmos. Revalue any one form to zero and the rest too must follow and share the same value.

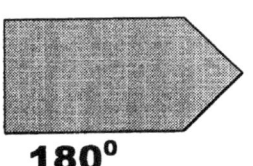

**$180^0$**

**$180^0$**

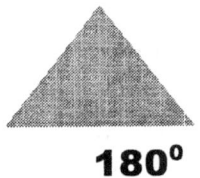

**$180^0$**

The only manner in which light can move one year apart form one another while each one is staying one year apart from the source it left is when the straight line that light use to move holds $180°$ true to the half circle they are apart ($180°$) and connects the two half circles in a triangle $180°$.

Put what I said just now into mathematical terminology it will be the same as saying there are 149 000 000 X 1 (multiplied by the kilometres) multiplied by what it is being measured which is 0 and what will the total come too... a full zero. 149 000 000 X 1 (km) x 0 (indicating what the km are made of) = 0 Mathematics says it. If there is something to be measured then the least value the measurement can have in relation to what is used in the measuring and as a factor that which is measured has to be one. It cannot be zero and be measured...and then we do measure outer space! It sounds as if something mentioned here is at fault. Yet I stand accused by Mainstream Physics of carrying the blame of being incorrect. It is not with my mentioning the inconsistency one should find fault but the fault is with the fact that the use of nothing is there and no one noticed! I am and neither is my work to blame just because I am mentioning it, but the blame must go where it belongs.

I should think that it is by now somehow understood although I imagine the implications of the statement using nothing to that affect is not nearly accepted by all. Going back to mathematical basics it is not possible that by adding a million of nothing to one nothing there will remain one nothing and that is still nothing. Nothing cannot accumulate therefore I cannot accept anything holding the vastness of space being able to constitute nothing as the major component. If that is true why try and construct the cosmos from nothing?

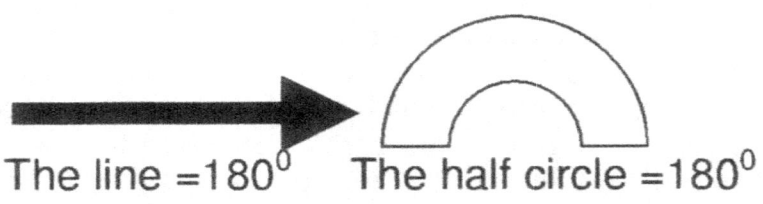

The line =$180^0$   The half circle =$180^0$

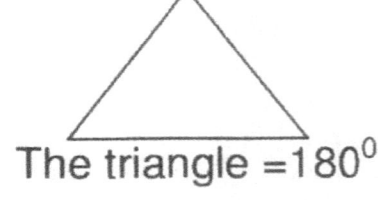

The triangle =$180^0$

Only if light exceeds mathematics and become part of pre-mathematics can light find validity. $\pi^3 = \pi^{2+1}$. What we find in this sketch is what we find in the law of Pythagoras.

The Law of Pythagoras is about angles in relation to lines and not one angle can represent zero because that will reduce all the lines also to zero. The measure of angles between stars at a distance uses parsec as the indicator, but the parsec between the stars indicating an angle has represent an angle whereby one may measure distance and such a distance cannot be zero because then the parsec will be equal to zero. Again it is multiplying the factor with the measure but if the measure is about a factor of zero, then the factor too becomes zero. That is as basic mathematics as I can present.

Let us dissect nothing from another angle. Mercury has $58 \times 10^6$ km and Pluto is $5900 \times 10^6$ km space between the Sun and the planet. The one measures about 10 x 10 times more than the other one does. The difference indicates a distance and a distance comprises of something, for if was nothing then both would have equal nothing and be next to the Sun or in the centre of the Sun. I repeat, the distance indicates something because nothing would place them both in the Sun and moreover in the centre of the Sun. Having nothing between Mercury and the Sun and between Pluto and the Sun places Mercury and Pluto at a same position within the Sun.

By saying Pluto has one hundred times more zero in between the Sun and Pluto makes such a statement laughable. Except if a learned Professor conducting a class does it and no student dare to laugh at his foolishness. If I would say Mercury holds one hundred times less nothing, such a statement will make me an idiot, but used as science makes such a statement plausible. That means the more zero or the more nothing one find between cosmic structures puts such structures further apart. There can only be distance with something concrete applying the distance between it. The problem is identifying something from nothing that defines the difference there is in science. I cannot see how nothing can become plural or more in some occasions.

When realising this I went in search of that which nothing is substituting. The issue I went in search of is what to substitute the nothing with and fill the nothing with that something, which is in place of the nothing and replacing nothing. Let us go on the interesting search of finding what prevents the Universe from tumbling in on itself. If the Universe was truly nothing, the nothing would not have the means to support the structure and the structure would disappear into the nothing that is not supporting it. The Universe is about lines forming angles and holding distance, that much we established so far. All that we know about what we now know started with Kepler's formula of $\mathbf{a^3 = k \cdot T^2}$ and that I may add was presented by the cosmos "in person" through calculating the orbit of the planets. It was rather incorrect of Newton to change this information by adding $4\Pi^2$ on the one side and G $(m + m_p)$ on the other side after all Kepler got his information straight from the "cosmic" horses mouth.

It was the Cosmos telling us humans through Kepler about the Cosmos. To change what the Cosmos told man by then telling the cosmos back what man is of the opinion of what the cosmos should be in the eyes of man is blatantly arrogant. That was what Newton's strongest characteristic was in any event and he even told the cosmos what he (Newton) as a person thought the cosmos should have told Kepler and what in his opinion was correct as how the cosmos should stand corrected when the cosmos gave the information to Kepler about itself. Newton told the cosmos what the cosmos should have been telling Kepler in the opinion of Newton. It can only be an act of utmost stupidity when man is telling the Cosmos what that person is of the opinion about what the cosmos should be telling man. This telling the cosmos instead of listening to the cosmos telling man is a trend still going on in our modern society. We run our lives by the cosmos laws every day and never notice the laws of the cosmos that we use to give us twenty first century comfort. Physics teach us that there is only energy with no differentiation between forms of energy.

The Big Bang was energy and the internal combustion engine is energy. Since there is only similarities one should then be able to find such similarities. In the fuel oil engine the engine heat up the atmospheric temperature in the combustion chamber and the fuel push the engine combustion chamber temperature to further levels where the equal of such levels we find on the rim of the outer edge of the Sun. Igniting the fuel enables the ignited fuel to be able to push the combustion chamber temperature to the same temperature as that what the Sun has on the outside edges of the Sun. That means we are pushing conditions inside the internal combustion engine to conditions that is applying currently on the skirts of the Sun. When the Sun broke free from outer space as the Sun atmosphere released into independent space in motion such developing came from outer space at the time as the temperature of outer space dropped through expanding. The Sun captured or freeze the limits of outer space as outer space went on further to overheat. The temperatures we find in the engine were the same temperatures of outer space when the Sun broke free from outer space at the time the Sun established individuality. What applied in the Universe some time ago we humans have to duplicate to get heat released as space from fuel. That is what enables us to use as energy. If it applies to the inside of the cylinder, which is merely a container it also must apply to space in the Universe, which is also merely a container. The question to answer is how can any container contain nothing that fills a Universe?

Let us dissect nothing from another angle, as we find nothing in the presence of the cosmos where science placed the nothings as the prime pillar supporting the structure of the Universe. We all know that the distance between the Sun and Pluto is roughly one hundred times more that what the distance is between Mercury and the Sun, but both has nothing between them and the Sun. This then means that one has one hundred times more of nothing between the Sun and Mercury than what the other one being Pluto and the Sun, has. The space filling the distance from the Sun to Mercury has nothing times hundred less than the space holding nothing between Pluto and the Sun. That means the distance between the Sun and Pluto is as equal in relevancy than the distance from the Sun and Pluto since both used the measure of nothing. If the one substituted the factor that we see as one represented by $Y^0$ or whatever we wish to use to represent a factor as one in the calculation with the nothing science at this stage place in place as a valid value of nothing, all laws of mathematics will go in disarray because when one multiply any number by zero it becomes zero placing both planets in the Sun. When there is nothing between the second whatever then whatever must be in the centre of the Sun.

The distance between the Sun and Pluto **is Pluto is 5900 X 10⁶** kilometres of space, but in that statement we take it that the one as a factor of a kilometre is present in such a multiplication. The one constitutes the presence of fact being a statement of a value. By saying the distance constitutes of nothing we have to substitute the one factor with a factor of zero. Then the calculation must read **Pluto is 5900 X $10^6$ X 0 = 0.** Including nothing as to state the presence of that part contained by the calculation delivers the total of zero. By excluding nothing from the equation space becomes something bringing in a value lying inside the realms of the infinite that must form singularity. Applying this logic to the Lagrangian system and interpreting that information to the law of Pythagoras a clear pattern come about.

When I try to point out such in discrepancies in the thinking of Mainstream science, which they are responsible for official information, those in power don't hesitate to show me the power of authority they enjoy. I would think that Academics would honestly be surprised when someone draws their attention to their possible being incorrect instead of they're getting all defensive and annoyed about the whole matter. They are in a powerful enough position where they can dismiss me by blaming incoherency on my part as the reason for doing so By the same margin as they declare my work being incoherent they use such a blame as an excuse whereby they then refuse me another opportunity defend myself when presenting my work again and a revision about my view by simply saying my arguments and my use of mathematics is grabbing straws. The Appreciated Academics of Important Standings may use any phrase to dismiss me and I do not have any chance to challenge them in further debating because according to them, I have had my chance.

As much as I have searched I could not once find any method any one used to prove mathematically that zero fills outer space but my effort to draw attention to that matter brought me the sum total of nothing. Please if there is proof out there to that effect I would love to see such mathematical proof. When using Kepler's mathematics without Newton's dubious changing thereof I can prove that the same substance produce distance in space and motion in time because of the infinite and precise control we find between neighbouring structures. While Academics blame me for being incoherent no one ever showed me where did I detour from accepted mathematical principle in doing that? But not once was my statements disproved mathematically by proving mathematically that nothing (zero) can become or form distance by which to measure. No…my correspondence is merely swept from the table or ignored on the grounds of a lack of importance which they find in the subject and then because of the lack of immediate importance, such lack of importance in my work disallowed their further involvement. Please for once prove mathematically where I can locate nothing between planets because that will mean that **$a^3$ = 0 and $T^2$ = 0 and k = 0.** That will bring about that all planets are on the very same plane being somewhere between nowhere and nothing. If space was zero or nothing then Kepler's principle formula will stand untrue and without substance everywhere in all of science. The fact that singularity spawns all natural weather phenomena is an indicator of such endless influence.

By reducing the line we come to the end of the mathematical equation of the circle but the circle does not end there. That is what Newton did not recognise from the figures the cosmos represented to Kepler. The circle only secures the final cosmic figure and the value to singularity where all things have equal value. At that point the half circle and the triangle and the line must start since all three having many different forms have equal value at $180^0$. Only after that point does mathematics begin where all factors in 1 have the value of 1 being $1^0$. In that conclusion one realises something must separate singularity from all other factors because singularity hosts all other factors but is by own initiative $\Pi$.

That will be the spot of origin. That will hold the eternal spot…the smallest spot ever because all spots that ever can be was secured in a position in the centre of that spot. Because of the progress singularity follows from the single dimension singularity only allow mathematics a start at $\Pi^0$ progressing further too $\Pi\Pi^0$ and from there the line is born as $\Pi\Pi^0\Pi^0$ or $\Pi2\Pi^0$ $\Pi3\Pi^0$ $\Pi4\Pi^0$ $\Pi5\Pi^0$ where $\Pi^0$ then may form the concept and value of r. But the line starts at $\Pi^0 = r^0$. Because cosmology is singularity based and the value is $\Pi\Pi^0$. This escaped the attention of the greatest mathematician about the work of the greatest cosmologist ever because Newton incorrectly introduced $4\Pi^2$. The introduction of $4\Pi^2$ exaggerated the value of time and removed space / time from the concept. Mathematics in cosmology does not apply pi, pi is the root value of all concepts in cosmology. The factor pi impersonates as much as it represents singularity. But we may ask why $\Pi$ will come about from singularity.

Let's go back once more and reduce the line by half every time. Then repeat the process until it can repeat no more. The reducing of the line by half every time will get to a point where all the ends land on the same position without any possibility if halving the two ends further. The points share one position and moving the points in any direction will lead too an increase of the line once more.

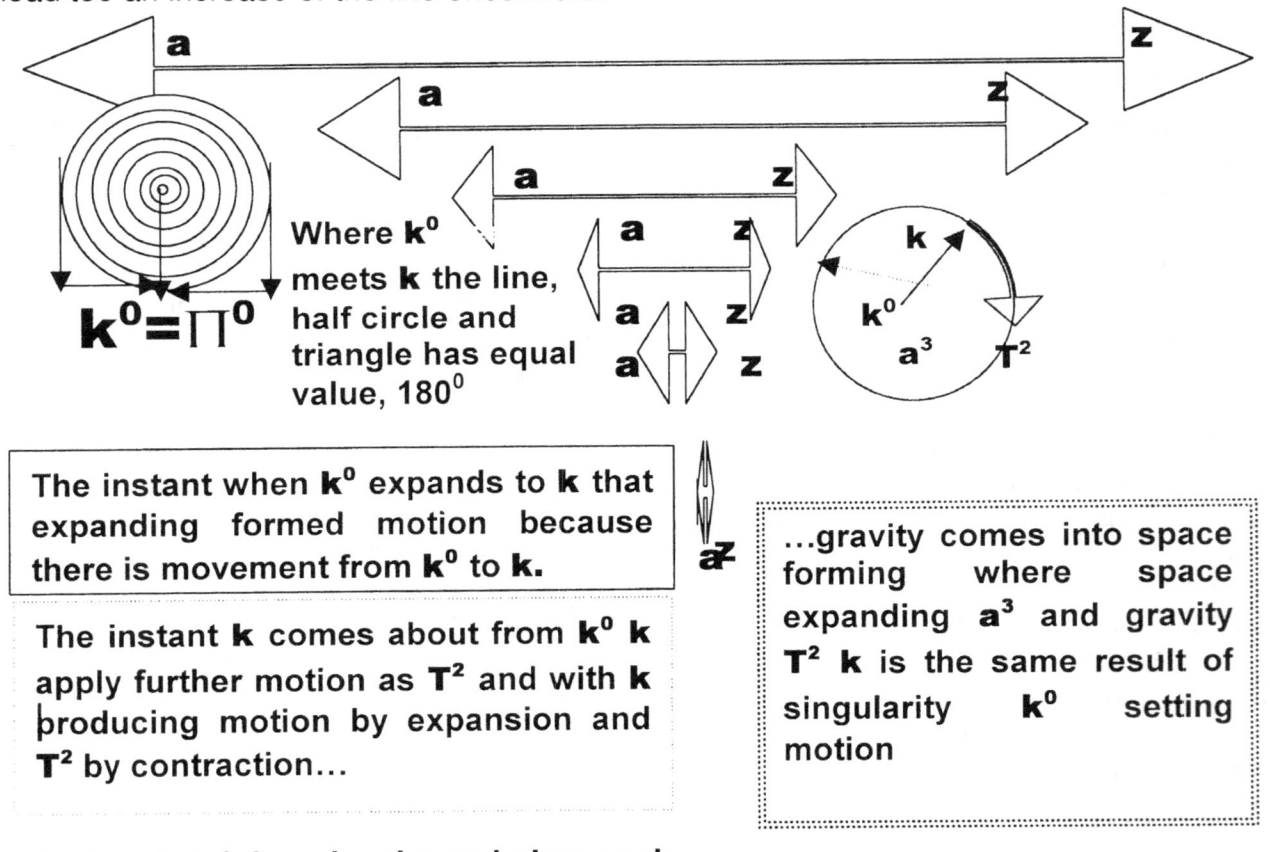

**Where $k^0$ meets k the line, half circle and triangle has equal value, $180^0$**

$k^0 = \Pi^0$

**The instant when $k^0$ expands to k that expanding formed motion because there is movement from $k^0$ to k.**

**The instant k comes about from $k^0$ k apply further motion as $T^2$ and with k producing motion by expansion and $T^2$ by contraction…**

…gravity comes into space forming where space expanding $a^3$ and gravity $T^2$ k is the same result of singularity $k^0$ setting motion

**In the sketch I made, shows below each of the lines reducing there is a space left open**

In the sketch I made shows below each of the lines reducing there is a space left open between the two ends of the line that is symbolising the end of the line in reducing. In the end the two ends will share one location even by having one single point holding each one. There is no chance that I can present any sketch reducing the line to a point where the points are sharing one location literally in the single dimension. The points are there and with the points being present they may not be dismissed as nothing. From there no reducing in a natural manner can lead to nothing without changing the rules of mathematics in such reducing. But the two ends has reached a position where any further effort of reducing must bring about the start of extending because every point possible share space with every other possible point at the point of singularity where all points share one common space. By moving any of the points such moving must then bring about an increase of space once more. This also applies to the circle because the circle uses a line to indicate size. By reducing the line and by reducing the circle the reducing will end up having the ends in the same position. It is this fact of the moving of any point from that spot holding singularity that such motion will introduce space as the space exceed the previous limits of singularity.

One must draw this statement of motion back to the point where singularity is getting sides. When there is singularity there can be no sides. It is 1 (one) from all angles there can be. That one fills a space. The space it fills does not really exist in the manner we humans see space to exist. It is a spot that is there without being there. It does not visually exist because it is not filling any substance and it cannot be recognised. Once one accepts the fact of singularity that accepting of singularity then is contradicting all the things we know by not being any of the things we can recognise. There is no space. In that space there can be no motion because there can be no space to have the motion within. It is a line that is so small it is not there and the only reason why we know it is there is because of the results it left as an imprint of its not being there. We cannot detect it on the merits of its absence because it is never absent. It cannot be absent. It cannot go absent but it can never be there where it should be in the third dimension if I wish to locate it. If it was absent then it was zero or nothing but since it is there it is not there and that makes it present. The centre spot we cannot see and that we cannot detect has no sides to any side and has no place it fills because it fills all the places we cannot detect.

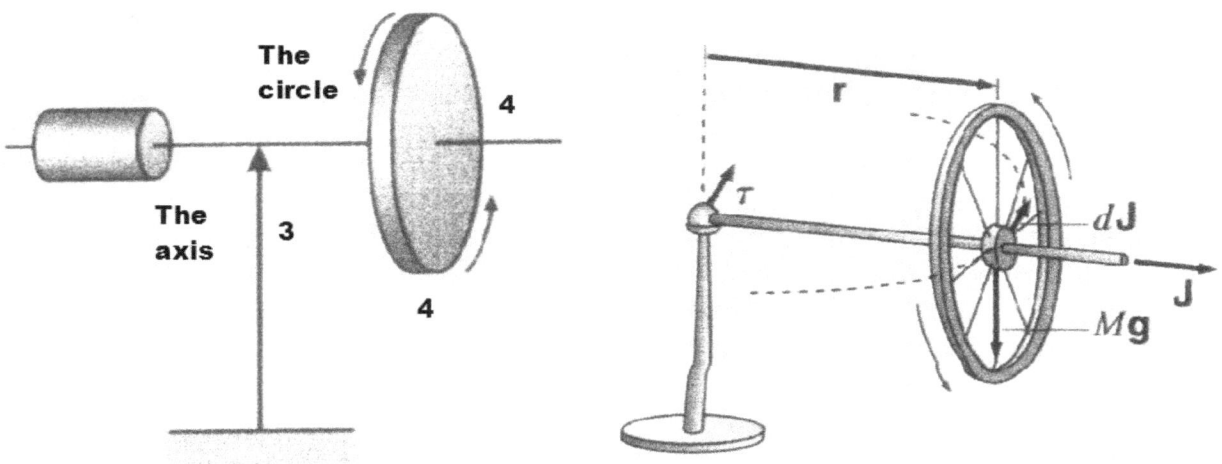

The only way such a spot can fill space is by doubling the space it fills to become more than one place to fill. But the very instant that happens it halves the space it fill because it then cuts the space it has into two parts. Any motion from such a point in singularity lands in and on the other side of the Universe. That brings about that the point of not being is doubling the not being and by doubling the not being into being it also cut the not being that became present into half. We have to find this spot as we find religion. It is something that we can only know is there because we cannot disprove it is there but we can never prove it to be there. It is something seen through intellect and not through the eyes.

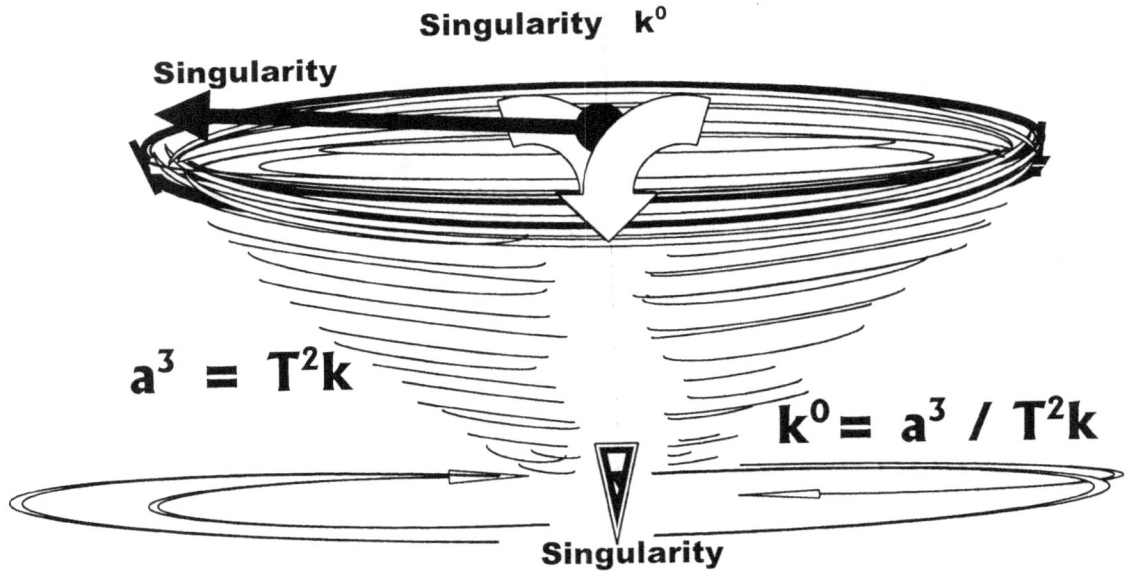

**Singularity  k⁰**

**Singularity**

$$a^3 = T^2k$$

$$k^0 = a^3 / T^2k$$

**Singularity**

From the smallest ever possible dot will grow a line in every imaginable direction relating to a prospect of Π because only Π will not favour one specific direction and that puts all directions at equilibrium meaning that any form of what ever might develop from such a spot will have the end and the start being in the same position, which will also have to be a sphere as the flow outward will be equal in all directions. This is why we humans show the incentive to acknowledge this fact as we the sphere being representing the form of the entire Universe.  The smallest spot in singularity is a sphere. Please think clearly, is that not precisely the commitment we find in gravity, where gravity is flowing from singularity outwards but never favouring any side?  The nature of gravity is to never end and never favour and where it seems to favour there is a valid explaining concerning singularity. This reasoning prompted me to look for singularity in such a spot because if the prime spot from which all came was a spot holding all, then the spot must hold the shortest line but more prominent it will hold the smallest form including the smallest circle or for that matter the smallest sphere. That leaves the door wide open for the advancing of any radius in all possible directions. With gravity always being in the centre of a sphere where the space is least available in the entire structure (there is not even space left to fill) one finds a flow of gravity from that centre spot outwards in all possible direction even-handedly. The fact that the original gravity will begin as a circle or will be a circle is the direction it will take when being the first spot created. All progress will be evenly in all direction because no direction will stand out or be in favour above any other direction at first.

Within the circle $k^0 = a^3 / (T^2k)$ holds gravity centred in the precise middle of the circle. By using mathematics in the way Kepler used it those rules and laws used correctly in the investigating of the formula that Kepler introduced must form the basis of cosmology. Also such intense investigation then must be without Newton interfering and telling Kepler what he (Kepler) should have found instead of Newton incorrectly correcting Kepler whereas instead Newton should have been looking at what he (Kepler) found because only then he (Newton) could have seen what gravity is. He (Kepler) said that the cosmos said that gravity is $a^3 = T^2 k$. The space is held in check by motion from a centre and that is gravity. It becomes more than clear that space $a^3$ is time by dimension $T^2$ and time is space $a^3$ without dimension $k$ Gravity is $a^3 / k$ but $k$ is an addition of motion $T^2$.

The spot forms a full circle, but the line running through the circle is forever present because that is the future radius of the circle that will one day develop the circle, which is equal to the present diameter. The fact of the presence of such a possible line in such a possible circle dividing the possible circle into two parts makes the centre line equal to the half circle. The line forms the half circle but not only that the line presents the half circle as much as the line is the half circle When referring to a circle I use the name of a circle because there is no other referring by name available but such a circle represents form and not yet space.

The Universe at the time was so small concepts we take fore granted today was eternities apart. In the centre of the form runs a dividing line that is the form eternally but also it is eternally smaller than the form because no measure other than eternal and infinite was available to use. Notwithstanding the concept that the line was the form but also what it came from was eternally smaller than the form this is because the line parted the form into two parts of equal halves of exact duplication. The line then is $180^0$ and the half circle forming on either side of the divide is split what formed into two separate halves leaving both half circles in equal value to each other and equal to the line parting the circle. The value in all cases is $180^0$ because in singularity the two factors are the same. Mathematics as we know it at that point in eternity has not yet developed because at that stage there were only space less relevancies between the same thing split into segments that was innumerable.

The same value is of course $k^0 = 1$. In this half circle of the future, which is no half circle as yet because of a lack of space there are three future points indicating the space less ness that will go on to become space filled with something. On top of such a circle to form must be a marker indicating an awaiting boundary or future border and at the bottom of the future circle there also must be a similar marker that is no marker as yet. Between the two possible points that are not there yet is a future line running that is not there yet. Then indicating the possibility of a position to come that will bring about the half circle being a future distance apart from the future line indicating a diameter that will one day be there a third such a marker must be established for the future. That forms a triangle with two more sides being connected by either a line being one or half pi being one. Crossing such a divide comes down to the very same as jumping over from the end of the Universe to the very end of the other side of the Universe. What was doubled so that whatever the Universe was before doubled by a not quantifiable margin in a not quantifiable number of possible jumps? Most important fact to observe is that the slightest notion took whatever was moving across the entire Universe. From singularity comes about that the line is the same as the half circle and is the same as the triangle and all has one value being $180^0$. Space and mathematics at that point was waiting for the future Creation to develop …therefore mathematics was in tern waiting to develop.

In any circle or sphere the size only depend on the fluctuation of r in the square as a component to the circle or sphere but that does not affect the form by indication of Π in any way there may be. The conclusion from this is that no line can start at zero because that will be a mathematical impossibility. This statement by itself excludes zero and with zero excluded one then begin to appreciate all the rest of the concepts governing corrected cosmology. A line or spot starting at zero would therefore be shorter than the shortest line possible. For obvious reasons can no line, or any line grow or extend from zero because such a line must then quit zero and become something, thus abandon its original value.

That would mean the start of the line has a different value to the end and a line holds conformity through out. When any line is starting from point zero it can never leave zero because of the influence of being zero disqualifies any possibility of growth. If the line then had to grow in all directions at the same pace the line must therefore be a circle or being three-dimensional, a sphere. Flowing from this fact is that in the Universe there can be no zero point or unfilled space. In the case of the growing sphere the value of the circle is Π, and that is where creation started. That gave me the clue where to start looking for singularity. One would find singularity in the value Π and the value Π will be in all things rotating in a circle. You might wonder how does that apply to the cosmos and moreover to gravity? You cannot fit nothing into outer space because it just will not fit. If any of the factors in Kepler's formulae represent nothing that is what you will get. The Universe will be nothing. $a^3 = 0$ $T^2 = 0$ $k = 0$. If the argument seems ridiculous it is not my mentioning such a fact that is ridiculous but the mere fact of the reasoning also becoming a recognising of an argument accepted by science making it as such ridiculous. It is the fact that one must argue about such a ridiculous matter that allows the ridiculous part enter the conversation because the trend reminds of arguing about fairies and little people and such nonsense. If space is nothing then it has a number to use indicating just that value being zero or the capitol O indicating zero. Try and indicate what is measured and calculated in space, but not by simply not thinking about the fact and therefore simply ignoring that what is measured forming the sole value of space, but put the value of nothing as part of the distance in calculation because that is what is measured. When stating the distance between the Earth and the Sun place on paper what will allow the kilometres measured to represent the factor that is being measured. If represented by one being the total of one by hundred and forty nine million kilometres of nothing put that language in the International language of mathematics that spans all dialects spoken on Earth.

**By reducing r indefinitely to the tune of half each time, r would become infinitely small, beyond human calculating means, however as mentioned in the case of the smallest dot holding one spot, r would become insignificant beyond human comprehension even, but never reaching zero and still Π would remain intact and dictating form. I believe one can begin too see where my suspicions are heading because the flaw comes about in the manner mathematics are practised for thousands of years. Before coming to the mathematics I would first like to bring your attention to the practical side. I am promoting a theory in which I am able to prove there is as much contraction going on in the cosmic universe as there is expansion and the contraction is as much part of the expansion. The universe rides on a balance and we have to locate such a balance. To prove my theory I firstly had to locate the centre of the universe.**

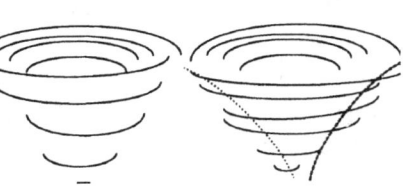

**To find the invisible I had to locate singularity. I realised that my effort to locate the point holding singularity enabled me to backtrack the exploding universe to its origins**

Even admitting to such a notion sounds like madness, but please give me a chance to explain in more detail. If I wish to achieve success that would depend on my ability to convince all that outer space comprises of material and as such we can locate such material even if we are unable to see such material. By applying some basic effort I have located the position from where all movement came and the direction it took moving forward in time.

We traced the line back to the spot and so we know where it all started there was a spot. The miracle part comes in the fact that eternity shows no change. Any change, even the slightest change represents the end of eternity. With everything motionless and locked in singularity everything remained eternal because even today singularity is locked in eternity. In the centre of all objects spinning there is a space forming a divide. It divides every aspect of space spinning into sectors of space having motion as much as changing direction of initial motion. However, since it was part of the cosmos, and although it is outside the cosmos, it still is part of the cosmos.

A spot holding what ever is and can be into a dimension so small it did not even have sides. Then for a reason, which at this time I do not wish to go into some miracle happened and the spot showed motion. The motion was deliberate but the motion was eternally small. The spot had one specific value, which it clung too…it had all sides on the same side and from that rotating diverting came about. The forming of $\Pi^0 = k^0 = 1$ had all possibilities available but it was only one possibility in the end. There was no space and then space expanded producing space $\Pi^0 \Rightarrow \Pi$ and from that motion comes gravity by the value of the motion creating contracting $\Pi^2$. The motion brought along $\Pi$ but even $\Pi$ was subject to relevancy. From this come the most basic principles in as much as forming the ground rules of the law of Pythagoras.

When drawing a line such a line then starts of with a dot serving the spot that holds all sides equal. That means the line serving as the future radius will be equal to the half circle which is then $\Pi$. The only aspect of the point that stands in for the end of the single line forming the radius of the circle is that we then mathematically reach the single dimension. We decreased the line to where a circle being $\Pi$ formed on the single dimension. This dimension also hold the circle dividing line because from there the radius must once again generate a value and by such a gesture that the extending would form the circle that forms the sphere that eventually lead to the formation of particles. This leaves a problem to investigate.

### The Roche limit is:

**The region surrounding each star in a binary system, within which any material is gravitationally bound to that particular star. The boundary of the Roche lobes is an equipotential surface, and the lobes touch at the inner Lagrangian point, $L_1$, through which mass transfer may occur if one of the components expands to fill its lobe. It names after the French mathematician Edouard Albert Roche (1820-83).**

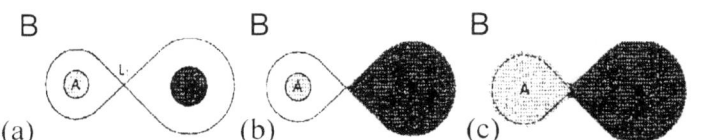

THE ROCHE LOBE: **In a binary system, the Roche lobes of components A and B meet at the $L_1$ Lagrangian point.**

**(a) In a detached system, neither star fills its Roche lobe. (b) In a semidetached system, one massive component, B, fills its Roche lobe. (c) In a contact binary, both components overfill their Roche lobes and share a common envelope. Lets explain the importance of this Roche limit and how the Universe used the Roche factor to produce the Big Bang. That is where it all started…**

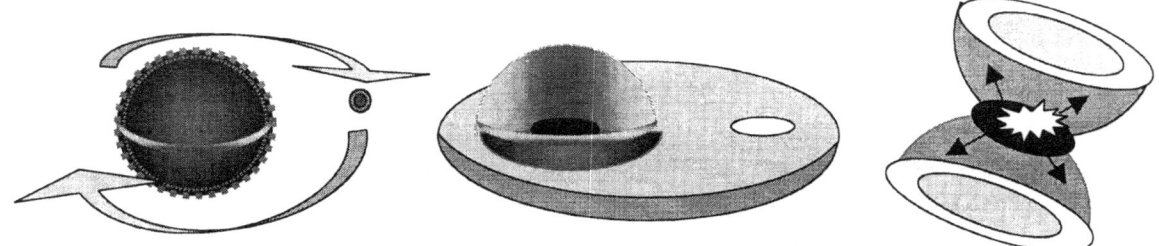

It is easy to see why it seemed as if there were carnage and destruction in a form of matter eating up antimatter and in a sense that is the case but the destruction came about because of the lack of space causing friction that brought about heat that turned to space. This took place because space is the motion thereof. $a^3 = T^2k.$ I guess one can say that if matter dissolves such dissolving matter can become antimatter since there were only two options available at the time but I prefer to use the term heat because by any other applying name confusion sets in again. There was material producing gravity by performing motion and then there was material performing anti gravity by producing heat.

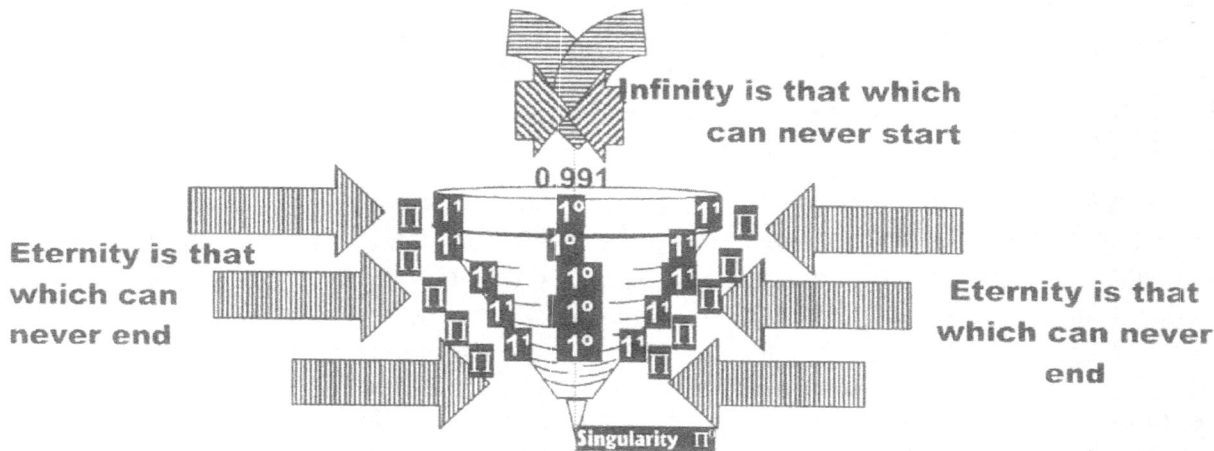

This was due to the restriction where by the particles was lacking the application of motion and with heating the heat created space that formed part of material destructed or then possibly became what science refer to as antimatter. But antimatter and plasma and all other names available does not for fill the function of establishing recognition about what one refers too. Using heat as the term is the least alien term available because in the last sense every one knows what heat is. Also in this process on the other side of another divide light became another product of heat and light is heat reflecting a sure connecting to dissolved singularity. The Roche limit is not only what we consider as the big and the large because the Roche limit had to be the driving force that eventually started the Big Bang. It is the Roche limit used to bridge the limits between the most volatile elements while expanding the space and reducing the time with electricity. The electricity both expands as heat and contract by locating heating placing the Roche limit in place and that start any nuclear explosion.

By heating the inside of the BOMB the material gets packed closer. With the increase in heat instead of the increasing the environment to accommodate such increase needed in space growing larger it brings the super volatile substance within the bridging of the set Roche limit and bang goes Nagasaki. Oppenheimer's dynamite did the very same in that first BOMB because the dynamite also produces excessive and uncontrolled heat that heated the volatile substance. It also is just what happened during and leading to the Big Bang. The growth of space placed some material within the limit of others while there was no other space to fill. The unlucky ones became time in space.

$1^1$

$1^0$ Infinity

$1^0$ Eternity $1^1$  +  $1^1$

$1^1$

How did this "Big Bang" take place? The best way to examine the reason is to see why anything in the universe expands. To get anything to expand one has to heat it. All matter expands when overheating. Science may come up with whatever brilliant theory, the fact of the matter is that when matter overheats it expands. The bigger the overheating, the bigger will the expansion be, it is as simple as that.

This means whatever leads to the forming of the "Big Bang", whatever preceded it, it had to come about from matter that overheated. With the event of the nuclear age, the proof came about that matter is heat in some frozen form. Unleashing heat from its frozen form brought about a jolt of heat, never yet experienced by man. By breaking matter from the frozen state, of which it is in, within the atom, heat produces light and heat. Where this process clearly shows how new space-time forms is where the releasing of heat caused winds that stun man's logic.

The nuclear explosion shows quite clearly what the "Big Bang" was, with the nuclear explosion being a very minute form. Yes, we have all heard the rubbish about matter and anti-matter. What can anti-matter is, since matter is heat, defined to a certain space occupied for that time. With matter being frozen heat, what would form anti-matter. Anti-matter means the opposite to matter, and if matter is frozen heat, anti-matter must then be overheating heat. This in itself is quite ridiculous. Anti-matter can only be matter with an opposite spin to that we think of as matter.

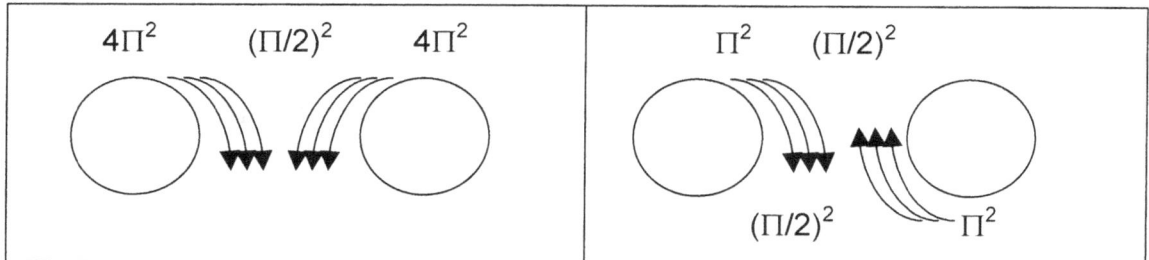

$4\Pi^2$   $(\Pi/2)^2$   $4\Pi^2$          $\Pi^2$   $(\Pi/2)^2$

$(\Pi/2)^2$   $\Pi^2$

In the second sketch $\Pi^2$ $(\Pi/2)$ / $(\Pi/2)^2$ $\Pi^2$ the Roche limit cancel each other and with that the space of the neutron effectively disappear. The two protons touch destroying each other and the neutron as well as the proton demolish and became heat $3^3$.

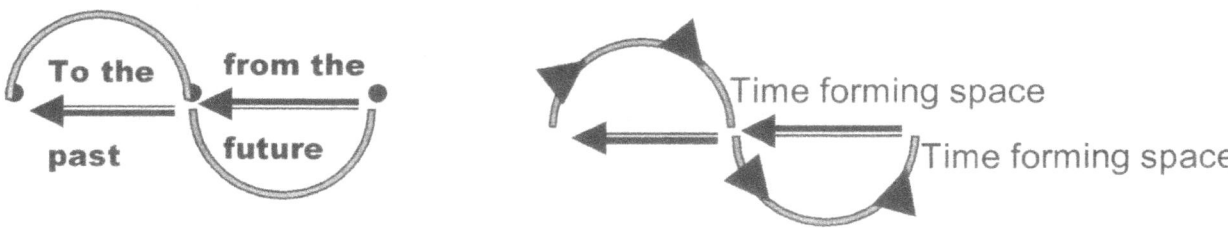

To the past    from the future        Time forming space
                                       Time forming space

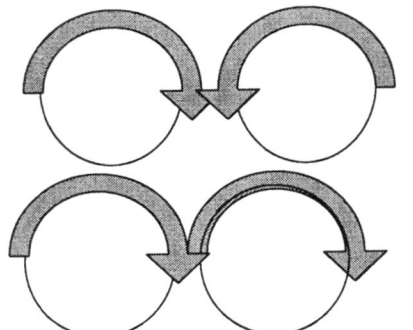

Due to the enormous absence in available space wherein to produce motion that creates the space when the first moment was born all those options coming about from relevancies applying would lead to friction. The scares ness of space between all the particles malformed some.

The deforming that allowed heat to rise brought about space and tried to apply motion that would accumulate into more friction. More friction resulted in producing more heat by more motion that produces more space. It served as a self-destructing devise in almost all cases but one. This cycle of heating affected all possible synchronised motion but one possibility remained viable to secure material. In all other cases the lack of space brought about heat leading onto more motion but this time the motion was directed in a single direction of securing expansion.

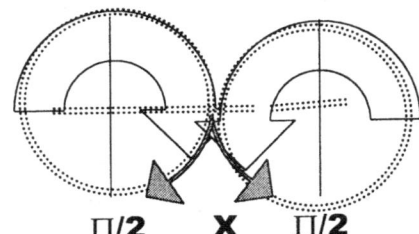

$$\Pi/2 \quad X \quad \Pi/2$$

Where synchronised directional motion is in harmony with the Universe but also is in opposite relevancy to the Universe the relevancy brings about form and creates space. Some particles were reducing space this way by motion in contraction. This then formed gravity and came in as gravity. If this form produces gravity any other motion not complying with this synchronised directional motion in opposing harmony must be producing anti gravity. This centre point is present since the very first eternal instant and is present in all of nature

In every one of the natural phenomenons the circular displacing of the Earth is turned into a linear motion and from the linear motion it receives rotational acceleration that generates a centre presenting singularity to take charge from that point. In all cases it is motion $T^2 k$ that sets space $a^3$ in relation to a specific centre $k^0$. Motion of many sorts establishes a generated singularity that activates space-time as it happens in the case of the spinning top.

There is the limit between all spinning material in a set time and that limit must not be crossed.

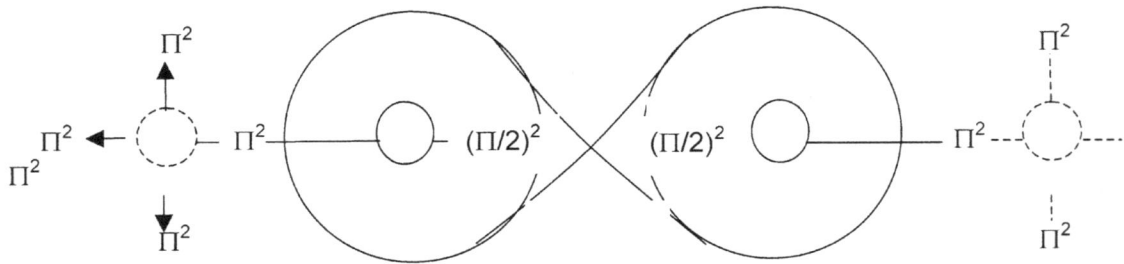

S₁

From the manner the Roche limit provides the Roche lobe with a gravitational fields we may assume that there is a link between points holding singularity provided by time in space

It is accepted but never yet proven that all things about and in the Universe connect. It has to since everything uses on common ancestral starting point but no one could ever prove where or what the point is. With one point connecting the lot everything is connected I a measure no human mind can ever anticipate and the connection is through motion providing time. Today we find that because everything started that way everything about nature comes by activating the Coanda effect in some manner where motion indicates an exciting of singularity positioned by applying motion. It is the value of the motion that brings about the time and the time "shrinks" space as the speed of motion will reduce or exaggerate the "size" of space claimed for use during that motion.

This sketch exemplifies all phenomenons in nature as we use nature to us with life's advantage

With the Universe being that small in the very beginning there were two options available for material to choose from. There was expanding because of overheating therefore becoming relatively becoming softer or remain relatively more solid and cool off by reforming through contracting heat released by other particles. It was gravity and antigravity. One must remember that in this time when all started the slightest motion, so slight we humans can never find

The rotating motion of liquids around a centre is what gravity is as the Earth rotates around its axis. Not for one minute should any one forget that the centre axis of the Earth is a liquid with all the heat gathered between the solid particles in the Earth centre. Take into consideration the fact that with the heat concentrated in the centre even the $iron_{56}$ in the centre is a liquid. By rotating objects in liquid (which is what the atmosphere is) the motion activates governing singularity a centre which is elected by all the atoms with the confinement of the Earth and intensifies but also localise gravity to a specific point shared by all space in that motion. It is using this principle where electricity is charged in this manner and where hurricanes generate space-time. Gravity will have a charge equal to electricity in the extreme centre of the rotating Earth. We must remember that gravity liquefies the atmosphere through compacting the space-time of the Earth from the gas it is in outer space to a liquid in the centre. The atmosphere in its natural form may display as a liquid in a natural form but can still also be a gas when water is in motion. The generating of singularity is displayed in electricity charging as it is in lightning, as it is in igniting fuel in turbine rocket engines, as it shows the ways the Coanda effect applies, as it is presented in weather and as it limits the space-time in the behaviour of the internal reflection where the flow of light is limited by the space-time supplied by the flowing water.

the ability to detect the space or the motion that went with the space, such motion took such space to another Universe or jumped to "the other side" of the Universe. In this there were two options to cause material destruction, which means producing space by heat and destroying singularity at the same time or performing gravity and preserving form with the preserving of material.

Coming back to that which Kepler saw, it was Kepler that gave the factors distinguishing symbols, but the symbols hold identical value but only holds dimensional differences.

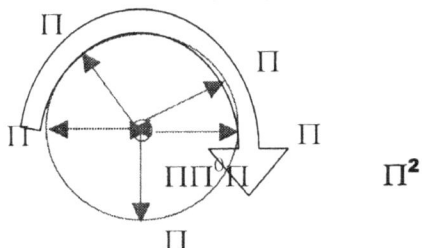 Singularity started with expanding $k^0$ to $k$ or for more precise valuing from $\Pi^0$ to $\Pi$. The expanding came about from the overheating of singularity and the motion coming about with space expanding. The motion is in itself creating space, which actually is time and not space, by extending $k$ four times since there are four relevancies. One should remember that since there was no 3D space yet the expanding was in the flat space less world of the proton at the double square of singularity forming time related space ($\Pi^2 + \Pi^2$). That was not yet the Universe we grew accustomed to. This was only formed by antigravity.

However, keeping $\Pi$ as one ($\Pi^0 = 1$) we keep the Universe in the first dimension.

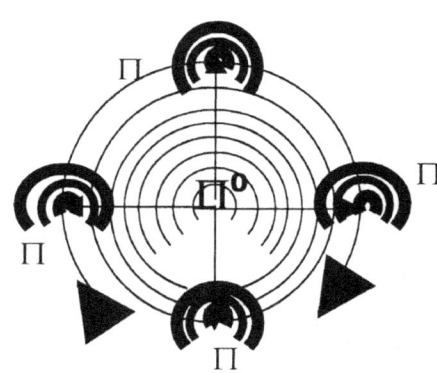 This point, which I now am referring to, is the point where $\Pi$ is a fully appreciated value while the diameter D still remains a dimensional factor of one. This is the dawn of the second dimension where space was there but space was sparsely shared in some cases. It was when $\Pi^0$ shifted to become $\Pi$ for the very fist time.

**However, the equation looks far more sensibility when using the value of singularity**

When singularity expanded for the first time evei and when heat parted from cold bringing about the Universe forming $1^0$ to $1^1$ from $\Pi^0$ to $\Pi$ a relevancy was born and that relevancy grew into what we now have as a Universe

Gravity in the centre formed time $\Pi^2$ by dismissing while the four time positions started the cosmic trend of duplicating.

**With every one of the four points taking form to the value of $\Pi$ at a measure of $\Pi/2$ each brought about the Roche value of $\Pi^2/4$ in relation to the developing centre. One has to remember that the star of today takes on the characteristics of the form of that era. Then did time form a component of 4 with space forming there in the time unit of three making it seven relating to ten**

In the past the scientific enormity of Kepler's statement passed Mainstream Physics because (I believe) that Newton at the time disregarded the importance of placing compositions in mathematical relevancies. Later on no one saw himself worthy of controlling what Newton may have missed and this blemish went unnoticed. But once a relevancy is established through investigating the relevancy from all possible sides the enormity of the concept becomes transparent. It links combinations that we now can see four hundred years after the fact.

In the Universe and throughout the Universe space contains particles by sustaining the six-sided space forming a balance. The relevancies are top opposing bottom where left is opposing rite and front is opposing back. When one of the six sides moves away, due to the direct contact with a sphere forming a cosmic structure the object then makes contact with singularity. By making contact with singularity the object falls from the sky towards the centre of the sphere. It is the same therefore it is dimensions repeating to form $a^6$ because $a^3 = a^{2+1}$ that becomes $k^0 = a^3/a^{2+1}$.

In the way Kepler presented space and time it is all the same thing that space is made of the motion forming time. This tendency Kepler confirmed Kepler introduced with mathematical equations that Mainstream science preferred to ignore. But all the further ignoring will only produce more ignorance because from using Kepler observations one can read so much more into cosmology. The future of understanding cosmology can only result from accepting what the cosmos told Kepler when the cosmos told Kepler that $a^3 = k \cdot T^2$. From this we see where singularity is $k^0 = a^3 / k\, T^2$

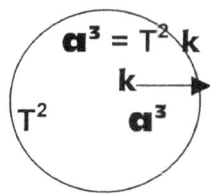

$a^3 = T^2 k$

$k$

$T^2$     $a^3$

Distance **k** is equal to the space $a^3$ it is in and the time $T^2$ it takes to move that space

$k = a^3 / T^2$

$k = a^{3-2}\,(T^2)$

$k = a^{3-2} = k^1$

$k = k^{3-2} = k^1$

$a^3 = T^2 k$

$a^3 = T^2 k^1$

$a^3 = T^{2+1}\,(k^1)$

$a^3 = a^{2+1} = a^3$

**and**

$T^2 = a^3 / k$

$T^2 = a^3 / k^1$

$T^2 = a^{3-1} = T^2$

$T^2 = T^{3-1} = T^2$

**It is all the**

From this we see where singularity is

$k^0 = a^3 / k\, T^2$

$a^3 = k \cdot T^2$ but Kepler also said

$a^3 / k = k \cdot T^2 / k$ and Kepler said

$a^3 / k^{0+1} = k^1 = k^{1-1} = k^0\, T^2$

$T^2 = a^3 / k^1$

$a^2 = a^{3-1}$ or $T^2 = k^{3-1}$  $k^2 = k^{3-1}$  $T^2 = T^{3-1}$

If ever there was one scientific blunder that puts science on its back exposing its under belly and brought along so many misconceptions, then it is Newton's ignoring of Kepler's brilliance in the face of facts

I believe there was much loss of life, not only from space flight but also from atmospheric flying in the past because of the blunder science inherited from Newton's incorrect presumptions of Kepler's work. When saying that I have to add that in the past my saying this brought about immediate dismissal of all I had to say, but I cannot ignore the truth.

We can see why motion is the culprit forcing us down towards the inside of the Earth in the process we blame as gravity. $k^0 / k = T^2 / a^3$  That is not what Kepler found $a^3 = T^2 k$.

Our instincts agree with Newton and we ignore Kepler because of what we feel with Newton. We feel we are pushed to the ground by some unexplainable force that is forever souring our lives. But as I have explained, it is because our motion is slower than the motion the Earth has that the Earth holds our space captured in the space we occupy. Because the Earth and we share space in the same space but is not the same space "we get dragged down".

However, keeping Π as one ($\Pi^0 = 1$) we keep the Universe in the first dimension.

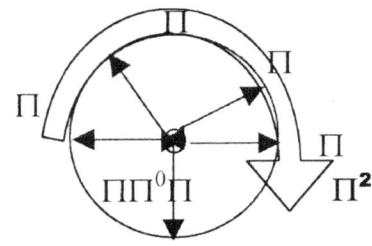

This point, which I now am referring to, is the point where Π is a fully appreciated value while the diameter D still remains a dimensional factor of one. This is the dawn of the second dimension where space was there but space was sparsely shared in some cases. It was when $\Pi^0$ shifted to become Π for the very fist time.

## However, the equation looks far more sensibility when using the value of singularity

$k^0 = a^3 / T^2k$ forms

$1 / k^0 = T^2 k / a^3$

$1 / (k^0 k) = T^2 k / (a^3 k)$

$1 / k = T^2 / a^3$

Expressing the equation by using the value singularity has instead of the symbols Kepler designated to the formula he introduced it makes far better sense expressed mathematically

$\Pi^0 = \Pi^3 / \Pi^2 \Pi$

$1 / \Pi^0 = \Pi^2 \Pi / \Pi^3$

$1 / (\Pi^0 \Pi) = \Pi^2 \Pi / (\Pi^3 \Pi)$

$1 / \Pi = \Pi^2 / \Pi^3$

By taking **k** into a negative the space will reduce the time because the space cannot sustain the demand of space growth.

$k^0 = a^3 / T^2k$

$1 / k^0 = T^2 k / a^3$

$a^3 / k = T^2$

$\Pi^0 = \Pi^3 / \Pi^2 \Pi$

$1 / \Pi^0 = \Pi^2 \Pi / \Pi^3$

$\Pi^3 / \Pi = \Pi^2$

In all my other work, I make exclusively use of the value of singularity Π since it makes a lot more sense, but when I use the value of singularity, which is Π then no one seems to have a remote idea to which I am referring.

$k^0 = a^3 / T^2k$ forms

$k^0 / k = T^2 / a^3$ that becomes

$k^{0-1} / a^3 = a^3 / T^2 / a^3$

$k / a^3 = 1 / T^2$

The replacing of the symbols Kepler used with the value of singularity the mathematic equation comes into practise.

$\Pi^0 = \Pi^3 / \Pi^2 \Pi$

$\Pi^0 / \Pi = \Pi^2 / \Pi^3$

$\Pi / \Pi^3 = 1 / \Pi^2$

Due to size differences the space we have has much slower space reproducing coming from the less space we hold compared to the space we share with the Earth. The duplication of space $a^3 = T^2k$ of the Earth is much more and therefore much faster in ratio of duplication $a^3/T^2k$ than we have. This discrepancy in space equality will tend to reduce our share of space $k^0 / k = T^2 / a^3$ because we lack the motion to equal the time thereof.

We have to remember that the Earth is at $k = a^3/T^2$ in relation to the Sun, but since we are captured properly of the Earth we are $k^1 = T^2/a^3$ in relation to the Earth. Reading the mathematical expression of the formula and from that translate what Kepler said we find that motion is gravity and motion produces space as motion causes space by duplication $a^3 = k. T^2$. This expression translated to $k = a^3 / T^2$, which proves the systematically expanding in time relativity which Hubble discovered that is in progress.

By every motion that came about since the start of the Big Bang the distance **k** expanded every time the cycle $a^3/T^2$ completed. Our atomic structure combined produced less **k** than the Earth atomic space combined because although the Earth does not have an atomic space structure yet, the atoms accumulate in the effort the produce and the accumulative effort combines as the Coanda effect, but on a massive scale. As the duplication of space is a product of gravity and our duplicating is on a much smaller scale than the Earth does its duplication and while we are sharing motion or time with the Earth in the space of the Earth the reproducing of our space will lag behind that of the duplication of space that the Earth has to complete. Since we reproduce less space we are confined to space reducing by the Earth. But since our atomic units are firmer than the reducing capabilities of the Earth gravity which is space reducing we resist the reducing and that resisting gives us the mass we find we have. This reducing in comparison is the result of the Earth producing an effort to **reduce and even bring on the** demise **of** our space by placing our **k** negative **in comparative relation** to the **k** the Earth holds. It is all about two factors that determine the question concerning the natural flow. It is the duplicating of space $a^3$ relating to the demise of space which is presented by the motion **k $T^2$** I delve much deeper into this aspect later on as this book progresses with information exchanged. The two factors we now examine was the very first motion that brought any and all forms of space into the Universe.

The point without movement, the point holding singularity must have a value of $\Pi$ being the eternal dot but since the dot has no dimension in having form the $\Pi$ that indicates the dot must be $\Pi^0$. From such a point there has to be to the side of the centre point be a point where space do start. That point will then receive a diameter but that point will have form only in being a circle. In that point there is a shift from in relevance from $\Pi$ to the centre $\Pi^0$ and for the first time it brought about two separate values for $\Pi$.

$a^3 = T^2k$ Space created from a specific centre is equal to the motion there of in time established by that centre. Applying the relevance value of singularity the formula reads as follows $\Pi^3 = \Pi^2\,\Pi$

$1/\,k = T^2/\,a^3$ reads the motion is in relevance to the superior space creation too slow to fill the space. When a relevancy is applying between two bodies and for the reason of not matching duplication in space by having independent motion the smaller body is unable to multiply space in ratio with the dominant space the dominant singularity will require. By the reducing of the distance factor **k** it tries to establish a point of equilibrium in motion setting an equal time applying that is duplicating space. The returning of a body towards the centre of the major body will reduce the requirement for duplicating space is unable to fore fill. By reducing the need for space duplication the individual space must find a level that will support the effort that such a space filled with those specific particles then can manage But then the space as it then is has to increase the time relevance. Applying the relevance in value of singularity controlling and singularity submitting contact and possible friction becomes a factor. The situation will be leading to the establishing of mass. Applying singularity as a factor in value the formula reads as follows $\Pi^{-1} = \Pi^2/\,\Pi^3$

$T^2 = a^3/\,k$ reads that time relevancy depends on the space the distance create in the relevancy. By increasing the motion the space will reduce by the margin of distance relating to time duration. By substituting the relevant value of singularity in the Kepler formula reads as follows $\Pi^2 = \Pi^3/\,\Pi$

$1 / T^2 = k / a^3$ reads when reducing the time as an object does when entering the atmosphere the space the object holds will reduce the distance the object maintains. Applying the relevance value of $\Pi^{-2} = \Pi / \Pi^3$. Again we see this mathematically proves that the flames we see that surround a body when the body enters the Earth atmosphere as the body is coming from outer space towards the Earth.

**That proves that the establishing of distance k will produce space $a^3$ and set space $a^3$ in motion $T^2$ where such motion is in opposition to singularity, which means gravity or contraction is the deliberate opposite of expanding $a^3 /k = T^2$. In the beginning the expanding then also involved three more points all just outside the border of singularity but within the atom exclusivity. It extends k while it introduce a returning relevancy back to singularity $k^0$ by creating motion in spin and duplicating space by reducing space.**

This is the gravity that takes place but the reducing as such must lead to heat amplifying once more because reducing space brings about increasing in heat and with that the reducing did not solve the problem of the overheating. The expanding was leading a movement flowing into the next-door neighbour territory. It brought the seven into the realms of the ten but with the reducing singularity already made a claim $T^2$ on the space it went into. That space belonged to another more overheating singularity without the reducing ability because of countless factors that lead to that singularity being in that state. By reducing the space it removed heat as particles from the neighbouring singularity by joining the gravity reducing singularity and accumulating heat which then compensates for space gained by plating the reducing singularity with more space / heat that then converts to material.

By reducing the space it removes heat from the density of the surrounding space and distributes that heat into particles. The motion is about removing heat from uncontrolled space situated in the neighbouring singularity and plating the removed heat onto the controlled singularity by having the removed heat joining the reducing singularity and allowing the accumulating of heat which then compensates for space gained by the plating process. The reducing singularity receives more space / heat it then converts to material. The plating would extend as the accumulated heat supplies a larger area to distribute the heat in. This is only effective in the very short term since the growth will not stop the overheating permanently. This would temporary cool the singularity. Then the expansion would once more come about, as the singularity will again start overheating due to the lack of motion, which then places some motion other than gravity in place on the space developed. I suppose in a way as the space in relevancy once more declined it would again force motion to create space once more. When the converting is done the overheating started once more and the process repeated again and again. That is time. That is space. That is space-time. The particles in a better position will slowly but evenly cannibalise the softer space-time by the growth it duplicates using the other less well place particle to develop. This action can be seen as antimatter because matter through gravity is eroding other material into more compact space. This process forms gravity in the one aspect that gravity has but there are nine aspects in all. It is motion of space moving towards a cannibalising centre of space.

Using this formulated method we can see that that is precisely what happens when Galileo's pendulum swing and the space decline because the time is being a constant. The space-time is enforced by the much larger growth in space of the Earth that reduces the substantially inferior space produced by the arm of the pendulum in the space covered by the swinging stroke of the arm. That is proof that space is reduced when time is a fixed measure as the Earth singularity will bring about and because of that we on Earth is stuck with mass.

Using singularity as a guide to find the position we are in being a subject of the Earth the formula applying to our position reads as follows $1 / \Pi^2 = \Pi / \Pi^3$ I suppose Kepler did not by own merit quite saw what he found. His interpretation of his findings at the time was not that conclusive but Kepler was many times over closer to the truth than was Newton with all his brilliance and the brilliance of all his followers.

In the overall view one may have the opinion that I am totally at odds with the antimatter theory which I m not. I only wish the Theorist would quantify and define the antimatter as a product belonging in the Universe. That the product he refers to is the same heat becoming eventually space that I refer to seem to be beyond any doubt in my mind. There was material producing gravity by performing motion and then there was material performing anti gravity by producing heat and with heat the heat created space that formed part of material destructed or that, which then became antimatter. But antimatter and plasma and all other names available does no for fill the function of establishing recognition about what one refers to because using heat as the term is the least alien. In the final sense every one knows heat. Also in this process on the other side of another divide light became another product of heat and light is heat reflecting a sure connecting to dissolved singularity. Singularity carries light because light is what remained from broken up and reduced singularity

Light is the highest form of antigravity in motion. Remember that gravity can only be if there first is antigravity. Space-time can only be if there is antigravity. When light became a presence as a part of the other or the anti side the motion that motion produced not space (I lack the incentive to give it yet another confusing useless name) but the spot holding $\Pi^0$ which was producing time, which in turn created three parts of space filled with heat. In the event singularity remained space less $\Pi^0 \times 3$ on three borders $3\Pi^0$ but even so the motion remained relative bringing about the Titius Bode ritual and the product coming from that is $\Pi^2$. Light is $3(a^0) \, T^2 k^0$ because $k$ never produced space and space never came from the motion $T^2$. In this the only aspect remaining was the three positions heat or space have as well as the motion the space produces $T^2$ in the contracting of $k^0$.

Kepler's formula of $(a^3) = T^2 k$ then produced another side $(a) \, 3\Pi^2 \, \Pi^0$
  $(a) = \Pi^0$ because $\Pi^0 = k^0 = 1$
$(a^0)3$ space in accordance with singularity

$\Pi^2 = T^2$ The duplication of space light holds

$\Pi^0 = k^0 = 1$ which is made up of $3\Pi^0 \, (a^0) \times \Pi^2 \, (T^2) \times \Pi^0 \, (k^0) = 29.6$
Light has a relevancy to one side of $3\Pi^2 = 29.6$ and to the other side it is $3^3 = 27$. The total sustainable space of light that has the ability of motion is therefore in relevance of 56.6. Two laws came about in relation to each other that produced the Universe and still secure the Universe. It is the forming of the Titius Bode law in conjunction with the Roche limit. Mass was never a factor!

Kepler said the space from the centre running to the space isolated by time is the same. The same space is isolated and the same space is orbiting using time. Kepler said $a^3 = T^2$

**k.** The time used is space moved because $a^3 = a^{2+1} = a^3$ as much as $T^2 = T^{3-1} = T^2$ as well as $k = k^{3-2} = k^1$. If any one is in dispute of my statement then please show me where did my mathematics fail me? That is what Kepler's formula says and if it does not say that I wish for once that someone can explain to me what is incoherent about my use of mathematical equations. Would some one for once explain to me when during my use Kepler's mathematical formula did I error and what do I do wrong? My work has been rejected by so many institutions in the past declaring my verbal reasoning about nothing not being a factor of properly expressed This they) (the Academics) say of my using logic which those then interpret as being incoherent on my part That they also say about the use of my arguments concerning the use of zero where I prove that a line cannot be able to advance from the original position. That is what I cannot understand of the Highly Educated Academics. No one tries to disprove my argument as every one dismisses my argument without disproving it first.

Kepler's formula suggests duplication by motion and that makes the duplication of space is a product of gravity and our duplicating is much lesser than the Earth while we are sharing motion or time with the Earth in the space of the Earth where the reproducing of our space will lag behind that of space the Earth is duplicating. Since we reproduce less space we are confined to lesser position in space where the duplication merits a reducing of our standings in relation to that of the Earth. But since our atomic units are firmer that the reducing capabilities of the Earth gravity which is space reducing we resist the reducing and that resisting gives us the mass we find we have. The atom form gravity is approximately forty times stronger than what the Earth present. Therefore in the conditions the Earth present, the atom form can withstand 40 times the gravity assault that the Earth gravity launch. This reducing in comparison is the result of the Earth producing an effort to demise our space by placing our **k** negative to the **k** the Earth holds.

The Earth is reducing our distance we have to the elected governing singularity and singularity allows such reducing of our standings in **k** because there is a normal discrepancy between our distance we have in value from the centre singularity and that which the Earth in its entire totality has. What I am referring to is all the atoms in the Earth that is relating to the material that constructs the Earth as a solid unit. Put another way might refer to the position the materials forming the structure have in relation to the social order the Earth insist on. In essence the balance ride on the dismissing of space-time by the group forming the unit in relation to the duplicating of space-time there is and more. The space we duplicate is less than the space the Earth duplicates and since the Earth is growing more the Earth is therefore extending the **k** applying to the Earth of the Earths growth in stretching **k** is placing the whole unit of growth of the Earth further away from singularity at $k^0$ centre than what we can manage in a comparable independent unit relating to the Earth effort. Through this we are naturally staying behind in the duplicating tempo and it is the staying behind that gives us mass. From our producing of space duplicating in comparison to that which the Earth duplicate in the same time factor and is duplicating at a greater pace than us our duplication involves less motion $(T^2 k)$ than does the Earth duplication by motion because we duplicate less space $a^3$ in relation to the motion but we attach to the same motion that the Earth allows us. Therefore our duplication in motion stands apart from the duplication that the Earth has $(T^2 k)$ to create to duplicate the relative more space $a^3$ our $a^3 / (T^2 k)$ is much less than the Earth's $a^3 / (T^2 k)$, which is giving us space.

When we find ourselves outside the atmosphere of the Earth $a^3 / (T^2 k)$ the odds turn much more in our favour. When we are outside the Earth boundary our synchronised singularity response has to match that of the Earth singularity growth and the body would then be a natural satellite being outside the Earth atmospheric boundaries. By being a satellite of the Earth we fall victim to the extending of k, which the Sun permits, and the growth of the Sun in relation to the space in influence allows us floating in space around the Earth much more leniency in response to the Sun's growing k. We then change the name we use for the applying gravity we experience there in outer space then to micro gravity because we then float instead of sink. It is not our mass that brings about the reason that we stay behind in the comparative growth. If our velocity as well as our speed in growth are in distance **k** and is dismissing $\mathbf{T}^2$ in synchronising is that of what the Earth produce we will grow in harmony and remain a satellite.

The reason for this is that we found a way where we fall outside the management of the Earth and under the equal management of the Sun. If we do not manage the balance we share with the Earth, we will eventually become either space debris or we will end up as particles that is part of the Earth. Every time gravity reduce what the Earth increases in space by measure of overheating the increases shows a difference between the **k** the Earth produce and the **k** all other bodies sharing space with the Earth does. But by not being of the Earth and still part of the Earth such increases in the difference between the extending **k** will slightly increase. Naturally that will influence changes to the mass and the volumetric inequality there are. It is also to this reason why Galileo found that there is no mass influences of heavier objects falling accompanied by lighter objects and the heavier objects will not fall faster when normal falling reduces the **k** there is between the Earth and the different objects falling. While falling no object has a mass because of the space the falling the object is in that is unrestricted and unrestrictedly allowing the decline of **k** on par with the Earth **k**.

If an object had a mass Galileo was wrong and Newton correct. Then the pulling of the larger mass would overshadow the pulling of the lesser mass onto the Earth's mass because the larger mass has then greater gravity it can produce. But since Galileo proved to be correct Galileo that in the same token proved Newton as being incorrect notwithstanding the corrupt arguments Academics make to vindicate Newton and his corrupting of Galileo's stance. I have almost heard all the corrupt arguments before and all the interpretation as how the Academics find ways to go around the inconsistency they create but in the end all that matters in the argument is that if mass had any role to play it would play the role while falling. Since it does not influence the falling mass has no role during the falling. If an object had mass while falling and the mass was effectively pulling on the mass of the Earth while the Earth was pulling back by mass inflicting force two objects with mass discrepancies would fall unequally. That is exactly what Galileo proved not to happen in spite of the many times I was very politely belittled very well mannered fashion by academics because according to their view I was "unable to understand" Newton.

Again I say as I cast all the clever and senseless arguments of the High and Mighty Academics aside that Galileo proved all mass is mass less while falling because all mass has the same mass while falling because all mass fall at the same rate which is $7(3\Pi^2)\Pi^0$. This is the factor of space displacement in the Earth atmosphere which particles has to overcome or equal in duplication or it will be the rate of their space-time demise. It is the rate that space-time displaces by dismissing related to duplicating. The duplication will grant the body in gravity confinement an independent duplicating rate as it applies a counteracting force resisting the diminishing of the space-time that the Earth produces by motion in gravity. When objects fall notwithstanding whatever mass differences there may be the **k** factor always remain the same in declining when the declining comes about from dropping to the Earth. This is because no mass is present in falling objects unless they on the ground find the ability using the **k** the Earth forcefully provides to duplicate the space created by the motion the Earth provide to the object on the Earth. This fact is the reason why dinosaur skeletons have grown that much bigger that they seem that much larger today than what the true size of dinosaurs were during their lifetime while they were walking on Earth. Once the skeleton is buried and forms a part of the ground in which it is buried, the skeleton takes on the same growth that the soil composition receives from the Earth expanding **k** by extending gravity because we have to remember the skeletons are referred to as skeletons but in truth that is the very last thing the stoned fossils are. As it became stone from bone it grew as part of the rock formation of the Earth while it caste the carbon qualities aside. The distance **k** is forever growing by increasing both time $\mathbf{T^2}$ and space $\mathbf{a^3}$. That makes the rock fossils still growing even after death. Being part of the Earth increases the relevancies applying.

From this relevancy we can see there is a clear issue to be made that at the point where singularity was presented by $\mathbf{k^0}$ it is where $\mathbf{k^0}$ breaks into space $\mathbf{a^3}$ through motion $\mathbf{(T^2k)}$ forming matter from which time $\mathbf{(T^2k)}$ develops. This is distantly linked to an eternal $\mathbf{k^0}$ but still is changing **k** all the while in dimensions through growth. That means time $\mathbf{(T^2k)}$ is as much a product of space $\mathbf{a^3}$ as space $\mathbf{a^3}$ is such a product $\mathbf{a^6}$ when incorporating singularity at such a point. $\mathbf{k^0 = a^3/(T^2k)}$ which then in our case only apply to all material which is within the limit space-time confinement in duplication thereof by the rate of 7/10 $\Pi^{6/}6$. The value that singularity applies to keep the Universe in a six -sided 3 dimensional Universe is 7/10 $(\pi^6)/6$ but I am getting to that explaining in a short while. From singularity $\mathbf{k^0}$ time $\mathbf{(T^2k)}$ is bringing motion $\mathbf{(T^2k)}$ at the release of space $\mathbf{a^3}$ from singularity $\mathbf{k^0}$. Singularity $\mathbf{k^0}$ is forming the six sides of space $\mathbf{a^6}$ or then 7/10 $(\pi^6)/6$. I also have indicated that even in the six dimensions $\mathbf{a^3}$ is claiming to be in, it's the one half of the six sides is the very same as what $\mathbf{T^2}$ represents or as that **k** represents. What will solve the problem is retracing **k** and from there we will see how time is aiding space and thereby is giving space the full compliment of what we find to be eternal space.

To trace **k** is to recognise **k** for what it is because **k** is $\mathbf{k^3}$ or $\mathbf{k^2}$ which is the same as $\mathbf{a^3}$ and $\mathbf{T^2}$ and those are just another form of **k**. **k** can be one line in six forming the cube with six sides as we all know but also **k** is the radius or the line running from the centre of any and all circles including all spheres to bring size to the circle or sphere. One has to bring **k** back to the value of $\mathbf{k^0}$ to see how space develops in time from singularity. $\mathbf{k^0 = a^3 / (T^2 k)}$. When reducing the circle in size one have to reduce the radius or the diameter because the pi is the indicator of the form as a circle. This divide is possible by reducing the r until there can be no further dividing Such dividing cannot end in zero because no matter how small, there will forever be a value in place. Our mass comes about from the fact that we have a negative relevancy $\mathbf{k^{-1}}$ (being a factor less than one) and falling into the 0.9 bracket of the comparing singularity to that of the Earth holding 1. One should not judge this by mathematical laws but rather see it in " less logical" terms of singularity applying cosmic laws.

It has never mathematically been proven that nothing is a factor in outer space notwithstanding Academics blaming me of incoherency when I suggest this fact should mathematically be proven. In overwhelming contrast I am about to prove the building blocks used to construct outer space. What Mainstream physics whish to convey is that in $F = G(M_1 X m_2)/ r^2$ the G factor as being the gravitational constant can be replaced by zero, but not one of them ever thought about that. Then that will bring about that $F = G(M_1 X m_2)/ r^2$ forms $F = \mathbf{0}(M_1 X m_2)/ r^2$ and the answer would remain the same. Any one with the least bit of knowledge will see the answer has o be zero. Take $(\mathbf{M_1 X m_2}) / r^2$ and substitute any of the factors with zero and the result coming about has to be zero. The factors in the equation have to have any and all the elements at a value of at least one.

Only if r was a factor of one can gravity bring about any mathematical equation that is developing from this argument and which is giving it coherency. That means the mass on both sides must have a factor of one being a limit, which does not allow such further reduction of r and any further reducing of r beyond the limit will not be tolerated. Only if r = 1 then $r^2$ can be 1 and mass can be apart. Like it or not but believing in the Big Bang must also bring about the accepting that the cosmos moved apart somewhat by some measure and using some means of control. The fact that r brought increase in the space separating the different particles the friction causing mass produces a problem that was solved already. About a century and a half ago Roche found just such a limit.

There is a limit to the radius that is tolerated between two independent cosmic objects. But before we try to find why the Roche limit intervene as it prevents objects crossing the limit singularity place on the boundary we have to replace the Mainstream idea of having a value of nothing in outer space as a mathematical fact. With such a zero in place in space it nullifies the Roche limit presence as a cosmic law by upholding the senseless idea of zero being a calculation applying in outer space. It hampers all my explaining about how the cosmos introduces space – time and when singularity brought about space-time by introduction of the Roche limit as a starting law by placing that at the point where the cosmos actually began.  Once again I were confronted by zero becoming growth. There is a huge hole that needs filling when bringing into a relation any forming of an alliance between a cosmos coming from nothing and filling with nothing and a cosmos growing spontaneously through balance shifting prominence.

Mathematically the fact of applying nothing as a value applying in the cosmos is not a strong and convincing argument. The minute one brings in zero as a multiplying factor forming a definite value working into the calculations of the cosmos, growth disappear. If growth was not a factor, the zero factors could be involved with some form of maintaining stability and where then further growth will accept the responsibility of zero. The closest encounter worth noting we ever had with this law in the modern age of news and Television was the Shoemaker-Levy 9 incident during the previous century, a bit more than a decade ago. At the time and even in the present no one drew any similarities but after completing this book the reader should find why I could draw such similarities, which there is between this incident and the Roche limit. Even the phenomenon called the Sound Barrier becomes clear when applying the Roche factor with the laws governing the influence of singularity.

### singularity $k^0$ by creating motion in spin and duplicating space by reducing space.

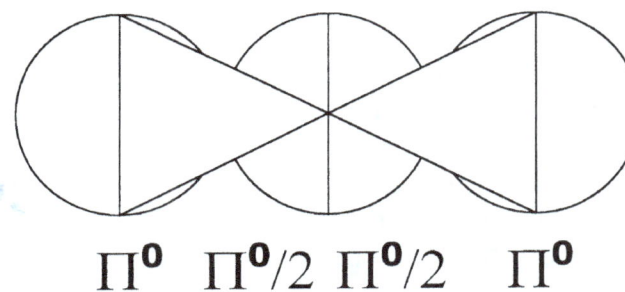

$$\Pi^0 \quad \Pi^0/2 \quad \Pi^0/2 \quad \Pi^0$$

**Let us start telling the story as it was.**

At the very first sign of any of the sides departing from the centre shared by all, all other points must also show signs of a willingness to depart. There will be one point where r still is one coming in as a factor but pi moves out from only being a factor of $\Pi^0 = 1$ and at that point, pi will become a full factor of $\Pi$.

At first there was singularity holding the entire cosmos with what ever is and will be in the Universe captured and contained in an area that too this day cannot fit inside the Universe we see and we appreciate. The best of all is that there was only one spot, which became innumerable dots but at first before the beginning began there was on spot with singularity in that spot. That spot had the great total of $\Pi^0$ but it could have been what ever you choose to use as a symbol as long as it is to the power of zero and not to the value of zero. There was no space therefore there was no motion but because there was no motion there could not have been space.

The single dimension is a dimension covering everything into the dynamic of one. This brings about that **k = 1 a = 1** and **T = 1**. That is the first dimension and the first dimension is a dynamic of one being the result of a dimensional 0. $\mathbf{k^0 = 1}$ $\mathbf{a^0 = 1}$ and $\mathbf{T^0 = 1}$. The factor **k** was at no stage zero. Only the dimensional factor is 0. The extending that **k** was capable of was zero but **k** as a factor was never zero. The factor of **k** was never zero. The factor **k** can never indicate zero as a point from zero or point to zero because just one zero will dump the entire Universe into zero. The Universe moved from the spot $\pi^0$ to a dot $\pi$ being a multiplication of dot $\pi$. That is why by having two $\pi$ it form $\pi^2$ and with time differentiating the value then becomes $\pi^2/4$. In four sectors of gravity applying gravity then became $\Pi^2/4$ **X 4 = $\Pi^2$**

Then motion became a factor. Motion created space as space gave room for motion and space gave room to motion, which is what the Universe is. The room it gave became liquid plasma or heat and the motion formed space growing or space in containing. With motion space is growing as space however the reducing of space is affectively securing and establishing the maintaining of singularity. That brings about growth by means of containing. Then through the relevancy of motion change came about but change involved gravity.

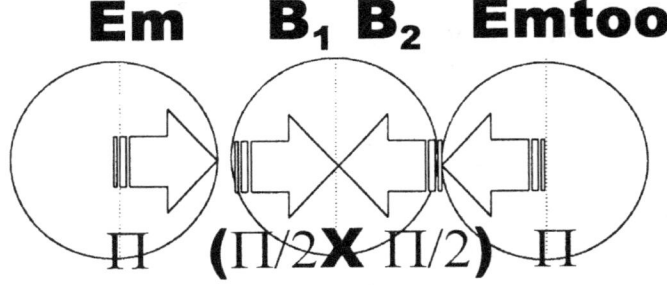

## Em   B₁ B₂   Emtoo

**By allowing the distance separating the particles to be zero, the particles melt into a unit. Again this is Mathematics and not my incoherency as some Academics chose to interpret my work or rather then to find grounds on which to dismiss my work without bothering them with the effort of reading my work carefully.**

We return to the fact we established before that Em and Emtoo was divided by r and then r had to be one since r could not be zero. Such a centre would then carry the same value as Em and Emtoo. That means whatever value Em and Emtoo receive has to go in equal measure to r with Em sharing half of the divide and Emtoo sharing the other half of the divide.

In some way I guess there is room for improvising slightly and find grounds to establish some basis too try and incorporate Newton's idea but that is only to find some connection and not to incorporate Newton. In an effort to begin explaining, we place potential opposing parties that came about from expanding in the relation to the very frozen instant it expanded by positioning the one mass or Em and across the space separating the two will be the other mass which is also a mass or then form Emtoo.

With no line possible to part the too because the two is taking up anything that may qualify as a line there had to be another dot that formed since the Universe has many dots that formed lines. But let us not to get confused and lost in the range of possible diversions but let us stick to two dots. One dot was next to the dot next to the dot, but as I said we stick to one dot next too the second dot. M X M / $r^2$ is the first step gravity began with. That leaves us with a huge problem in as much as when r = 0 then $r^0 = 0$ and 0 dividing any value will leave 0 as the answer. If the particles were inseparable at the start it must bring about that gravity would not be forming since the distance will not permit any dividing.

By allowing the distance separating the particles to be zero, the particles melt into a unit. Again this is Mathematics and not my incoherency as some Academics chose to interpret my work or rather then to find grounds on which to dismissed my work without bothering them with the effort of reading my work carefully. Let me run through the argument one more time because I have been insulted by Academics in the past telling me I am bending mathematic rules with my applying double values to try and produce some argument. While they are blaming me it is they whom are being guilty of blatant misrepresenting mathematical laws but by deflecting blame onto me as being incoherent, that incoherency they pass on to me to use to blame me. Please judge my arguments in line with finding correctness in the argument and not just to establish a loophole through which to find an escape root as to avoid reading the work. Please think carefully while you examine the next couple of sentence because this is what max Planck with all his mind-boggling brilliance missed.

At first the two particles moved apart from the spot, as they became dots, which by then they became neighbours as much as associates and formed from at first being an inseparable unit to the then later divided associates after the split separated their sharing of a spot. We know that at least two dots formed because there are many more than just two dots that remained to become our all inclusive as well as the visual Universe. Let us name the dots because that is what humans do best if they do not know what to do with what they have to do. Let us call the one dot Em and the other one dots next to Em we then call Emtoo. Between Em and Emtoo there were not nothing because Em and Emtoo were separable.

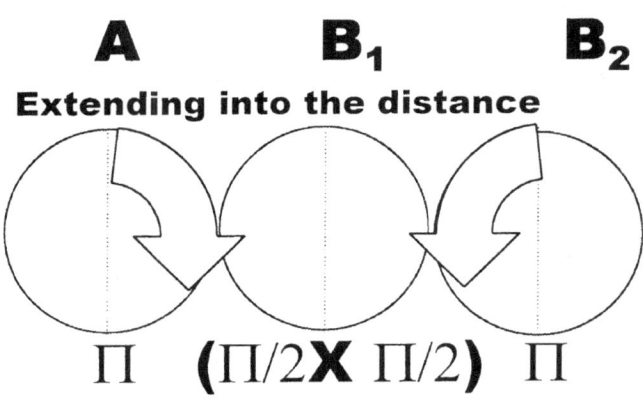

**A       B$_1$       B$_2$**

**Extending into the distance**

$$\Pi \quad (\Pi/2 \textbf{X} \Pi/2) \quad \Pi$$

We have established the fact that Em and Emtoo was divided by r and then r had to be one since r could not be zero. Such a centre would then carry the same value as Em and Emtoo. That means whatever value Em and Emtoo receives has to go in equal measure to r with Em sharing half of the divide and Emtoo sharing the other half of the divide. This is the only manner by which the division could have taken place

**Because the three points existed on equal terms in singularity sharing a same spot the coming out of singularity forming dots will enforce that equal value comes to all.**

By their being separable, but still remaining infinitely close, we would naturally be inclined to think that the separation value should be nothing or at least zero. But putting zero in that place is a mathematical excluding procedure leaving future mathematics excluded. With m multiplying m$_2$ and then dividing ÷ r with r being zero (r=0) such a procedure will leave the lot at zero and with that much of a total forming nothing then that much nothing (or is it many nothing), the nothing is going nowhere. That means although we think the space between the two parts are nothing it is our concept whereby we are placing such a value in that location because with our flawed position we are looking at the cosmos we rely on corrupting the statements that we make about the non-existing space. We know well that the space has to be at least one to be a future factor.

Because the three points existed on equal terms in singularity sharing a same spot the coming out of singularity forming dots will enforce that equal value comes to all. The expansion had to produce $\Pi$ due to the reasons given in a similar explanation elsewhere in this book. That means the circle gets to become $\Pi$, the diameter becomes $\Pi$ and the distance setting the structures apart will also become $\Pi$. This is what the coming from one point brings along. Only when being part of the second dimension can there start being separate values. While the form was still being in the single dimension from the one side of the form the dots had to establish identities apart but not separated yet. The one circle had a factor of $\Pi^0 = 1$ and the centre had to have a value of $(\Pi^0 / 2)$ extending past the very next object but also cutting such an object into a square double half value that was going to come about as soon as the other dimensions came into form. In the relation at present Em is extending toward Emtoo by means of establishing a valid r and Emtoo is establishing a valid extension to Em by using r and this leads to two valid values for r being $((7+7) /10)$ and $(10/7)$. The values I give here I shall explain later on because the full understanding in the correct context requires much more introducing with many more facts introduced and established.

The only definite place one will locate zero is in between the starting point of the lines going in opposing direction in the position the lines hold before there was the least of directions applied, but that is only because there is no such a position, not because any line is coming from there. As I have indicated and positioned Em and Emtoo the two points may share a position but separation is forever a possibility and for that reason should there then be no other reason we then have to put a dividing possibility at a value of one $(r^0 = 1)$. By dividing it means the shared dimension is $r^0 /2$ which then is half ½. The two lines are still one holding the opportunity of parting as an option but have not yet parted and therefore are on the very precise same spot from which came duplicating the next dots.

Being on the same spot does not mean being inseparable or being the same. Everything that is now in the Universe was part of Creation before Creation started because when the dot moved from the spot all of singularity that now is present in the innumerable dots was present in the single spot as well as being present in our Universe. Every dot that was were and is because nothing can remove from the Universe but the rest which was in the spot formed a dot and every dot is part of our Universe. It only means that everything was sharing a spot. The line coming from every dot being there is already there because it already has the choice of going in any and all opposing directions in spin direction and when it starts running it will place filled space in that location not yet present but also holding a factor of one since it will become filled in the future. This is because the space at present is filled with a line and the line is sharing an equal value that does not double the value yet. Where this space is, is now already filled with a line without the line exactly being there in accordance to every detail of the three dimensional standards we apply at present. However it may be the line has to have had a start. The starting became the line running and by running the line is filling space. That means with the line there it filled the possibility that a line could form and not with a line not being in place at all. It is again taking the r separating Em and Emtoo on its factor value of one and not our human visual accepting value of zero. One may not discard any future possibilities of growth by giving those possibilities a value of zero. A line might form or space may form where the line later may form. But by using a value of zero in that spot one then remove such a spot and all the future potential values that may fill such a spot in the future. We humans tend to dish out a value of zero where ever we do not visually are able to find a value at that precise second and in doing that we also place such a value as a running obstacle into the future. But our habit of doing that is proving to be a human shortfall because with our shortsightedness we think of the here and the now in excluding possibilities while we should think of the future by including possibilities. With myself personally not being on the moon and with the great likeliness there is at the moment that I may never be on the moon does not remove all potential future possibilities and with that my personal possible future ability to reach the moon whereby putting my future chances at zero such issuing of zero is removing of all chances there is. The zero will have me never ever able to be on the moon. While I am alive there may be an infinite but still applicable possibility that by some extremely remote chance that might occur however infinitely slim such a chance is still a possibility however remote such a possibility may ever may be,

By disregarding a positional value as zero we exclude such a position from ever being possible. On the other hand when giving it a factor of one we include such a position as a future possibility. When reversing a line we might find a better idea of what is in place and where it is in place. Gravity is according to official sources a force without limits going past and through borders and has an unlimited reach. It seems to remain even and this is conflicting with the flow of perceptions about mathematics. In as much as showing that r is serving in a factor value, as one such a value has to form a limit where Em and Emtoo has divided sides of r and aligning the discovered singularity produces the Roche limit as such a dominant factor in the cosmos. With my effort in retracing a simple line that helped me to find an explanation about the Roche limit, a feat not yet done in science. The Roche factor is next to singularity the second most basic foundation in cosmology and is the starting point where singularity spawned into dimensions. As it is fundamental in all cosmic development and with that it denounces the gravity principle introduced by Newton as $F = G \, (M. \, m) \, / \, r^2$

The formula $\mathbf{F = G \, (M_1.m_2)/ \, r^2}$ is unable to explain the principle discovered by Titius and later by Bode and in contrary to all statements to that effect made by Accepted Science policy makers the Titius Bode principle is not coincidental. In fact it is one of the four most adhered and important cosmic pillars holding the cosmos structural in place. From the two examples mentioned above comes gravity. In past few pages I proved how one could arrive at the facts that prove how the Titius Bode Principle leads us in the direction of the origins of the solar system. But before we can accept the influence of the Titius Bode Principle we have to return to the aspect of nothing being what is what the Universe is built of. We first have to deal with "Nothing" and as such dismiss nothing from science. I wish to introduce the fact that the reader should entertain the idea that the Universe are made up of pockets of space built by gravity where gravity cements the layers in blocks of 7/10 and 10/7 in relation to the Roche factor of $\Pi^2/4$.

This measure cannot be possible in a Universe where the manufacturing material is nothing and lumps of nothing onto the bargain. "Nothing" in the Universe is coincidental; "nothing" in the Universe does not apply. Where mathematics places objects and objects meat with lines nothing disappears. Nature subscribes what applies in the cosmos. Should any principle not match that which an accepted theory translates too or nature principles has too change to fit what the Philosopher introduce in support of the new theory in order too get it accepted, then the theory does not apply. If one cannot place what is in nature today in the cosmos at the very first instant of cosmic birth, the cosmos then had to change later on to accept that which the theory introduces and frankly, that is not possible. The cosmos is unchangeable. To imply that this or that was formed later on, or this that developed from, such presumptions are totally irrelevant.

The content of my work holds a new view about Cosmology, which I have been working on for the past twenty-seven years and exclusively for the past seven years. I always had a problem with the idea that space constituted of nothing, while I came to realise that lines mathematically couldn't start at zero because there is no evidence of zero as a factor in mathematics. Should you disagree with my statement about outer space being mainly formed by nothing the question in need of answering is this: What will the length of the shortest hypothetical line imaginable be and moreover, what would the total overall length be in that case? I once again come back to this idea after I introduced my idea about how the Roche limit came about and why the Roche limit forms the absolute partition in the cosmos between objects with variably similar dimensions. But seeing this can only become clear if the notion of nothing forming outer space is demeaned to the nothing it represents. By coming to the point where the line cannot possibly reduce more than it already reduced such a point that is holding the shortest line is precisely where the Universe started.

At that point the Roche factor came into prominence. But so to did the Bode law come into prominence because the one intertwines with the other. But to get to that we first have to abolish nothing as a valid concept. If one cannot trace zero at the start then the start must be filling the Universe and those in doubt whom are persisting on having zero must prove where zero later found the opportunity to enter the cosmos. The shortest possible line (hypothetically) must be so short it must have an initial and the eventual ultimate point sharing the same spot. The two points must be one with all the other potential separations being in the future and only then can further reducing of any line not occur. As I said before I say again: if any or all lines used zero as a start, the zero part would not count, because with the slightest growth the zero will either continue extending its current value of zero. Using zero is the same as going nowhere all the way into eternity. It is either that, or if otherwise the composition must change into some other value at a specific point. But that is invalid because what inexplicable reason, would come about to bring such doing. It cannot then remove zero.

The line will continue repeating the make of it in the same way as what it was before without change and that fact will bring about the line will then only start. By changing such composition the line may regard what it was before as something and this is still totally unacceptable to science. There can be no changes to the line from what it is when it is flowing from what it was at first. At a point past zero where the line then will start forming an infinitely small spot. Even by that measure such arguing eliminates zero as a valid factor. I press this point about the dot moving from the spot in urging the reader the understanding this concept because there is such a point from where the Universe is ruled, but I have to get acceptance first about such a fact in my attempt to underline the fact, I have to per sway the reader to abolish four or five thousand years of accepted and practised mathematical culture and that is no easy feat. Notwithstanding more than five thousand years of thinking we have to eliminate zero as a valid factor from mathematics.

In applying the most basic method of taking the line back as far as possible bring us to a dot $\Pi$ and going past the dot $\Pi$ till you reach the spot $\Pi^0$, which is actually still a spot $\Pi^0$ that is infinitely smaller than the dot $\Pi$ but since that is only a notion of something being outside our Universe we will have to call it a dot $\Pi$ since the next value carrying is the spot $\Pi^0$ is already taken by singularity $\Pi^0$ and we are passing that to a spot where $\Pi^0$ is single and is infinite in the extreme infinite $\Pi^0$. Because of the equilibrium that will stem from such a position the dot is the most balanced form there can ever be. The spot $\Pi^0$ is in infinity plus one but, however small it then might be, it still is not zero. Zero dismisses the position zero claims by vacating all possibilities of such a position being filled leaving the place full of nothing. Zero ultimately means not existing and then that point, which miss zero by one single eternity, holds nothing in such a start from such a point that does not exist.

Taking the line down the line by reducing the line proves that the line must still have a start and an end that disqualify zero as any of the two points or all the ends from front to back must be zero at all times. There is no other option as to have no line at a value of zero or a line starting from infinity, which is a valid number and running all along the line in a uniform rate to the line's end.

The smallest line has a beginning and an end at the very same spot located in infinity, and infinity may be beyond any possible human scope, though infinity is still not zero. Infinity puts the start and the end at the same spot, but in that it does not remove the line and all possibilities of a future line from the spot holding the line, as zero does when using zero as such a starting position. Infinity may constitute of something we do not yet understand, but we may not define our human misunderstanding of being zero just because infinity is not present in our minds and therefore by not sensing a value we disregard such a value as nothing whereas if it is visibility nothing then that does not mean it is nothing but in being potentially there it qualifies the point that it is one. It is the same as a person hearing a dog bark and investigate. When not sensing what the dog was barking at, the person turns around and disregards the barking as the dog is going on about nothing. The dog's reaction was not the indicator of the nothing because ultimately the dog sensed something. The man's wits let him down and his wits produce the nothing. The dog will not bark about nothing because then the dog will not bark at all. The fact that the dog barked produces a possibility of something being out there, which the dogs is getting annoyed about. The man's inability to detect what it is that the dog is sensing becomes the nothing but that nothing does not exclude the possibility of something out there being worthwhile to investigate by someone with better senses.

From the onset my approach to cosmology prove to be somewhat unconventional but through the abandoning of the accepted, it enabled me in locating the precise location of singularity that forms the connecting basis of the Universe (and this I say with some degree of confidence). There **are two locations** but I shall **first concentrate** my explaining effort on **the prime singularity**. Singularity did not vanish into the unknown after the completion of the Big Bang development but is in a place science incorrectly valued and classified incorrectly and in that, there is something hiding which we named as nothing and through that became an obstacle that is hampering our recognising of what is the truth. If singularity was or is where the beginning is we have to go back and see just where such a beginning was.

It is also true that where infinity hides singularity is at a place where we can only detect nothing as we cannot detect the position or location of singularity by using our meagre senses we have. But since we are the only part of creation that is (presumed to be) blessed with wits we have to locate such position with our human side and not with our animal side. We must bring in something that the atheist cannot find because that which is in control of the Universe by appointing value to singularity from outside a point in the Universe understanding such a concept goes beyond the animal and will also exceed that which the atheist can relate to. Understanding that the concept that the control of the Universe comes from points that is not part of what we perceive as the touchable Universe is far beyond the capabilities atheist have. If they had such abilities they would not be atheist in the first place.

I cannot accept that the Universe started at zero and neither does anything else in the Universe start at zero. My excluding the possibility of zero includes that the Universe is not filled to the top with nothing and neither is nothing part of outer space. The Universe is about lines allowing light to flow from one point to another point and in following that line it has to continue in the line as the line has to represent something. The Universe is all in relation about lines indicating distances between cosmic structures.

The cosmos is in short about lines connecting points in space being apart. It is about a line starting and continuing from such a start. But science advocates their opinion that such a start of a line flowing between any and all objects that can hold zero because according to them the Universe are full of nothing. If the Universe in as much as outer space is a container filled with nothing at the present moment, and there is no place to place anything that was part of outer space previously to substitute for the overflowing something could release to and there was no emptying of what ever filled it before, then it could not get rid of what was in the outer space when it started with what it started off with. We must then accept from what is not in the Universe meaning that that is absent at the present time was not in the Universe at the time during the start. Everything, which we find that is at that in this present in accordance with our observation which is present in this present time according to science part of the present because it then still must contain the same nothing and must have that same filling from the start present. If it was nothing it still must be nothing and that same substance being nothing is what it also used to grow and by using nothing to grow brings conflict in the conception that form because how can nothing accumulate it as it grew because it filled outer space with nothing growing from and growing to nothing. Is that true? If such a presumption is true then the filling of the Universe could not go anywhere if one has to presume it started off from nothing and from there it kept filling with nothing since what ever was in the Universe at the start had no place to escape to or no place through which to escape.

That is only applying if it is nothing filling the Universe at large. Can nothing grow as much as a line is growing from a start of nothing? The answer is that such lines not only indicate a distance but since the Universe came from such a small space as science propagate with the theory of the Big Bang then all particles in the Big Bang Universe were rather cramped for space when the Universe started from that small line between particles and is now the same line but is now so big. In the past everything seemed being so small and showing that the space between particles seemed then to be awfully short at the time during the star of the cosmic concept. It was short but how short was it? Did it start off as nothing? Is the line starting at nothing as science wishes us to believe? If it does then all lines must start from nothing so we better investigate this trend with the start of a line. In this following I show my argument with which I hope to prove the counter part of what science believes. Later on in this book I am about to prove that which science sees as nothing in space and in material is the very location of singularity. But lets return to the start before the confusion came about with space being nothing. The start has to start with Kepler because science in the new era started with Kepler and not with Newton.

I have to belabour the nothing for the last and final time because from what I am about to present cannot be presented if Academics dismiss my presenting of my work with they're bluffing everyone with nothing being used in outer space. If there is one still persisting on nothing being used in outer space such a person should either by now be convinced about my reasoning and if not that person will find no benefit in any further reading of this book. In that case please donate the book to someone more intellectual and therefore more presentable to the obvious. Kepler's finding cannot stand true if the cosmos is nothing and by persisting to the accepting of nothing then Kepler becomes a discoverer of nothing, which he absolutely was not.

The value of Kepler's space he indicated as a third dimension $a^3$ does not depend on indicating a structure $a^3$ that is in rotation $T^2$ but only needs one position having a constant of some sorts. Any point where $k$ may indicate a position one will find a value matching $a^3$ and the matching location will fit $T^2$ at that point. That is the relation there is in the solar system between all planets and the Sun. The Sun always indicates the centre and the planets always indicate the rotation. But $a^3 = T^2 k$ is only producing a relevancy of three dimensions that is equal to two plus one dimension.

Let us take it from a point where the Sun provides a centre $k$ then that centre $k$ will provide a line from the centre and the line $k$ will provide three spots in a formation that produces a structure by the square $T^2$ of the dimension. That means every single point that $k$ indicates there are three positions $a^3$ implicating sides of a double dimension. $k = a^3 / T^2$ That is what Kepler said. There are three dimensions $a^3$ between any two points $T^2$ flowing as time from the centre of the Sun, which is indicated by the line $k$.

The implication of the relevancy produced by the use of the formula $k = a^3 / T^2$ brings about that when dividing $T^2$ into $a^3$ there is $k$ left. The fact is that $a^3$ is a three dimension ($^3$) of single $k$ ($^1$) showing one or $T^2$ is two dimensions of $k$ being the one dimension it means that $k$ is a part of space $a^3$ or $T^2$ which is time. It is the same thing in a double dimension or space being a triple of $k$ then $k$ is one factor and $k$ cannot show a position of zero. If $k = 0$ then there is no possibility of $k = a^3 / T^2$ because $k = 0$ then $0^3 / 0^2 = 0$. That does not make sense.

Mathematically space cannot be zero because those being of the opinion of space being zero or nothing must first prove mathematically that space is zero. I have tried to convince the Super _educated by using that line which is more than correct and being more than correct brought me nowhere. Those in charge of serving Academic policy decided that nothing in outer space is a proven and established fact proven and accepted by those with prominence and who ever comes afterwards has no reputation to produce whatever truth there might be to produce. Moreover they then must prove mathematically how zero can grow through the Hubble constant. That too says nothing. Those in charge does not have to prove anything about nothing being accepted as fact and that makes all my principle efforts not applying with nothing being the norm. I cannot prove anything notwithstanding that Kepler proved the lot there is to prove. Kepler said space could only be space if space is in motion and therefore how can nothing move? If $k$ cannot be zero then $k$ could not start from zero. With $k = a^3 / T^2$ no point can be zero because $k$ shows space $a^3$ in the duration of the time $T^2$. Then the next thing I know is that through the inspiration of Newton, Kepler is not accepted and Kepler's formula is not even disputed, it is blatantly ignored by misrepresentation. Even if $a^3 = T^2 k$ is about motion we know that nothing cannot move and therefore only if $a^3 = 0$, $T^2 = 0$ and $k = 0$ can outer space be worth nothing.

We use nothing not as a value to measure by but to avoid what we wish to disregard when the effort to trace and determine become s to stringent and tiresome to further investigate and not to valuate. In this aspect lies the difference there is between arithmetic and mathematical science where arithmetic can have position such as zero since arithmetic excludes the cosmos calculating numbers only. The nothing we see we made we made that the nothing we find but the fact that there is a visible and measurable distance between the structures which we may appreciate as being there proves that the distance is there, it is separating structures and by that is bringing in the factor of one which we might not be able to explain but as such we are able to see. It is the way we try to disguise our inability to detect which produce the nothing we then use as a value, but still we substitute the nothing in applying arithmetic with the name as nothing and then the names used becomes the factor of one. Cosmology is not about numbers because no one can calculate the number of stars in spite of ridiculous Critical density attempt Cosmology is all about lines and angles positioning objects, and in those lines there features no zero. The cosmos is about better or in other cases lesser development by extending of **k** as the barometer of singularity development in space-time.

No line can be zero long and forming a position of zero degrees in relation to another object. Doing that shows we use our culture to hide our inadequacies behind just one more misconception. Let us find a place where zero does apply. A man may have that many oxen or so many sheep and even this amount of wives, (in Africa) or not have any therefore having then a total of nothing, but there cannot be nothing between the Sun and its orbiting structures. The having and have-nots are part of arithmetic. Light will indicate a line flowing between the Sun and whatever planet, following dot after dot from infinity crossing infinity to reach the next infinity and thereby proving the existing of the possibility of something going about by a straight line. Any straight line in is relation to other straight lines and will be valid under the law of Pythagoras where the law is in as much as obeying the rules of trigonometry. At any and every given or imaginable point between the two points forming the line, the line can be interpreted by something just larger than the epitome of the infinite small line and up to the size just larger than the size used by the line.

Regarding the possibility of zero there is no possibility of a straight line not forming in space. If there is space, there can be a straight line. Kepler said space is the motion thereof. That means space is immediately following another straight line by motion thereof. The mere fact of two spots having different positions in space gives the two dots different values. If the line has the length of zero and is the line separating the points that it is not present, which puts everything represented by the two objects and the separation between the two points holding the ends of the line apart, outside the Universe we have to use. We gave a name to something we identify as being representative of something we gave a name to such as nothing and not the fact of zero as such. The nothing is a name and what the name nothing represents. Then the triangle where one angle is the zero it then is no triangle because all other angles are dismissed at the same time.

Mathematics converts the values of integrating lines according to Pythagoras and arithmetic is about numbers to be added or subtracted. By mathematically excluding zero from cosmology a new Universe opens to the human mind. With the distance between the Sun and Pluto being roughly one hundred times more than the distance between Mercury and the Sun, the distance must hold something more than pure vacuum filled with nothing except one atom hear and there occupying the vacuum between whatever object we speak about and the Sun. If space supposedly comprises of nothing how can nothing then become plural forming more or be multiplied by a number as to indicate a growth in something not even existing. As the one becomes one hundred the one cannot substitute a value of nothing but then must be part of something. If the one substituted the nothing, all laws of mathematics will go in disarray because when one multiply any number by zero it becomes zero placing both planets in the Sun. If Pluto was one hundred times closer than it is at present was it then one hundred times nothing closer? In Mathematical term using a mathematical expressed manner the words used then translates to being mathematically expressed as $100 \times 0 + 0 = 0$. That is the expressed factor what we read into mathematics! By allowing the three hundred a value carrying the value of nothing then nothing must form one making that which is between Pluto and the Sun not nothing but representative of something we gave a name of nothing to as we would name someone George or Jack.

The nothing is a name and what the name nothing then represents, such representation has to be something as would George or Jack. Allowing this concept to apply this argument then follows mathematics to the book and in precise detail. With Pluto and the Sun being apart that being apart has to have one of something a in place of a value forming the being apart from each other's cosmic position one time where the one is a factor multiplied by the many ones we find in that space standing relative to other space regarding whatever the space becomes what is that we think is between the Sun and Pluto. That factor cannot stand in for the value of "not being" one, which is the same as nothing. As that is because one cannot take the place in the position that zero secures. By excluding nothing from the equation space becomes something bringing in a value lying inside the realms of the infinite that must form singularity. As the zero becomes a dot, something else becomes clear about the dot. Looking at the night sky we find darkness overwhelming the space in relation to the stars bringing across light. In another one of my books I show that we are unable to see darkness because we consider the darkness we see as nothing which represents no visibility but we can see darkness and we do so see darkness very well therefore the darkness we see must be light we see. That excludes nothing on another term.

From the onset my approach to cosmology prove to be somewhat unconventional but through the abandoning of the accepted, it enabled me in locating the precise location of singularity that forms the connecting basis of the Universe (and this I say with some degree of confidence). There **are two locations** but I shall **first concentrate** my explaining effort on **the prime singularity**. Singularity did not vanish into the unknown after the completion of the Big Bang development but is in a place science incorrectly valued and classified incorrectly and in that, there is something hiding which we named as nothing and through that became an obstacle that is hampering our recognising of what is the truth. If singularity was or is where the beginning is we have to go back and see just where such a beginning was. I cannot accept that the Universe started at zero and neither does anything else in the Universe start at zero. My excluding the possibility of zero includes that the Universe is not filled to the top with nothing and neither is nothing part of outer space. The Universe is about lines allowing light to flow from one point to another point and in following that line it has to continue in the line as the line has to represent something.

The Universe is all in relation about lines indicating distances between cosmic structures. The cosmos is in short about lines connecting points in space being apart. It is about a line starting and continuing from such a start. But science advocates their opinion that such a start of a line flowing between any and all objects that can hold zero because according to them the Universe are full of nothing. If the Universe in as much as outer space is a container filled with nothing at the present moment, and there is no place to place anything that was part of outer space previously to substitute for the overflowing something could release to and there was no emptying of what ever filled it before, then it could not get rid of what was in the outer space when it started with what it started off with. We must then accept from what is not in the Universe meaning that that is absent at the present time was not in the Universe at the time during the start.

Everything, which we find that is at that in this present in accordance with our observation which is present in this present time according to science part of the present because it then still must contain the same nothing and must have that same filling from the start present. If it was nothing it still must be nothing and that same substance being nothing is what it also used to grow and by using nothing to grow brings conflict in the conception that form because how can nothing accumulate it as it grew because it filled outer space with nothing growing from and growing to nothing. Is that true? If such a presumption is true then the filling of the Universe could not go anywhere if one has to presume it started off from nothing and from there it kept filling with nothing since what ever was in the Universe at the start had no place to escape to or no place through which to escape.

That is only applying if it is nothing filling the Universe at large. Can nothing grow as much as a line is growing from a start of nothing? The answer is that such lines not only indicate a distance but since the Universe came from such a small space as science propagate with the theory of the Big Bang then all particles in the Big Bang Universe were rather cramped for space when the Universe started from that small line between particles and is now the same line but is now so big. In the past everything seemed being so small and showing that the space between particles seemed then to be awfully short at the time during the star of the cosmic concept. It was short but how short was it? Did it start off as nothing? Is the line starting at nothing as science wishes us to believe? If it does then all lines must start from nothing so we better investigate this trend with the start of a line. In this following I show my argument with which I hope to prove the counter part of what science believes. Later on in this book I am about to prove that which science sees as nothing in space and in material is the very location of singularity. But lets return to the start before the confusion came about with space being nothing. The start has to start with Kepler because science in the new era started with Kepler and not with Newton.

I have to belabour the nothing for the last and final time because from what I am about to present cannot be presented if Academics dismiss my presenting of my work with they're bluffing everyone with nothing being used in outer space. If there is one still persisting on nothing being used in outer space such a person should either by now be convinced about my reasoning and if not that person will find no benefit in any further reading of this book. In that case please donate the book to someone more intellectual and therefore more presentable to the obvious. Kepler's finding cannot stand true if the cosmos is nothing and by persisting to the accepting of nothing then Kepler becomes a discoverer of nothing, which he absolutely was not.

The ether might now be now regarded as unnecessary, since it is recognized that electromagnetic radiation can propagate through empty space but it is the empty space that replaced the hypothetical ether that brought along the misconceptions I am fighting as hard as I can. When cosmology was in an infancy Mainstream Science then realised there had to be a conductor to conduct gravity as a force. Ether is or was a hypothetical medium found in space that was presumed to act as a conductor of the force of gravity. Electromagnetism was the presumed conducting force through which the force flowed. Some how back then Mainstream Science had the sense to foresee that electro magnetism was part of electricity and electricity needs conducting to flow. If magnetism did not need a conductor, electricity would either flow without challenge of resistance or electricity would not find the ability to flow at all. Ether is a hypothetical medium once thought to permeate all space, through which electromagnetic radiation supposedly travelled; formerly spelt aether. On the basis of this supposition, the Earth should move with respect to the ether, and it was predicted that the speed of light would vary when measured in different directions.

This presumption is based on another presumption the time in which we are and that apply to us every day is a cosmic standard time. The presumption was and is that the time applying on Earth as years, days, seconds or whatnot is equal everywhere from Mars To Magellan's cloud and every where thought may reach. The tests were based on the idea that time was similar through out the Universe. Please I whish to put one thing straight and remove whatever doubt my arguments might provoke, but I am not trying to re-establish or reinstitution ether. My fight is on the incorrect view of the standard unified Universal time that we can set our clocks by from here to whatever stellar system there is fifty billion light years away. The way the experiments were done in the 19th century (e.g. the Michelson-Morley experiment) which failed to detect any such variation in speed was as flawed as the ether theory by its own merit. The ether tests using the time as was done is by indication as if time could be measured by a clock watch simply because the standards of measuring used in the principle experiment was the true indicator applying to time wherever time was to be measured. The way in which the measuring was done typifies the cripple manner in which science does not understanding the concept of space-time being space which cannot be if not in motion through time doubling as the second space component.

The ether is now regarded as unnecessary, since it is recognized that electromagnetic radiation can propagate through empty space but that statement alone mesmerises the calculations that electricity requires. If nothing is able to conduct the flow of electrical current the current resistance must be zero because nothing simply cannot resist a flow. That makes the actual flow of current as measured by the heat displaced from one point to the other point beyond any limitation or measure of any sorts. The volts must be eternal which is just another name for a concept if we use an infinitely large measure to try and pass on an idea of the flow of current so big it goes beyond understanding. Using human terms available belittles the measure possible for the current that might come about when there is no resistance containing the flow of electrical current. The volts will burn holes into a Black hole while the amps will fry the Black hole to smoke without resistance coming about to the flow of electricity. Again it boils down to the fact that science trash my work as incoherent and not worth reading but in the mean while every time I return to the nothing issue I find words lacking of what there is to express how incoherent the argument about outer space and nothing is. The first mistake was the presuming of space being empty. Space can be less dense but space cannot be empty.

Mainstream Science is forever so concerned about mathematical proof, except when it suits those in charge not to insist on such proof. As much as I searched I never once could see how any person brought about calculations to prove that outer space is empty or outer space contains nothing. I challenge all concerned to show mathematically that there is nothing in outer space. I do not promote the idea of either as it was first called or I do not think either is part of the cosmos. I do think there is something in outer space or outer space constitutes of some form of material that we cannot detect because such material is the basis, the new ingredient, which formed all material including heat and space. The Big Bang was the ultimate nuclear explosion which all War Lords of the Warring Empires in the world today dreams of.

The War Lords I refer to are those who bombs nations to dust because they promote peace and democracy and Human kindness and Brotherly love in such a manner while they go robbing them blind of their oil. They kill those children they say they save from brutal dictators while the dictators killed far less innocent woman and children in a lifetime than the big saviours do in a month. We all know who they are because all commercial oil funding flow through the banks of those nations. Let's get back to cosmology. Then through some investigating testing proof came about that ether does not play a part in the conducting of the flow of electricity. Those tests had to reflect on the flow of light as well. Light is also just electricity because electricity can produce light as much as light can produce electricity and light flows through space. In the events that followed the disproving of ether being present in outer space led to a belief that a vacuum came about. The vacuum was present in the minds of those explaining what was present in outer space and the vacuum then transformed to the void and from there the void went out to outer space, which then formed space. But the only void I can detect was in the heads of those placing the void in outer space. The critical density proved that no void could be in outer space just because a factor indicating the possibility of a critical density was detected. However there was a realising that the required density was not enough to prove Newton correct but it did not prove that much nothing was filling the vacant space between the thinly spread material either. In outer space there is less density of space than in the more compact atmospheric space.

The density I refer to is that of material that would be obtained if all the matter contained in galaxies were smoothed out across the Universe. Although stars and planets have densities greater than the density of water (about 1 g/cm$^3$), the cosmological mean density is extremely low (less than 10$^{-29}$ g/cm$^3$), or 10$^{-5}$ atoms/cm$^3$) because the Universe consists mostly of virtually empty space between galaxies. The mean density of matter determines whether the Universe will continue to expand. By accepting the density parameter, as the ratio of the mean density of matter in the Universe, which is standing in regard to the critical density of matter. That which accompanies the material is what proves the density factor. That is released heat that becomes space and the space, requires another associated material of compromised singularity to bring about the eventual collapse of the Universe. A value of $\Omega$ (where $\Omega$ symbolises the density parameter) will bring the accepted ratio to ensure that the collapse is immanent and Newton can rest in peace in his grave as he then was ultimately eternally proven correct beyond question by his followers. It is if the Newtonian presumption is presumed that the Universe must collapse to vindicate all who believed in Newton and with Newton being correct and gravity is contracting the Universe. The value of one or more will ensure science that the Newtonian concept they use will apply and will bring such a gravity collapse. It is also anticipated (correctly) by those Super-Educated that the opposite apply. On the other hand will a value of less than one bring about expansion forever and the Universe will drift apart with space forever growing.

Off course the whole concept I just now mentioned is flawed and corrupt beyond any effort of saving and not even Einstein's best efforts could manage to bring a rescue about to save the face of all the Newtonians involved. Notwithstanding that not even Einstein was able to correct Newton's incorrectness.

Most surprising with all the genius they have, it is still the way the mathematicians go about arguing. In an effort to prove Newton above all logic those bent on proving Newton say there is space and somehow the space contributes in the effort of gravity. The space with least protons contributes by producing the gravitational constant. What will bring about the gravitational constant where the gravity is somehow elastic because Kepler's **k** does flex from season to season? Why then would nothing flex? Why does the Universe not expand into the oblivious since nothing means not having any restriction at all present as much as there being nothing to bring on friction tension? Why would Kepler's **a**$^3$ apply in combating such unrestricted expanding and what prevails when material is absent? The fact that any density of any sorts is present which secures space proves the fact that nothing cannot ever be present. The nothing is like all other things. The nothing is a relevancy brought in to fit the needs of man and to support some or other concept that those in charge of producing official policies wish to promote in such an advent. Not one of those in charge of producing official policies ever tried to prove mathematically that space is nothing because in space there is density and density of whatever kind destroys the concept of nothing.

We can fill or empty a container the container does not mean we are filling the container with nothing. IN cosmology emptying any container can go back to the era where the Sun started forming a cosmic independence because it is in any persons range to reduce the stroke or **k** back to what the outer space was when the Sun started to burn space. We have used this method for ages on end to tap into energy sources. By our using of fossil fuel of whatever means must involve a process of pushing time in a container back by reducing a relevant **k** that will alter the space **a³** and that will produce a time running to earlier applying conditions in space –time by prolonging the duration of time**T²**.

The internal combustion engine works on just that principle and so too does the turbine engine. By reducing space the space becomes hotter and the heat forms the workable energy that we humans found a way to tap into. In our quest to harvest energy we take the environment within the cylinder walls in the engine back the conditions as it applied in outer space when the Sun established that very first thin border line that parted outer space from the Sun inner space as the Sun declared independence from the Milky Way That is when the Sun became a star. We use this cosmic development technique by retaining such development. The manner in which we do it is to place air in a large area, which we reduce by motion into a very small area thus changing space and converting the space into usable heat, We then introduce into the reduced space fossil fuel and mix it well with stored heat that we pushed back to a time that dates when the solar system awoke from the Milky Way heat blanket. That was long ago.

We place this either solid or liquid fossil fuel and place it in a very small container that is terribly hot and well secluded from the container we think of as the Earth atmosphere. The container that is burning our mixture is small. By method of harvesting a combusting mixture it is pushed even smaller. By the reducing of the stroke or the **k** factor of combusting the fuel we have a process where we turn the fuel into heat and with that much heat the space can duplicate space in favour of antigravity. In that manner we push time back billions of years in this very small container. By returning time to where the Sun started its quest to individuality from the Milky Way we then can harvest energy created millions of years ago and which nature stored in solid fuel. By reducing the space **a³** contained to predating the arriving of the solar system and introducing fossil fuel by method of igniting it, the process enables us to create motion we can use in the present day to the advance of life's wishes.

**The piston in the exhaust/ intake position stroke**

**Compression / ignition**

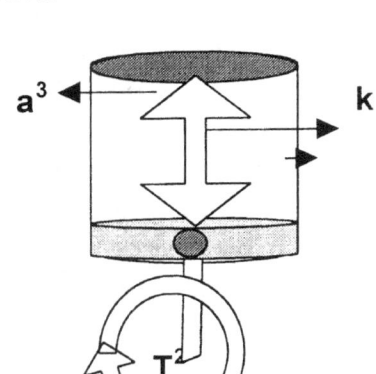

The internal as well as the external combusting engines drive on the same principle as the Coanda effect and it represents Kepler's formula precisely. In analysing, the engine we find the principle of space which Kepler introduced applying in the same way.

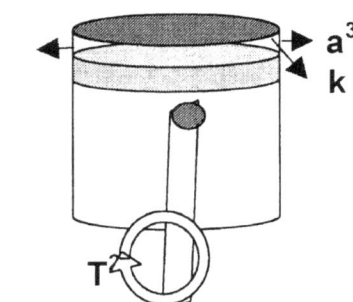

**By decreasing the stroke, which will be the distance, k the area a³ diminishes considerably while roaring action of the crankshaft T² produces the time motion keeping the sequence in ratio.**

**By the same measure that the distance reduces, space demise in ratio but the heat concentration represented by T²**

The reducing of the stroke of the connecting rod will have precisely the effect on the space where the reducing will come into effect as the reducing of **k** produces a smaller combusting area. The cylinder with piston and rod turning will also adhere to the same formula Kepler proved. The area of the cylinder **a³** depends on the connecting rod length

**k** and the concentration of heat $T^2$ that is representing the gravity aspect applying increases in ratio. It once again proves $a^3 = T^2 k$.

The only difference there is, is that we turn the prominence of the different factors around to suit our needs in accomplishing the harvesting. But the factor most important is that the motion increase the heat density as the space reduces. $T^2$ increases by doubling or tripling as space declines in volume. $a^3 = T^2 k$ is an undividable unit wherever space, time and distance form an interlinking action. The focus must be on the temperatures rising as the stroke reduces the space. This implication is what space-time are all about and that is the strongest force preventing any space travel.

In this the vital part is the release of heat in producing space that establish a linear motion, which produces a circular motion. If not for repeating this cosmic action no motion would have come from the whole episode. It is heat producing linear motion that turns to circular motion that produces the energy we tap. Whatever way we look at the working of artificial or manmade machines, the Coanda effect is ever present. There is a centre point we call a driveline. There is a distance **k** we call a stroke and there is a rotation factor $T^2$, which we call the revolutions or the machine speed. The rotation personify the cyclic rotation we find present in the completion of a circle and all rotation from an engine and an electric generator / electric motor down to the mathematical expressed graph uses Kepler's symbolic space-time. It further lays waste to Newton's view that by rotating a value of zero work is completed

When realising the error of science in they're accepting that a value as zero is presented and is legitimate in mathematics, we then have to take it further. One can establish from that that the circle does not employ zero as a value after the completion of one rotation therefore $F = G (M_1 \times M_2)/r^2$ is invalid, one has to return to Kepler's $a^3 = T^2 k$ and establish a value from that.

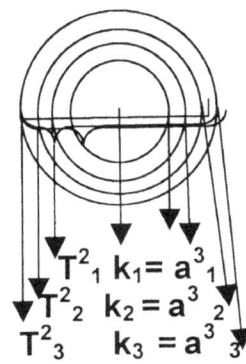

$T^2{}_1 \; k_1 = a^3{}_1$
$T^2{}_2 \; k_2 = a^3{}_2$
$T^2{}_3 \; k_3 = a^3{}_3$

**The space on point from k towards time is space. Space in the six sides is time directed from space by one dimension. The motion of time provides the three positions another three to form the universe as we see it. By motion at the point of singularity the time forms the space doubling in value from 3 to six. Take away time and space collapses and takes away space and time disappears. It is not possible to have time without space or space without time. By destroying one the other will disappear. Where singularity meets space and k changes from one ($k^0$) to $k^1$ space as well as time comes about and time is the compliment of space where space is result of time coming from singularity.**

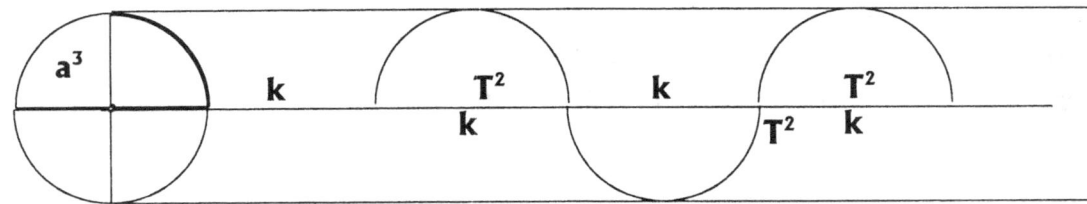

**From the graph, one can establish the link in the circle's rotation around a conforming unit being singularity.**

Saying that one therefore has to admit that the smallest spot has to hold space because the most insignificant dot can transmit light and being able to accomplish that, one must accept it then too has to carry a value of something. If that spot had the value of nothing that would mean that such a spot was not there to begin with. If it is holding space-time then one should return to the original formula indicating space-time in as much as $a^3 = T^2 k$ where $a^3 = \Pi^3$ and $T^2 = \Pi^2$ as well as $k = \Pi$. Being space-time and time to space it has to alternate positions and that can therefore only apply to $k$ where $\Pi$ will indicate a relation to the space-time in question or the relevancy to singularity being $k^0 = \Pi^0 = 1$

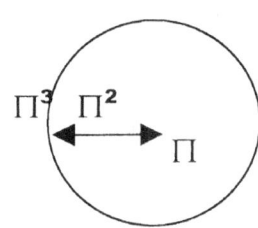

**Time is always a displacement of space in relation to the implication of singularity, and comes about between two points in space relating to the centre of singularity as positioned by k, either too the value of k or too $k^0$. There is no larger k or smaller k but when one factor changes other factors has to compromise and it is this compromising that places cosmology above normal Mathematics since nothing in the Universe is static enough to use general mathematics.**

$\Pi^3 / \Pi^2 = \Pi$ **or** $a^3 / T^2 k = k^0$**. With this fact established we then must return to the value as indicated by singularity being** $\Pi$**. In this we find that** $\Pi^3 / \Pi^2 = \Pi$**, and** $\Pi$ **is a stand in for** $\Pi^0$**. This brings about the value relating to space-time relevancies as a formula consisting of** $\Pi^3 / \Pi^2 = \Pi$ **or** $\Pi^0 = \Pi^3 / \Pi^2$ $\Pi$ **in various forms and relations. One also must keep in mind that there are always four time factors relating to the universe from any point holding singularity, and since every point in the universe contains singularity in what ever form, every spot in the Universe comprises of four time points initially extending to the next spot by means of** $\Pi^2/4$**, which we know as the Roche factor. By rotating space-time the atom forms as an identifiable independent cosmic structure. It is always about relevancies applying differently because the mass of the atom can be higher on Jupiter than it is on the Earth or lower on Mercury than it is on the Earth. However the atomic relevancy will always remain in place except in stars where the star already abandoned the implication of certain factors that no longer can apply since the motion, which the star established exceeds the limitation that those factors carry.**

**From such a relevancy there then must be four different values relating to singularity and since the atom has a relevancy of** $(\Pi^2 + \Pi^2) \Pi^2 \times \Pi \times 3$ **that then also must be true.**

---

$\Pi^3 / \Pi^2 = \Pi$ or $a^3 / T^2 k = k^0$. With this fact established we then must return to the value as indicated by singularity being $\Pi$. In this we find that $\Pi^3 / \Pi^2 = \Pi$, and $\Pi$ is a stand in for $\Pi^0$. This brings about the value relating to space-time relevancies as a formula consisting of $\Pi^3 / \Pi^2 = \Pi$ or $\Pi^0 = \Pi^3 / \Pi^2$ $\Pi$ in various forms and relations. One also must keep in mind that there are always four time factors relating to the Universe from any point holding singularity, and since every point in the Universe contains singularity in what ever form, every spot in the Universe comprises of four time points initially extending to the next spot by means of $\Pi^2/4$, which we know as the Roche factor. By rotating space-time the atom forms as an identifiable independent cosmic structure.

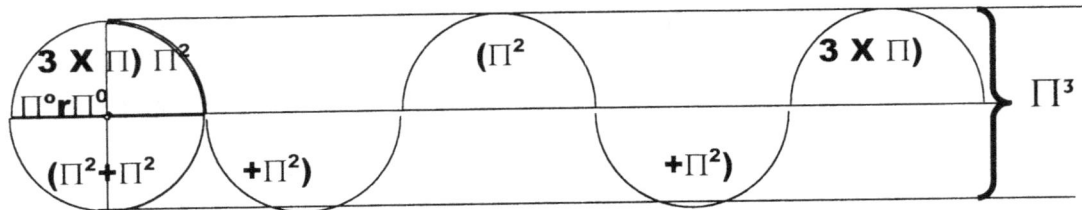

$\Pi$ X 3  $\Pi^2 (\Pi^2 + \Pi^2) = 1836.$

In our ability as we found a way to reduce space and energise singularity the possibilities we can achieve seems to be endless. Our generating energy or tapping energy goes by the means of taking singularity back one stage to a time that goes back as far as we may go.

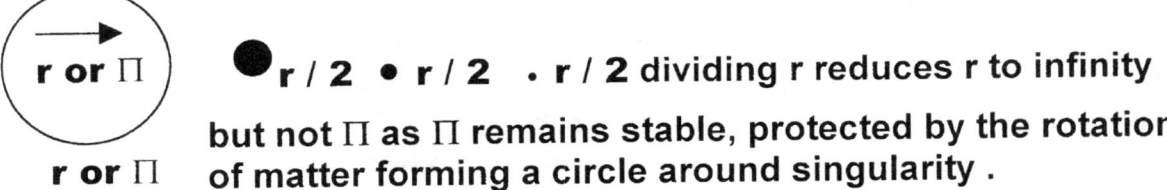

**r or $\Pi$**  ●$r / 2$ ● $r / 2$ ● $r / 2$ dividing r reduces r to infinity

**r or $\Pi$**  but not $\Pi$ as $\Pi$ remains stable, protected by the rotation of matter forming a circle around singularity .

**When one starts reducing singularity from where we stand in the Universe such reducing is physically endless.**

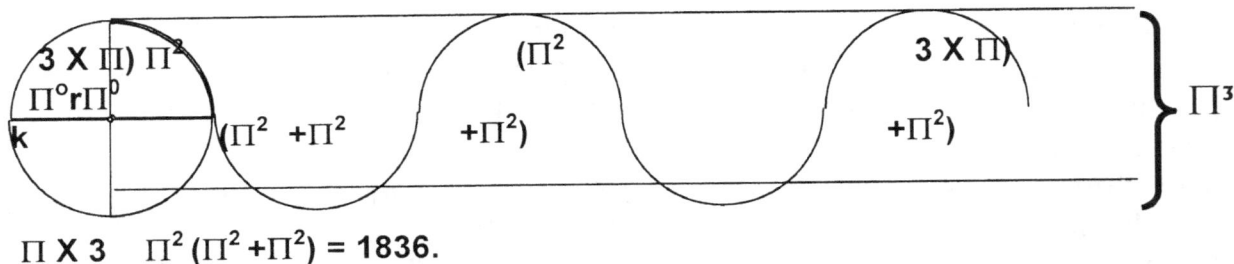

$\Pi$ X 3  $\Pi^2 (\Pi^2 + \Pi^2) = 1836.$

It is always about relevancies applying differently because the mass of the atom can be higher on Jupiter than it is on the Earth or lower on Mercury than it is on the Earth. However the atomic relevancy will always remain in place except in stars where the star already abandoned the implication of certain factors that no longer can apply since the motion, which the star established exceeds the limitation that those factors carry.

From such a relevancy there then must be four different values relating to singularity and since the atom has a relevancy of $(\Pi^2 + \Pi^2) \Pi^2$ X $\Pi$ X 3 that then also must be true.

━━━ ━━━ ━━ ━━ ━━ ━━  ──────

**0.9 $\Rightarrow$  0.09 $\Rightarrow$ 0.009 $\Rightarrow$ 0.0009 $\Rightarrow$ 0.00009 $\Rightarrow$ 0.000009 $\Rightarrow$0.0000009**

In our ability as we found a way to reduce space and energise singularity the possibilities we can achieve seems to be endless. Our generating energy or tapping energy goes by the means of taking singularity back one stage to a time that goes back as far as we may go.

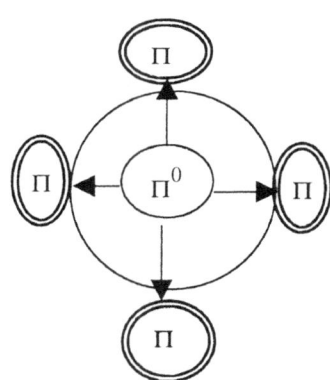

**By moving from $1^0$ to $1^1$ and from $1^0\Pi^0$ to $1^1\Pi$ requires space. Yet, such moving does not leave the realm or the domain of singularity. The motion is still within singularity because moving involves forming a relevancy between heat and cold and between infinity and eternity, between space and time and most of all producing what will in the far future develop into a Universe that can even be a host for life albeit on a very small spot for a very short while in relation to the vastness space has and the duration cosmic time has. This where time started and time remains at this edge of forming space by motion from singularity that cannot move because it has no space.**

The one Mathematician will use an r to indicate the radius and the other use a D to indicate the diameter, which is double the radius and therefore needs to be divided by a four to eliminate the Newtonian inverse square law amounting to the difference there will be between the two. The one using the radius is $\Pi r^2$ and the other formula is using the diameter is $\Pi D^2 / 4$. The factor that mathematics normally allocate to the circle which is carrying the square is given to the radius or the diameter which indicates points in relation to singularity running from singularity to the edge or the border of the circle and that circle is implicating the factor of time in cosmology. Every mathematician during so many centuries has missed the chance to observe that Kepler's $T^2$ proved to be time and time is the gravity that is applying at that point in space-time. In that the process goes beyond what mathematics may deliver. This misunderstanding is about time not being a fact of human perception but a factor containing the entirety there ever can be as a relevancy. The fact it is a factor containing the Universe. This brought the about incorrectness in the understanding about spreading the confusion. Any and all drivelines (even photons that travel) are about securing motion through singularity and in the process displacing space-time. That is what energy amounts to. It is finding heat release and attaches it to a drive through a line held by singularity and put space-time by duplication in motion of the space-time.

Time predates mathematics. Time is motion of space ($a^3 = T^2 k$) and therefore time can never go single. Science put time at a single value as $t = \sqrt{(1 - (C^2 - V^2))}$ but having time as **t** is the same principle as showing a photograph of an event that happened in the past as time taking the onlooker back in the past where the image froze time to the single dimension and the image has to rely on some other person using that persons imagination to interpret the photo. However we look at it the interpretation takes on the role of t and not photograph that claims space-timer in the third dimension. The photograph holds space $a^3$ in the motion of time $T^2$ because the photograph is constructed from material that is part of the Universe while only the image we find in the ink is presenting the part we consider as time in the past by the image we have. Not even the ink, but only the picture printed with ink is what Einstein's time formula t represents. By motion of space the image we find to hold our perception of time was destroyed one moment after it was established.

$0.9 \Rightarrow 0.09 \Rightarrow 0.009 \Rightarrow 0.0009 \Rightarrow 0.00009 \Rightarrow 0.000009 \Rightarrow 0.0000009 \Rightarrow$

Taking into account the behaving rules of singularity it is important to recognise that notwithstanding the size of a line, there eternally is another line (or dot) eternally bigger as well as eternally smaller than the line in question. This is gauged from our perspective because we can never achieve singularity. We can never grasp the size of a line that forms the utmost or the least of possibilities in size and therefore size belongs to the human mind forming conceptions of big and small, but it has no place in the cosmos at large. This concept not only applies to size, but also to all limits and divides we wish to create that is forming borders, which we can appreciate. When looking at the circle in the conventional manner, we persist with errors brought about in culture and not by applying some significant modern logic. The reversing of the circle radius is not alien to nature at all.

An observation coming instinctively to mind one may recognise is that the form reminds rather explicitly of natural phenomenon as hurricanes, water whirls and even the shape most commonly favoured to express the cosmic object referred too as a Black Hole. The similarity may be more than coincidental. Let us consider the statement in the reverse. In our calculating of a circle we apply two formula methods. The reducing of the line can go as far back in history to where the line does predate mathematics because the line came well before mathematics arrived and it predates the numbers we use to apply in mathematics as it even predates positional relevancies we use in trigonometry.

This means it even predates directions in space therefore it totally predates human perspectives and goes further back than what human perspective can go. It goes back to the point where only the most basic mathematics being the size the squares and the shapes can take the human mind and that I am afraid is further than the human mind can go. There is no person that can truly explain more than just accept that a straight line and a half circle and a triangle could all carry the same value of $180^0$. This fact makes the mathematics one are able to use in cosmology rather different from the mathematics one may use to design an aircraft wing or al large hanger to store the huge aircraft wing. Mathematics going around in normal everyday use does not apply that straight forward into the field of cosmology since cosmology carries laws mathematics never heard of and was in place long before mathematics was concluded This description was possibly what confuses everyone to this day.

Time is duration; it is motion of space filled with particles running from one specific position to another position or from one point to another point. It takes time it fills space with motion presenting time. There is no one cosmic Central African Time in cosmology where an American clock in Washington sets all time through out the Universe. Every time **k** changes through motion time changes. This makes that every time space changes through motion time establishes many different cosmic time zones. Time is created by motion and motion is heat spinning through space. Any dragster car driver will tell you that the time he or she experiences while driving is much longer in relation to what the clock tells. The motion brings about time discrepancies Academics are unable to explain.

The time exaggeration the driver endures is not physiological but is physics in the truth as science do not yet appreciate. The acceleration enhances the space duplication that produces less space per occupied volume in more time. The sound barrier is proof of this. The Coanda effect is proof of this. Momentum gained or lost is proof of this. Momentum is the increase of linear gravity affecting the mass by accumulating space occupied in time duration that is increasing through motion applied to an individual body compared to what they can manage under normal conditions when the Earth secures time to what is valid on the surface of the Earth.

Let's reflect for one the moment on the truth of every aspect about the Universe at the time in the Universe before the Universe was formed. What ever you may think of as being part of and in the present in the cosmos in the present cosmos now being seen to be a part of the overall and all-inclusive cosmos was not yet in the present term thought about but still very much under development in the far future. It all was locked in a spot, which was one eternity away from the first dot. Whatever is was, was not yet but it was confined to a spot that was so small it did not as it still do not fit into the Universe we now know. The growth that came about depended on the resistance to growth that the growth responded by not resisting. It means the longer the waiting was the more severe the outcome was and what wait can be longer than eternity which was the length of the wait. However drastic the need was to grow there was no space to grow in so with no space the confinement was the natural while the growth was the unnatural.

**The motion is still within singularity because moving involves forming a relevancy between heat and cold and between infinity and eternity, between space and time and most of all producing what will in the far future develop into a Universe that can even be a host for life albeit on a very small spot for a very short while in relation to the vastness space has and the duration cosmic time has. This where time started and time remains at this edge of forming space by motion from singularity that cannot move because it has no space.**

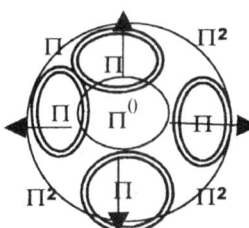

**By moving from $1^0$ to $1^1$ and from $1^0\Pi^0$ to $1^1\Pi$ requires space. Yet, when form came into form such moving did not leave the realm or the domain of singularity. That is still with us since the principle has nowhere to go but to remain in the universe. The motion brought about $\Pi$ as the motion brought about $\Pi^2$ using the same motion. It is the motion that moved $1^0$ from $1^1$ or $1^0\Pi^0$ to $1^1\Pi$ that became time in the square and the motion including time became space $1^0\Pi^1\Pi^2 = \Pi^3$**

In the very beginning before there were $\mathbf{a^3}$ dimensional Universe the motion was only accomplished by the surging of heat, which gained more points in singularity and by the surge of overheating new singularity activated which started and produced relevancies in relation to existing singularity. It still apply in the same manner in the modern Universe by still using the Coanda principle to the measure found in the atom that is the manifesting of occupied space as material.

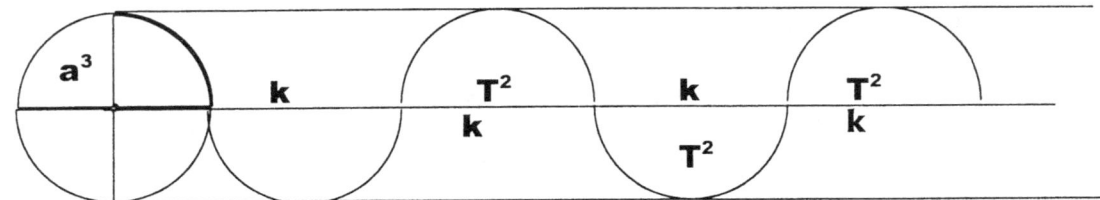

**The graft is a basis on which the entire Universe was formed when only form was available. In that scenario the atom was set without using space...yet when the atom developed before the Big Bang.**

That is very much in contrast to our thinking now in the present where the growth in outer space is the natural flow of space-time or gravity and the resisting by containing that produce mass, which is the force to resist and remain independent and any force is part of the unnatural. Then before the release of space into the Universe in waiting on independence forming, it was no option while at present fighting for independence is a struggle all matter will eventually lose in the end but until such time all matter is fighting the struggle for independence rigorously. Just before space came about a time had to come when the need for growth became as strong as the need to remain confined. The need to remain confined was helped by the fact that there was no space to allow space to form space. This may sound as if I wish to become poetic but it is not because with every mention of space it refers to yet another condition through which space formed and those conditions we will never be able to imagine.

We must view the Pre Big bang era as multiplying by developing heat. Then the graph came to a conclusion where space occupied formed within space unoccupied and that space became a time factor. The Big Bang era is about accumulating time into the material and in that diminishing time once more.

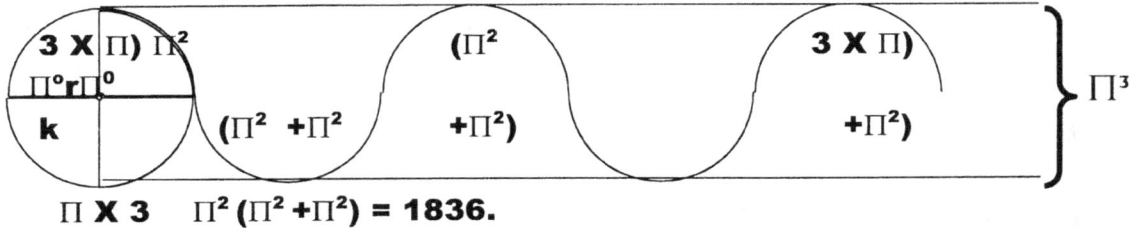

$$\Pi \times 3 \quad \Pi^2 (\Pi^2 + \Pi^2) = 1836.$$

Finally there is the star era where time and space is removed from the atom by diminishing occupational three dimensional qualities and we as Earthlings with life has as little a notion about the true mystique that goes on in that era as we have about the pre Big Bang era of multiplying without dimensional intervention.

Giving it complicated names would be the normal way of acting but I do not whish to confuse myself with adding more names to the congested naming already in place. The only solution was to invent something new such as space in order to break the dead lock balancing the Universe by creating space and starting Creation. Remember eternity persisted therefore everything was in equilibrium. There had to be some introduction to change the status quo because eternity and time standing still had the Universe in a state of not changing for many eternities. Motion became the tool and what has more motion then light? Light is motion where all motion faster than light presents the killing off of motion. But motion is the altering of relevancies by changing locations and positions. By motion the material strike independence from whatever it is moving away from. The independence coming from motion applying brings about space differentiation in space differences between independently active areas. The important issue is to realise that with cosmic birth came a change about that was contradicting what was natural.

Let's reflect on the moment the birth concerning every aspect about the Universe at the moment of birth. We are discussing a time when what ever you may recollect in your memory is something that still is something yet to be. The Universe was locked in a spot that at the time mentioned still had no space and no place to have space where that place or space forms part of our Universe in 3D which we now know so intimately. Then the dot came about as $\Pi^0$ enlisted $\Pi$ to perform duty. This brought along growth but the growth depended on the resistance there was to growth. The growth was equal to the resisting of growth and that brought about gravity as cooling coming from retracting and heat was material releasing and expanding. What brought on such resistance must have been the fact that there was no space to grow. There was no space to grow into or to progress the distance between points in singularity in relation with one another.

That means there was no space yet to become space and that is a concept we must accept as much as we accept that a line is represented by $180^0$ as is a half circle and a triangle notwithstanding every thing we see as huge differences. Everything we now see originated at that time. The only solution was to invent something that will break the deadlock the Universe found it in. Creation came to a solution by creating motion, which is not yet space but is a change in relations between particles where the altering of the position meant the creating of space during time.

Motion became the tool to use in all further cosmic development from that moment on. All that motion actually is the altering of relevancies by changing the manner in which they apply from moment of change to moment of change. Motion became the change in direction, which then progressed into forming space and the duration, or the distance it took became the time where the formed in a relation where both cannot be without one another. The motion established independence but not yet space because the independence created space-time and that is what it still is. The space-time is only created differences in relation to changes about relevancies that apply through motion acting as such.

Newton saw the apple falling and that made him conclude the dynamics of gravity. The dynamics of an apple falling from a tree might have inspired Newton to realise gravity but the totality of gravity as Kepler introduced gravity by motion passed Newton by completely. It still passes Newtonians by to this day.

The Unit forming the apple is in motion but since the unit can perform the motion as the apple makes the apple in motion through the act of motion equal to the biggest star as much as it is equal to the smallest piece of material humans have not yet discovered. The motion confirms the independence and that is what makes the apple able to move. The bonding of the material that forms an exclusive including unit by a group of atoms show that the bonding of the group is provided by the combining effort of all the atoms allowing the apple to have the motion while the motion is also very much in defiance of gravity. The falling is one small part while the maintaining of the composition forming an apple is the largest dynamic of gravity.

The fact that movement can take place produces duplication that performs as the composing of the Universe. While duplicating, as motion is a process of time duplicating space, space is vacated by space filling vacated space. This goes on to the smallest detail that the largest computer is unable to calculate. Finally the control of the motion will dismiss all motion and also the atom as the atom within the star gives up the independence it fights for to the star.

## Motion increases space as motion brings about space .

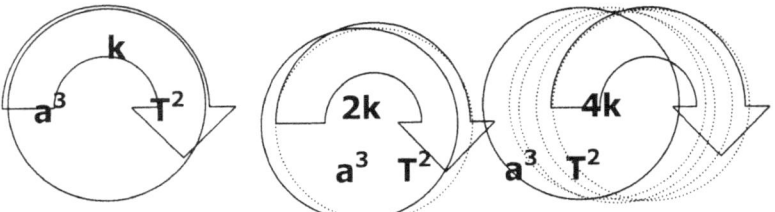

The object seems to be motionless but in reality, the space occupied is evenly distributed in relation to the centre. As the motion changes the positions of points in space-time and although it seems motionless, it is changing by motion. That brings about the

By motion coming about the motion doubles the positions space would occupy when at a reduced motion and the space halves during occupying duration while holding the position because the space occupied is reproducing twice instead of only once. The space then is reduced by duplication as **k** increases and by that space, $a^3$ occupied captures more space $a^3$ through motion in less time $T^2$ bringing about more space distributed in lesser time duration.

By increasing motion, the space that the object captured is increasing the validity of singularity connected to the space-time in question. By moving the object remains in time duration but space-time increases. Since one may either place the relevancy on space occupied or time experienced with the increase of **k** it will increase space-time $a^3$ $/T^2$. Since our observation stands related with space remaining the same and time at a constant, we find nothing changing. However, we on Earth do not experience the true cosmic effect of gravity in balance.

This effort totalled in material capturing more space through motion, which in truth is the other half of gravity that brings about space by singularity fighting for independence. Mainstream Science Academics refused to acknowledge that gravity is space applied in time through motion of space. By failing to salute the truth Science went about making such a mess of the concept behind cosmic gravity. By not looking at Kepler they opted instead not to recognise gravity stated as such by Kepler and refused to put it down as part of gravity. Instead they use another name in identifying the second part of gravity as being momentum. Any motion and all motion alter space and change time.

Time and space is interlinking and is the very same but the one is following the other.. It is what Kepler said. If mathematicians cannot read mathematics it surely says much about those holding the profession. $a^3 = T^2 k$. It is precisely what Kepler said when he translated what the cosmos told him using the language of mathematics. Space will increase as time duration slows down $k = a^3 / T^2$. Space and time is so much interlinked they are the same. There is no time constant as much as there can be no space constant.

There can be no constant.

As space expanded the Universe came about. However, one look at that confirms that the expression constitutes to what we find when we use the term momentum. .First space and time freed from eternity slotting in a position being next to eternity in a place that was just between infinity and eternity and that is where we still are. Therefore that is why such space and time seems to us as eternity, which is coming from infinity. Then matter occupied space through the duplication of space by using time. Space excluded heat from space by having excluding occupied space forming separate time. Some particle overheated and created space and others formed gravity and conserved space by freezing space.

Take a circle and reduce such a circle constantly to where it no longer can reduce. Reduce it to a point where only form remains part of the circle because the radius has gone beyond human measure and becomes so small it is not noticeable with what ever tools man may use, then what remains is pi since pi does not indicate size but indicate form, and form is all that then will remain. I believe one can begin too see where my suspicions are heading because the flaw comes about in the manner mathematics are practised for thousands of years. Space is though to be nothing because that means man thinks about space as a standard fit all issued everywhere that came about when time came about. Before civilisation taught man to read and write, even wind was part of magic. Today we know wind is part of heat forming space in motion. Then it was thought that space cannot increase and winds were ghost blowing their breath.

If there is any suggestion of this thought being ridiculous then how ridiculous is it to pronounce space as nothing. Nothing and ghosts are more or less similar therefore scientifically amongst the wise and informed of the day, little has changed since then and now. More seriously wind is space holding more heat than other space holds heat in relevancy.

All the four pillars of the Universe depend heavily on the form of $\Pi$ and if not for the liquid of the neutron and the form of a double $\Pi$ (to the square which indicate gravity, the Coanda effect would not have been able to establish the atom in the form we now find the atom,. The importance of the Coanda effect in duplicating space-time while producing a diminishing thereof in the form of $\Pi$ is an indication that $\Pi$ serves as mould and form to the entire Universe.

Winds are as much antigravity returning reduced space back to the ranks of increased space. Before coming to the mathematics I would first like to bring your attention to the practical side. I am promoting a theory in which I am able to prove there is as much contraction going on in the cosmic Universe as there is expansion and the contraction is as much part of the expansion. The two factors are inseparable the same. The Universe rides on a balance and we have to locate such a balance.

To prove my theory I firstly had to locate the centre of the Universe. The failure of Newtonian science to locate the centre of the Universe is an obstacle they never noticed. It should be facts they have to predominately first establish where to locate the centre of the Universe. With gravity pulling everything in that general direction and finding such a point should show where the lot is heading to where all contraction will eventually lead. Identifying that precise location is a far greater problem to investigate than is the critical mass density factors a devastating problem.

This inconsistency to point where the contracting should be heading proves to be the Waterloo of science because science has no idea where to position such a centre. If we backtrack instead of fast track the contraction of the Universe we should be able to find the point of the beginning of everything. Its because of where science position the end of the Universe some thirty odd billion light years from where we now are that I concluded the centre of contraction must be allocated. Closer to home we must search for the point of gravity where the gravity is the strongest as it must be in the centre of the Earth.

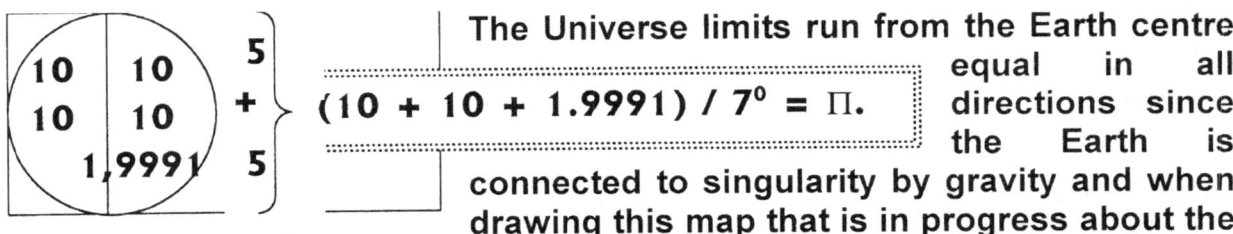

The Universe limits run from the Earth centre equal in all directions since the Earth is connected to singularity by gravity and when drawing this map that is in progress about the cosmos the allocated centre must be where the Earth now is. That was what inspired me to locate my centre of my Universe. Even admitting to such a notion sounds like madness, but please give me a chance to explain in more detail. I realised that my effort to locate the point holding singularity enabled me to backtrack the exploding universe to its origins. By applying some basic effort I have located the position from where all movement came and the direction it took moving forward in time...and yes, even time as such. Gravity is the dimensional changing of space holding r as reference in the cube as to the sphere holding $\Pi$ as the reference. In order to generate spin that is producing time in matter occupying space, therefore creating dimensional change, $\Pi$ has to be a factor indicating the possibility of spin because by implementing $\Pi$ the circle sides will follow one another without establishing separation. As soon as motion takes gravity straight, singularity will reposition the direction changing the direction of motion by $7^0$. It is this turning of motion by redirecting the continuing of motion that sets the critical time within the proton connecting to singularity. Instead of r being a line gravity will inevitably be $\Pi$ which is the form value of singularity. That is this $7^0$ redirecting in the square of space of space, which is ten on both sides of singularity and time is that what we find to be the Titius Bode law of 7 / 10 and 10 / 7 in relation to the Roche limit of $\Pi^2/4$ which is producing the gravity of $\Pi^2$. However the reducing in it is going from ten that is on one side and is crossing over the figure of 1.9991, (which is singularity on both sides of the Universe) and coming into contact with another 10 while turning $7^0$ that we find to form $\Pi$. In all being the total forming on both sides of the Universe it is $(10 + 10 + 1.9991) / 7^0 = \Pi$. The answer must be in finding $\Pi$, and thereby locating singularity. If singularity is in affect the original point of the cosmos birth, the reducing path we should follow will indicate the whereabouts such a point must be. That is where cosmology diverts from mathematics.

In the normal applied mathematics there are two standard formulas used to calculate a circle. The one use an r to indicate the radius and the other use a D to indicate the diameter, which is double the radius and therefore needs to be divided by a four to eliminate the Newtonian inverse square law amounting to the difference there will be between the two. The one using the radius is $\Pi r^2$ and the other formula using the diameter is $\Pi D^2 / 4$. By implementing either neither produces results therefore such a lead will bring one no further than the understanding that person has. However one looks at the mathematical expressions and Kepler's formulating of space-time and we find there is an exceptional difference between the two scientific uses. When investigating Kepler's formula one do find it appreciably differs from the normal Mathematical equation such as we find the normal allocations to be $a^2 = r^2\Pi$ and $a^3 = 4/3\ \Pi r^3$. In the normally used mathematical expressions such equations tend to concentrate on the volumetric aspect. In the case of Kepler's expression it is something else that wants to surface. It is totally another idea that is coming to mind. In Kepler's formula $a^3$ stands to symbolise the third dimension and such a third dimension becomes equal to two other dimensions grouping and sharing value to equal $a^3$ efforts. It is not the circle of the rotation because with such a normal circle the radius is in the square and $\Pi$ evaluates the form.

Here in the calculations Kepler received from the cosmos there is no mention of a factor $\Pi$, which one would expect to be somewhere applying since the circle is $\Pi$ and $\Pi$ is the circle and the two are inseparable. But not in Kepler's $a^3$, where there is no mention of $\Pi$ at all. The fact that there is a radius of $\Pi$ used to replace r of some sorts used to indicate a position, which cannot hold the square as it normally does in the case of the normal equations. In the mathematical equation the factor indicating the position of the circle edge has the square value being called the radius or in some cases the radius doubles and which then is the diameter, and the circle indicator is $\Pi$.

But in this event the formula value will bring about a square value to the answer one receives. It will bring a value to the surface of the circle. In Kepler's formula it specifically does not. I am not the first one that brought Newton into disrepute. Before I did the cosmos did. The comets with they're not colliding did, and so did Roche and Lagrangian principles. Hubble was another one and it becomes apparent that every one that made a study about matters in the cosmos was in some disagreement about Newton. However no one in the past had the audacity to confess they're being in disagreement with Newton.

By Newton's effort to improvise on behalf of Kepler Newton made a statement that Kepler never made. In all honesty nature reacted strongly against the claims Newton made on behalf of Kepler and not about Kepler's work but about Newton's modifying of Kepler's work. In short: how can a comet sail past the Sun time after time without colliding and still apply a contraction in the manner which Newton suggested by the one claiming a freezing grip on the other? This strongly contradicts $F = G\ (M.m)\ /\ r^2$ How can five structures as the LAGRANGIAN POINT form around a centre structure while the centre structure keeps the five in position at equilibrium? It is so clear that all of the cosmos is rejecting Newton's improvising and rejects just as strongly the contradicting of the formula used to incorporate what we think of as cosmology in $F = G\ (M.m)\ /\ r^2$

In the event where I refer to r, more terminology of indicating a line than it is referring to a radius of a circle.

We find this proof in what we see in the Roche limit. In the Roche limit the extending of the radius does not commit r in any way but produce singularity by form of $\Pi$. It is the sphere reducing the other sphere by one applying material robbing through gravity contraction. The one sphere heats the singularity of the second sphere by some form of electrical crossing space-time without acknowledging a radius of sorts. It does implicate singularity by measure of $\Pi^2/4$, but that refers to singularity bridging tie and it has no hold on a radius. The mistake science made in the past in their studies of the Roche limit was their trying to implicate r as a radii factor.

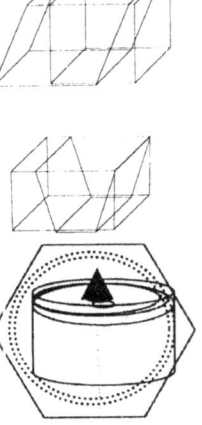

Anything occupying space in the cube will apply r and by r I mean just a distance not using $\Pi$ because $\Pi$ serves as a form indication while the collective product of r will determine form as well as accumulative dimension total. Notwithstanding the name used confirming the shape or r named as length width or height, it is all just a straight line bringing about the cube with all its other names that may find attachment to specific form but nevertheless still remains only a six-sided cube with connecting lines applying different angles changing in some cases. The normal perception is that any circle growing spontaneous would grow by the radius, which is r. In mathematics that may be true but it is not true in nature. In nature that cannot be the case because, r is an indication of a straight line. By growing with the aid of a straight line from the centre to circle the influence that that would have on the circle would result in many circles following one another and not a continuous growth.

Applying the Roche limit brings about a divide to the value of $\Pi^2/4$. This places singularity at a relevancy in a divide with singularity . When there is insufficient dynamics to place the correct space-time in order to bring about the minimum relation in space-time dominance, such dominance comes about where the one would secure a proper **k** factor with the minimum $a^3$ space and the $T^2$ time of the dominator will capture and increase the gravity charged by the dominant which then turn out to be antigravity to the lesser body. In the manifesting of the Roche factor the growth of gravity of the dominating singularity turns the space-time of the dominated partner into liquid heat, which is equal to antigravity of the sub structure. This process repeats the very first cosmic action there ever took place but since this action was part of the cosmos it will remain part of the cosmos until the end. In order to establish space-time singularity has to give up the privileged position of capturing and securing what ever may follow from such a release From that singularity the dominant factor will release space but it will release time as it intensify the heat density in the released space.

The heat becomes the time factor but then forms a sector of space-time and will lose the inclusive qualities singularity has. Before such a release as which I just mentioned singularity space in heat is forming time as a unit. We will not be able to understand this relevancy and such a ratio because such a unit falls outside the Universe or domain we have to secure and  use to form our concept about our Universe. That position singularity holds is very much not part of our Universe though it controls our Universe by laws and whatever forms or represents singularity. That makes our concept of singularity fruitless and singularity is very much beyond our explaining. But the space-time released forms a limit we find running between specific borders in law as underwritten by singularity. There is a heat or hot which is liquid but is still material representing the heat and then holding the liquid is space representing the cold that places the "border on the other side" . We humans regard the heat, which we measure because we cannot measure the space holding the heat that forms the limit in the cold the space-time can reach. Space forms the cold and what space contains is the heat, which limits the cold from expanding by accommodating the heat into the space. The representing of the two and the establishing of space-time is also representing the demise of what forms singularity. In singularity within the star we find that singularity reunites heat and space into something we are not aware of but that we can see diminished all space as it took heat to whatever limits heat can achieve.

When we look at the Sun we look as if searching through another persons waste to see how the person is fairing and living. Police are famous for this work method and so are anthropologists, but that is at least while they are keeping in mind that they are investigating the waste and not the person. They look at the skull and realise the person is dead and not just smelly because it hasn't had a bath for many years. The work method actually involves the looking at what the person's are not because what they are not they discard. That which the person discards as waste is what does not represent the person's life style. Yet in applying cosmology we stand outside the domain of the Sun and see the Sun discards the amount of heat it does and then by what the Sun discards we categorise what the Sun is. It is as if standing outside the biggest freezer ever built and feeling the heat the that the cooling pumps extract and discharge and then by that evidence declare the freezer as the hottest place ever. Why would the freezer discard heat if not for maintaining a cold within? To discard heat it has to be cold at another place where the heat is coming from. This we can only realise while of course getting rid of the senseless and meaning less idea that the Sun is a hydrogen burning coal stove of medieval proportions that would some day run low on coal and die! Think about what became of all the gravity in such a cold lump of rock.

For God sake I think science has evolved somewhat past the burning boiler concept where stars burn fuel and then when the fuel is finish, then the star die off by getting cold. For Newton in Newtonian age steam kettles and boilers working off massive coal stoves were the cutting edge of technology but at present that cutting has rather worn thin and the edge has gone very blunt. Today we know about nuclear reaction and fusion producing a work principle in stars. We must realise that when heat is being vacated from an area the use of such a process will insist on a compromise of having cold somewhere in another area and when there is no compromise required for absorbing heat it is because no compromise then is required as is the case in outer space. There is no limit to the possible heat outer space can absorb because it has all the heat that is available already contracted in that area. It is then as hot as it will get. When an object discharge heat, then the inside must be cold because no one can flame a fire on the inside of the Sun while the Sun is freezing hydrogen to liquid at a temperature of $6500^0$. Being $6550^0$ must have a meaning to a human because a human will not stand such a heat, but in the Sun where we see rivers of heat flowing and ejecting like a liquid sprouts things look pretty much different to what the needs of life is.

On the other hand we see an area, which are absorbing all heat that it possibly can absorb without affecting the area by heat rising in the slightest. Yet because the heat cannot raise any further we declare such an area being as cold as it can get. If something is so hot the heat cannot rise further notwithstanding what effort is achieved in raising the heat level, then that area absorbed all the heat there is going around and is as hot as anything can get. Outer space takes what ever it is thrown with in terms of heat and then shows no sign of change. That has to be because the heat level is so high. It is as if the sea can take all the water all the rivers throw into it and while seeing the water levels not rising from that, we declare the sea as bone dry because water coming in does not affect the sea level upward. We cannot declare something of value with value looking at what the something is not. Outer space cannot rise from heat because it is as hot as it gets while the Sun cannot discharge that much heat without cooling to a freezing.

This we find in place in outer space and because of the contraction we are unaware of, we find outer space to be cold while outer space is in reality so hot it is exploding and has been exploding since the Big Bang commenced. The exploding process we named after Edwin Hubble. In contrast to this we find in outer space heat at the very ebb in space in outer space at the highest pinnacle space can achieve with the time that developed as far as time did develop this far. Hot and cold loses unity when singularity releases the two aspects and singularity unravelled as it has in outer space again join the two aspects where singularity is totally unravelled in becoming space, which is representing heat. Before the unravelling space and heat was one but when Creation started it was in favour of heat whereas outer space now is space and heat still contained in a unit that is demising all the time but is favouring the heat aspect. The separating of the two and the establishing of space-time forms the demise of singularity but as heat levels demise space levels increase.

By space-time expanding singularity is in the demise. On the other side in forming material singularity vested the counter of this action by forming motion that demises space to increase heat and reinstall heat into singularity eventually where such singularity is maintained by material in gravity. The motion of expanding in outer space $k = a^3 / T^2$ establish singularity favouring recouping of heat by increasing space producing more space and the motion in the atom $k^{-1} = T^2 / a^3$ favours the recouping of heat by destroying space But in all cases it is singularity in control of cosmic law to secure the maintaining of singularity.

If we wish to believe in the Big Bang and we wish to accept the factor of singularity then we have to accept that there was a period where there was no space anywhere at all. We can backtrack the space to a point where the space is no longer space on the precondition that the space is coming about only from the motion of space. If more heat comes to such a centre the centre will produce more motion. The motion will produce more duplication of space and the duplication of space is the gravity we experience as a contracting direction of motion where as heating is the expanding of space through motion. But it had to have started with a space less motionless dimension-less Universe wrapped in singularity. The differentiation coming from motion is a dimensional barrier that changes many aspects in cosmology. The dimensions came about as the Universe came about and each had its individual introduction period. Space and time parted at $(\Pi^3)^2 = 961$, material formed identities at $\Pi x \Pi^2 x \Pi^3 / 5 = 192$ and $\Pi^2 x \Pi^2 x \Pi^2 / 5 = 192$ where space either had material or had heat without material and space separated from heat and matter at space holding $10/7\pi^2/2(\pi^2 + \pi^2) = 139$ material $7(\pi^2 + \pi^2) = 138$ and space having liquid within $7/10 \, \pi^2/2(\pi^2 + \pi^2) = 136$ This is suggesting that these are meaning this was the first time liquid became part of the cosmos while all were still part of the same unit as the Roche principle would suggest $(\pi^2/2)$ as well as the $(7/10)$ and the $(10/7)$.

By my mentioning this I am not only once again jumping the gun but it is more a case of my being half way down the race track before the gun fired the start to the race. This, which I mention is ahead and almost at the end in the direction of where I am heading... I am on the one hand forced to do this because I am unable to find a connection between what I try to introduce and connecting that to what is available in science for me to connect too. Let me try and once more somehow locate a connection between what I suggest and what Mainstream science find acceptable.

Material seems to get glued to the Earth by some force, which holds the name of gravity. Moving such a gravitated particle needs some drag by motion. The secret of lessening the effort in applying motion to the object in need of shifting is reducing the drag that is not being dragged at all but is all about motion that is not in motion and hiding behind the name of mass. Let us find this drag in nature and work from there to find a better natural understanding of being in a solid state on ground or a liquid flowing down into the ground or a gas floating above the ground. It is accepted by science that water can rub together and form static electricity. Those Super-Educated teach us that that is how lightning is generated and the best of all are that those Super-Educated then sit back and feel pleased in sharing their vast insight and wisdom. Can you believe respectable men say this while still feeling blameless about their view and acting so absolutely shameless!

The scientist that thought this one up had some or other big problem with his hair and thought that rubbing his hair with a plastic comb will be the same as water rubbing against water. I cannot believe that science can indorse such shit and shit it is! How can water form static electricity in vapour in the atmosphere by rubbing water to water because lightning is the product of heat that has expanded to gas then by motion of wind and cloud again concentrate such heat that before this expanded into gas. When the water vapour crystallise the vapour then again condense and form water drops but to condense the water release the heat that kept the vapour apart. The heat in turn then also condense back to heat being more intense in a smaller space as the heat is going into liquid (the most common and widely present example of heat to form liquid heat is in the form of electricity) and comes down by gravity as lightning in the turbulence of heat. The liquid heat forms lightning and as water does, the heat condensation we see as lightning flows down to the Earth in the form of transmitting electricity. The electricity we refer to as lightning has more in common with wind than it has with electricity but that I explain on another day in another book… Let's remove gravity and find mass.

It is well documented that heat turns to space and space becomes more than what it was before the space which the matter is occupying before the event came about when the heating formed additional space by using a method we gave the name to as exploding … While the action is a well documented fact for many, many centuries Mainstream science to this very day never realised space creating and exploding is directly or even indirectly connected!! By heating material there is an introducing of space because the space needed after heating becomes more than what the space would have required before the heating started. By heating the material with a sudden burst will bring about so much space available it brings along the destruction of matter in the position and form it holds. This destruction we know as an explosion. Because culture gathered through many centuries of steady science development left us name we use for the process of exploding as an inheritance package. When the explosion occurs there is abundance of heat released and by turning the heat into space (which they then call shock waves as if that serves as all the explain the action requires).

The advancing of the newly formed space reconstructs the position layout the matter holds in all the immediate surrounding space. With the knowledge and countless demonstrations brought about by war and other destruction, this knowledge is edged into our minds to the same extent as getting dressed or eating. With everything having an opposite and a counter action the opposite must also apply. There has to be a removing of heat from the condensed space by the condensing of space In that case the reduction of space bringing about the forming of condensed heat, concentrated by a removal of space as the reconstruction of matter. The removing of material from one side and replacing it onto the other side By removing space from space in concentrating space such space that is removed materialises the condensed space to liquid heat and then further into solid material by way of the electron condensing space further than what the atom cam achieve.

The electron serves the liquid neutron as a gateway into the Universe of the atom. Where matter removes heat from uncontrolled space to reconstruct its element worth and element position in value, the reconstruction is the direct opposition to the deconstruction of matter by explosion, therefore the relevancy changes and with the relevancy changing the result therefore must become reversal to the explosion. If the Big Bang was antigravity in exploding of uncontrolled or compromised singularity, then gravity is the constructing of un-compromised singularity confirming the compromised singularity onto the realm of the un-compromised singularity and extending the un-compromised singularity to the devastation of the compromised singularity. In such a manner space-time does not waste but is re—affirmed by control.

In outer space an object floats. On the moon nothing solid will float but there is no liquid air either. Those not familiar with this statement must think about what will be the difference there is in outer space of space in outer space and space in the atmosphere of the Earth and why objects entering the atmosphere suddenly acquire the ability to heat up and burn out. The reason why outer space is a gas is because the density applying in the space in outer space and material holds very little liquid material in comparison to what we are use to in the Earth atmosphere. Outer space is not colder but hotter, much hotter. In space all objects are very loosely connected and move quite freely about. When this occur it reminds one of a gas because in a liquid there is much more density in the matter relation and when solid the matter is as close as can be found.

Density comes as a result of a cold environment (not a cold atmosphere) Therefore the conditions in outer space form a gas and a gas is the hottest of the three conditions there are available to substance material. Comparing the likeness with anything we can compare too with our vision of what is on Earth, we must move to something we all consider to be a natural in all three forms, one being solid ice (very cold), two being liquid water (less cold) and three being gaseous steam (very hot). Conditions in outer space come down to steam because there is much space between the particles bringing about more space and less in the density in the space bringing about that there is a lot material. In the following example I use water as the subject because of every person's familiarity with water however the example rings true with any substance we may choose to inspect. By introducing heat to water, water changes from being a solid we call ice where there is much more material in the ratio between space not filled and material filling space whereas with a liquid substance such as we call water there is slightly more space unfilled between the solids than there normally is in other solids being apart but much less space as there is in the substance we call gas and the gas form of water we call steam or vapour. The scenario does not apply directly that much to water but I hope I'll manage to bring the point I wish to make across being what fills the space which fills and what constitutes what the solid / liquid / gas relevancy there is. The state in which material is in form being between material molecules is not written in rock but can change as situations change. It is space that is filling the entire overall space in relevancy to denser or less dense space filled at that point sharing space-time.

By introducing heat to water we change water from a solid (cold) to a liquid (less cold and more hot) and with the introducing of much more heat we get the heat to become a gas such as it is in outer space. By introducing even more heat to water we get clouds forming. By introducing even more air heat to air we get clouds moving, and with more heat added it is moving excessively, where the movement in fact displays a density increase. The motion provides a density release of sorts.

There is a density increase in the atmosphere during the storm and therefore the wind can then uproot large trees. Hail falling in whatever intensity can strip leaves but it cannot uproot trees and rain falling cannot break tree trunks. Yet those responsible for deciding what the scientific know-how must be and other superior members in Academic circles holds the opinion that air born particles such as oxygen, nitrogen, hydrogen, helium which are all extremely mass less particles, can have a density of superior such means that by wind blowing the particles the particles will collide with the trees and the intensity of such collisions between the air born and air driven particles will gather sufficient momentum as to remove the tree in totality of trunk, branches, bark and all from the soil holding the roots of the tree.

To suggest that something as light as say oxygen and nitrogen can blow down a tree with the quantities present in such a density as one find in the space we call our atmosphere proves how little science are able to think! With more increase of density in the wind we find spiral motion adding to the in lateral movement. With wind circling it has terrific density because in such a form it not only uproots trees but also takes on houses and much of what man can build. The wind in access blowing extensively produces the same qualities of producing destruction damage than water in a river cause the flooding that the water in the river can match. When the density increase by adding motion much of the increase goes along with vapour, that is a form of air that is thick with water, (I distinguish between the terminology because why not only use one word, either steam or vapour. After all it is the same thing!)

In clouds we find lots of vapour but we find little water. The difference between water and vapour is that vapour has more unoccupied space and less space filled with water material in ratio. The thick density is there, but the air is so thick the vapour and the air combines to form a gaseous liquid we can see as a cloud. Remove the heat in the cloud (which the cloud needs to be if it wishes to be being vapour) the water returns as rain or as hail. On the other side the heat separating the particle excels in motion that becomes space also turns liquid by motion of space and the space forms a blanket that (by motion forming space) it may become so dense as to rather remove the tree with all the tree holds than to allow space between the particles to part. We also see such dense liquid space remove as a liquid and produce a form of heat we named lightning,

The heat that I refer to is different to the idea we have about the heat that makes you sweat but a much more concentrated in spin value such as electricity has liquid heat. It is all part of Kepler's formula where space density is the product of the motion and space-time is all about spinning motion. The liquid forms when the heat concentrates with a much more formidable spin that produce a higher electrical charge and comes as pure electricity in the manner it is produced. As the liquid heat we call static electricity contracts in becoming denser in measure that uses less space the static receives motion and the motion becomes lightning. The heat is forming static electricity because of the lack of motion present and that static is in between the vapour that liquefies forming electricity which is just plain old liquid heat separated from the vapour water and by letting the vapour form a denser water drop and allow the water to separate from the from the liquid, the water forms a specific density because the motion that lets the water vapour crystallise to water. Then in that case and being next to more vapour the water forms the solid although in our mind set water is a liquid but still... it then is a solid in the form of water we named rain. The liquid which was the water did not vanish but became liquid air which we call lightning. The question about density increase always comes from material being more prominent and more abundant in such a space. With the increase of density it always accompanies the increase of heat and the discharging of heat.

If one would think that it is vapour in the wind that increased to such extend that the wind can uproot trees, then why does hail with such a lot of solid water not uproot the trees. In order to judge the logic try to imagine how much water is needed that is converted to spray and is forming a steam or vapour. Then think how much of that is needed to form a substance with the veracity that is required to uproot a large tree? There is a world of difference between windstorms and hailstorms because of the abundance of electricity or more bluntly phrased heat in spinning motion. Windstorms having the ability too uproot the trees have a very sticky substance between the molecules and the more sticky evidence there are, the bigger the ability to cause damage. This sticky substance can only be liquid air or liquid heat. The air substance shows a bigger resistance to part or create space than the tree shows its willingness to remain secured to the soil by its roots. The substance can be sticky to the point where it breaks braches that will require an effort of many hundreds of Newton meter to break.

Even spraying the tree with water which man created artificially by pressurising the water would hardly break the branches and even less hardly uproot during windstorms the tree. If it is done, the water flow will be enormous but I have seen many times trees uprooted without one drop of water visible. The only logical remaining supplier of such a density increase must be the heat in the wind becoming denser. In this there are changing relevancy dynamics, which I then introduced as equal to the substance found in atoms. Let's look at relevancies applying according to singularity distinction.

By accelerating gravity such acceleration can reach a point where space-time diminish to the size of a photon as time accelerate to the speed of light. Establishing a Coanda effect basis for singularity becoming established in an iron confinement where copper dismiss the space –time electricity changes as gravity is in the atmosphere of the Earth to what gravity is where the Earth dismiss space-time in the inner core.

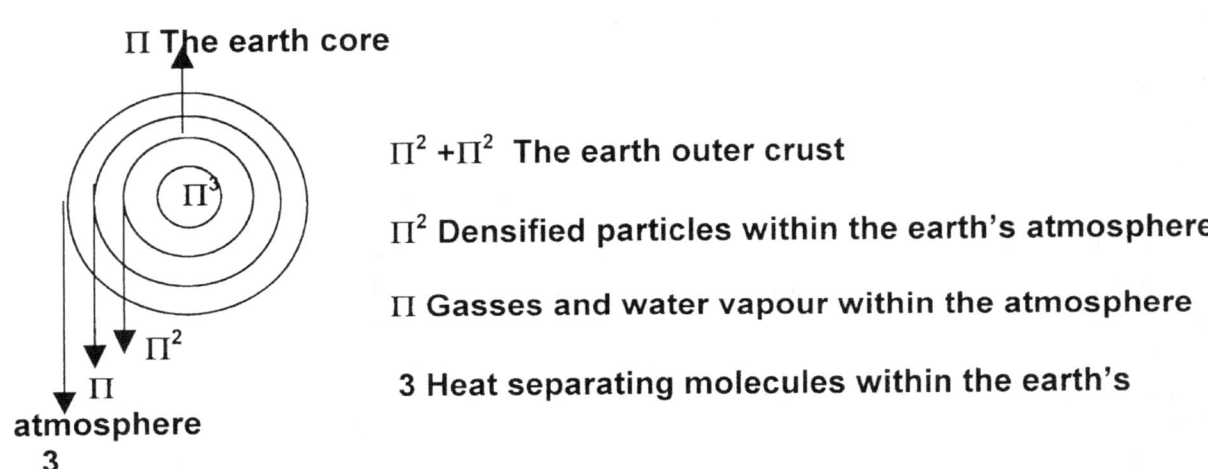

$\Pi$ The earth core

$\Pi^2 + \Pi^2$ The earth outer crust

$\Pi^2$ Densified particles within the earth's atmosphere

$\Pi$ Gasses and water vapour within the atmosphere

3 Heat separating molecules within the earth's

$\Pi^2$

$\Pi$ atmosphere

3

When a structure is part of a solid it holds the value of $(\Pi^2 + \Pi^2)$ which is the proton number and that number I dedicated to the solidness of the soil and everything within the soil structure. The formation of $(\Pi^2 + \Pi^2)$ is uninterrupted from the centre base of $(\Pi^3)$ which we find in the very centre of the Earth through all the way to the solid top of the Earth soil. Everything attached but not being part of the Earth where the Earth then holds $(\Pi^2 + \Pi^2)$ and forms another separate value but is still resting on the Earth soil nevertheless I gave the bottom value of the neutron in as much as forming $(\Pi^2)$.

The Earth holds a proton value of $(\Pi^2 + \Pi^2)$ and the structure that is solid but has an individual and independent solid form that is unattached to the Earth carries $(\Pi^2)$. Being $(\Pi^2)$ the object is in the realm of the solid but which through possible motion still has find independence from the Earth through achieving separate motion and construction. This would be a rock being loose from a mountain or a tree or even life being on the ground or then the proton part valued at $(\Pi^2 + \Pi^2)$. When it is not struck to the ground and has an ability to apply motion independent of that of the Earth, which makes it float in the atmosphere I gave the second neutron value of $(\Pi)$.

Water vapour in a cloud formation will be $\Pi$ but when forming rain in condensing to a semi solid the water drops will then become$\Pi^2$ by becoming relatively solid to what it was before. As $\Pi^2$ it will move to $(\Pi^2 +\Pi^2)$ which is the ground. This has all to do with space-time duplicating versus space-time dismissing. When the object is in total suspense and away from the Earth all the time being part of the atmosphere such a particle is valued at the electron value of 3.

Water forming vapour will have a relevancy of 3 to $(\Pi)$ where 3 would carry the relevancy value of the electric charged air and $(\Pi)$ would then take the relevant value of the vapour. When the vapour condenses the heat will turn to liquid as lightning and change the relevancy of heat from 3 to 3 $(\Pi^2)$ where then the water will become a solid form of liquid being $(\Pi\Pi^2)$. One should take note that relevancies can change as quick as motion would allow such change from where liquids can be as dense as solids $(\Pi^2)$.or being a liquid $(\Pi)$ or a gas (3). It indicates position and is not a tool to use as calculated formula or as measuring devises.

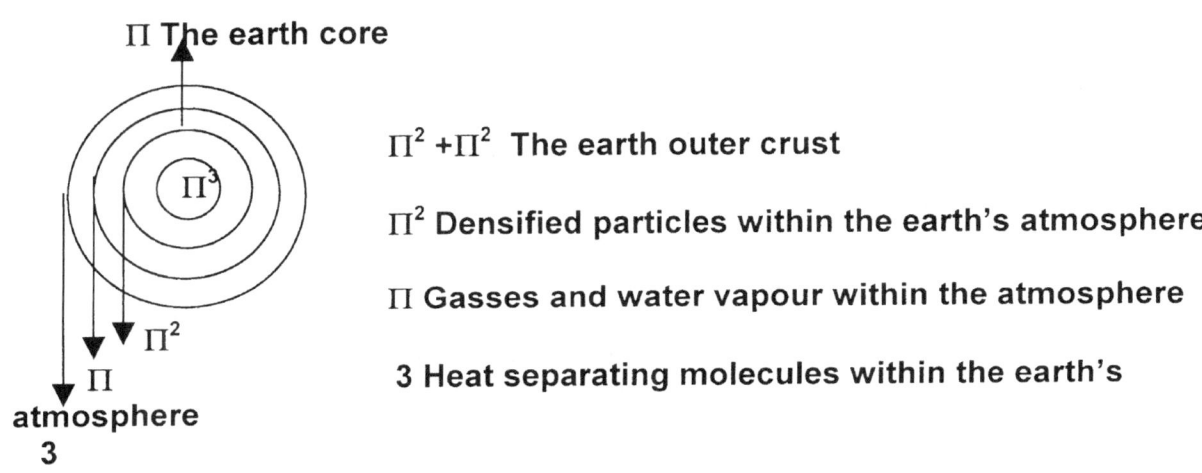

$\Pi$ The earth core

$\Pi^2 +\Pi^2$ The earth outer crust

$\Pi^2$ Densified particles within the earth's atmosphere

$\Pi$ Gasses and water vapour within the atmosphere

3 Heat separating molecules within the earth's

$\Pi^2$

$\Pi$ atmosphere

3

Changing from solid to liquid or to gas is re-establishing a relation between particles such a relation is also between a solid proton $\Pi^2 +\Pi^2$ or a liquid neutron $\Pi^2\Pi$ or gas 3. In each dimension, the light displaces space-time that is more concentrated and therefore influences the projection path of the light's future position in space-time to unoccupied-, occupied-, and densified space-time. Motion is what causes gravity. Motion is the contributor to gravity or the replacement of gravity and motion comes about when heat concentrates in space. In every development it is possible to see what implications in changes came about where the Roche principle was either $(\Pi / 2)^2$ or $(\Pi^2 / 2)$. The Roche limit is therefore a joining space or a separated space.

The very first of every aspect that became the Universe started when the very first action came about and that action we now call the Roche limit. As singularity expanded motion brought about a relation in singularity that reflects a value of space shared by motion singularity shifted from one $\Pi^0$ to innumerable formations of $\Pi$ just by applying motion by overheating and gravity resulted in the motion of contracting to form $\Pi^2$where neighbouring $\Pi$ starting a rotating relation and all along $\Pi$ stood in for the values coming about. But because the space shared is about seven points standing in regard to ten points over the Roche factor the square of $\Pi$ formed the divide as well as the joining of such a divide But at the time and even the represent we find where singularity forms space –time space is very much at a premium therefore singularity is halved.

The discoverers discovered all the singularity that forms a unit also produce a singularity taking charge of the combined effort and this yet again is named after the discoverer because no one at the time had the foggiest notion of what. In the case I just mentioned the discoverer was an engineer by the name of Henri Coanda. Henri Coanda discovered without him or any other person eve realising that what he discovered is the effort singularity in single produce can remove and form a unit in an allocated centre from where the motion of gravity takes charge to establish a space with boundaries set by motion. In this the Earth as just another cosmic structure is no different and all the atoms in the confining space of the Earth elects a centre wherefrom gravity takes charge as space is dismissed. At such a point singularity-forming gravity shifts the collective singularity to govern by producing gravity at $\Pi^2 = 9.8$.

This position is what Newton saw as cosmic gravity but cosmic gravity it is not. It is a bit of localised gravity that has no cosmic importance at all. But that is not where the centre of the Universe is at…no it is in the centre of any product of singularity that charge space-time and by charging space-time it calls upon the use of the Roche limit as it introduce contraction in $\Pi^2$. But since gravity is motion only half of such gravity takes charge on any particular side of the Universe and since the Universe we discuss at this point is the accumulating effort of all the centres of the Universe that group together to form the Earth. When a body is within the atmosphere of the Earth half of gravity applies as $\Pi^2/2$ and between separate bodies the value is gravity which is time divided by the factor forming time which is four $\Pi^2/4$.

There is another system called the Lagrangian system where a centre aligns with 5 other points to form a six-pointed relevancy.

LAGRANGIAN POINT is another result flowing from the singularity position and most directly coming from Pythagoras principal and connecting that most basic mathematical in conformation of Kepler's formula.

As singularity parted two very much opposing factors rose from that action where each formed a relevancy in relation to the other as well as a relation to the third factor which was singularity controlling the positions. The two particles that came about in a fight for space was round and the space required a square to fill. In one case the round could remain round as stay solid by applying motion in moving about an axis and in the other case the other one had to apply motion by expanding and relatively grow softer as it deformed and lost shape. There were many more than the two that stood against each other forming the required motion but only two points in singularity apply a relevancy matching each other. Each one in such a confrontational position had to form a match between two particles where the two were one Universe away as two separate Universes matched.

The one was in a position to apply gravity and kept the form remaining true to structure and shape at the expense of the other where the other miss formed by overheating. The one factor, which remained solid, we now call material and the other got the name of plasma but to spare confusion I use the term heat, because all things are either one of two forms of heat. The one form is heat frozen solid in a unit of heat and the other is liquid in a fluid heat. But in all events it is heat-matching space that is holding the heat that is in a unit, in another unit. The softer compound became cannibalised by the solid firmer structure and demise came to the liquid singularity in order to save the solid singularity a sure fate of destruction. There are today two factors, which still is in opposition still as relevant as the first motion. There is heat and there is matter, which is a Universe apart although they keep a position next to one another. The crucial factor is the relation every element of every atomic grouping has is the form it holds by number of protons grouping in clusters. Being secured by an electron outside, a neutron fluid and a proton solid centre is the characteristic all atomic elements hold.

With the electron, neutrons and electrons in equilibrium there is a matching of representation in the unit it is in holding. The cluster forming is by itself the unit into another unit that is much more securing to form. The form Lagrangian layout plays a cardinal role but is far too time consuming to explain at this point however I do explain it in my book Matter' Space In time : The Hypothesis. The relation the atom form in cluster grouping of the protons absolutely dictates the relation the atom accepts with the heat surrounding the atomic structure. The Lagrangian proton layout will eventually dictate heat/ material relation applying in every case of every structure of every atom. Since any star is a cluster of atoms, and any cluster of stars are several groups of atom clusters, the groups of heavenly bodies will assume the characteristics enforced by the characteristics of individual atoms.

Kepler brought in a formula that came through the dedication of two lifetime studies and the formula read that space is equal to time in motion. Mathematically it reads that $a^3 = T^2 k$. This formula brought Newton's claims into dispute because what this formula said was that geodesic space is not confined space and while laws apply in the confined space of Earth it does not apply in geodesic space. From Kepler one can see the precise moment the Cosmos started. It is Mainstream science that is hindering the accepting of the formula that is dampening the human understanding of the Cosmos. Kepler said without saying that the first moment was when singularity $k^0 = a^3 / T^2 k$ produced space through motion and we know motion is the product of heat releasing space. That is what Kepler interpreted when he declared that $a^3 / T^2 k$. Understanding the cosmos largely includes our realising that heat and cold is the very same thing. It happened at creation when heat brought about creation. The moment, the instant heat came about cold also was invented and the maximum heat there was, was also at the same instant the minimum cold there was. By creating heat it created cold that separated cold from heat. The birth of heat produced the birth of cold because the same thing was born at that instant. Heat and cold is the same thing bringing borders we humans' whish to create but such creating only furthers our inability in performing worthwhile understanding. But like planets heat and cold are human creation and have little interpretation in actual cosmology. Our being part of the Universe depict us with a realising of borders and boundaries but also prevent us from seeing the full picture as observers of the Universe should be.

When did the cosmos start is the question every one is in search of since Biblical days. Even at present the remaining unanswered question is this question every race and every culture is in pursuit of. I would say it had to have started with space but science has the cosmic time linked with time. The truth is that there could not have been space without time and there was no time without space. The first moment came when $k$ moved a way from singularity $k^0$ to establish space. Creation was when $k^0$ introduced individual entities $a^3 / T^2 k$ and went about moving the different entities apart. It was when the confinement was broken and particles appeared for the first time. Understanding this comes hand in hand with accepting that there is two forms of structures other that elements confined in an atom.

Elements being in liquid is volatile because of the motion such elements are capable of. They are volatile because of the high capability they have in duplicating. Being volatile is requiring space to move about. The heat links directly with the density and the density links directly with the motion or capability of movement. To move extravagant requires space in which to move veraciously. Being solid is being frozen which means all the heat separating the elements has gone vacant. There is no more heat left simply because there is something inside the atoms of the elements which is removing the heat from the outside faster than the outside can replenish the heat on the outside which is lost to the heat absorbing inside. It is when looking at the scenario in that manner that the elements start to make sense

| | | |
|---|---|---|
| **Hydrogen 1** | melts at $-259^0$ C, | boils at $-252^0$ C, |
| **Helium 2** | melts at $-269$ $^0$ C | boils at -268,$9^0$ C |
| **LITHIUM 3** | melts $180^0$ C | boils at $1300^0$ |
| **BERYLLIUM 4** | melts at $1287^0$C | boils at $2770^0$C |
| **BORON 5** | melts at $2030^0$ C | boils $2550^0$ C |
| **Carbon 6** | melts at 804 $^0$C | boils at $3470^0$ C |
| **Nitrogen 7** | melts at -$210^0$C | boils at $-195.8^0$ C |
| **Oxygen 8** | melts at $-218.8$ $^0$C | boils at -$183^0$ C |
| **Fluorine 9** | melts at $-219.6^0$ C | boils at $-188.2^0$ C |
| **Neon10** | melts at $-248.59^0$ C | boils at $-246^0$ C |

Melts (meaning that the element becomes a liquid) at $-259^0$ C. **Hydrogen boils (meaning that the element then becomes a gas) at $-252^0$ C, This does not concern the basic atomic element but only applies to the space on the outside of the atom**

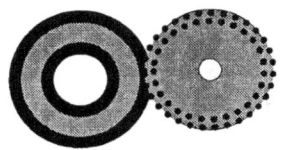

**The heat surrounding elements is not a match that fit all and is a separate issue in every case with every element. The heat /space /material relation gives elements their characteristics they are recognised by. Such characteristics changes as the heat /space/ material relation changes and alters the presumed characteristics**

This is the matter and the antimatter that science is referring to when science claims the devouring of matter by anti-matter, according to my opinion. Some particles became matter and others became antimatter or plasma or heat or just what you wish to call the by-product that came about from the friction that brought about the heat that led to the expanding of the space. Those forming units we call elements had established characteristics relating to space-time where each one holds an individual identity according to the number of protons the element has in one cluster unit.   There are elements. The elements presume in the role of being solid and form as units the solidity of solid materials. Then there are liquids. Amongst those elements mentioned, we regard some of them as being natural "liquids" and others being natural "gasses", however they are solids notwithstanding what our culture call them. There are those we consider to be mainly gas or liquids and only then there are the state of being "frozen gas" (as we regard outer space to be) but such presuming underlines our mistaken culture we have.

The two forms I have just mentioned being gas and liquid are the same and the only difference there is, is the state in which elements can find form being between the two forms of having liquid/gas between elements or not one of the two which in that case we then call the state in which the elements are being a state of solids. The form of being liquid or gas or even solids alternates and the changes come about with more or less heat being part of the density factor. The form of heat being more (gas) or less (liquid) or absent (solid) establishes and reforms the state in which the element in cluster together. Nevertheless liquid and gas is the opposite of the very same thing that came about before the Big Bang event took place and represents a period at the time the cosmos was still forming the second form other than atomic solid elements.

The forming of liquids came from the friction between solid elements that were in a position without unoccupied space-time. This forming of a form other than being atomic clusters took elements into having the opportunity of choosing between another forms, other than solid atoms forming clusters we think of as elements. This is a very much identifiable period as it developed in the cosmos as the cosmos developed and went through a specific development period. This period went on up and until a time when the Big Bang took place. The Big Bang is a period after the forming of space-time came about and normal cosmic growth started influencing the development direction of the cosmos.

The Big Bang commenced when some particles went liquid (or softer) which took liquid space and turned a space-less liquid space into a spacious gaseous space b y adding more time per cycle to the relevancy. But elements found in nature are not naturally solid, liquid or gas as the element table wish to teach us. They do not form unbreakable columns that one can take anywhere and it will match everywhere. The element table is a human concept and not a cosmic law that is valid on Earth and then again when thought about it realistically it is not even valid everywhere on Earth.  The heat or space that brings form is an ultra blanket that wraps the element into an enclosure to suit the conditions applying at that point.  We then identify the conditions atomic elements may be in, in such a thin margin that it only apply under certain conditions which is mostly artificially created by humans in laboratory conditions at a very precise pre-arranged heat range under very specifically stipulated environment.. These results will change notably and have a varying outcome when tested in different climate conditions through out locations on the Earth.

One cannot put the characteristics these elements will have and prescribe the same characteristics to conditions applying in the Sun or on Jupiter. All elements are all forms and it depends on the conditions permitting what form will apply at that point. Water boils at very low temperatures on Mount K2 and at very high temperatures down at the bottom of Stilfontein Gold Mine, which are several kilometres straight down vertical. To the Human mindset we form, water must boil at 100° C or 210° F because that boiling temperature suits man and human grew accustomed to that idea. It is as far from the truth as thinking about outer space being cold or the Sun being a hot gas structure.

Forget about the example we find in water because it seems if any one talks about the three materials all concerned immediately thinks about water as one structure that can ice and can boil when it is not flowing. That is not even an example because water is not a true substance but is a compounded combination of volatile elements forming the most outrageous concept the cosmos could think up. By combining some of the most volatile elements the cosmos created the least volatile substance known to man, which is water, that in the combination $CO_2$, comes to from in that combination, a substance killing fire. But the three forms water personifies is the structure compiling the Universe and prove stages of development. Water on the moon is ice. Water on Venus is not even a thought that will ring true and water  has never been on the Sun. Water is to us a vital means having an ability to sustain life, which to our concept is to sustain the most critical of just about all substances.  We even think of water as only in the way it is used as fuel to get us away from the future possible planet we might be visiting As long as we create a Universe with life as the centre of all focus we will find a miss fitting Universe we can never understand and what ever we think will never apply anywhere but on Earth. We have to start our focus on    singularity in charge of whatever space-time we investigate. In the beginning there was singularity. That makes singularity the centre from before the beginning of all times. Singularity is still the centre on which our focus must be. Singularity came first but also staged the centre.

From singularity came material as a solid substance. Some rubbed against others forming heat in some cases. From elements disposing form comes liquid, a softer substance that will allow some form to match other less flexible forms. Then comes gas when singularity loose control and turns space to liquid, which loses density as a result. Again we can see a pattern coming about. But it involves Kepler more than any Mathematics. It puts distance $k$ in space at point $a^3$ between $k$ and $k^0$, which provide distance $k$ with two points bringing about time $T^2$. $k^0 = a^3 / k\, T^2$. Once Kepler showed relevance existed no one could denounce such a relevancy. Not even Newton could because relevancies in mathematics show clearly an interaction that changes and reapplies. Yet Newton did by putting $k$ at a value of $T^2$ and put that at zero.

A circle has no point serving as the start and therefore there is never an end and cannot form zero. Singularity brought about $\Pi$ from the instant when the point took up more space than what space was available. The space less ness brought $k$ into an equality of $T^2$, which was then equal to $a^3$, and that equality still performs as 180°. The result was that $\pi$ formed and $\Pi$ has no particular star or end. It cannot show a start because there is no ending spot to singularity. In all cosmic wheels there is no specific point that indicate where the circle motion first started at any given point. Therefore after one cycle it may end precisely at any point where the immediate start starts. All starting points or ending points are human concepts but has no relevance on the truth of the state of affairs where we wish to divide our life in seasonal changes. There are only relevance's reapplying as motion contribute to cyclic reoccurring. Therefore there never can be a cyclic zero as Newton believed. All singularity would form a duplication of another singularity that forms a ring as relevancy to the inside and a relevancy to the outside but the value remains $\Pi$ at all points in all stages .

Singularity forms a divide $\Pi$ that is sharing value since there will always be a possibility of yet another line in the realms of singularity lying between the two lines in question. By reducing the size infinitely to either side of the divide is what we humans create. Boundaries therefore are human and as man made substances it does not belong to the cosmos outside the influence of man and must be discarded. The understanding of this insists on one vital precondition. We have to abolish our instincts about things being big and small, high and low, hot and cold, tall and short. All the human measures are truly unfitting to the cosmos. The biggest of whatever there is in the cosmos is beyond human detecting and will be classified according to human standards as nothing because in human standards we are unaware of the largest there is. The same goes for the smallest there is because man will never even almost have any ability to detect the smallest in the cosmos.

The biggest and the smallest are human inventions. Those are measures made by man. In the cosmos the tiniest object is the key to that particular Universe. By going smaller one is going bigger. By going within the atom we go outside and meet the Universe. In singularity we find the cosmos because every singularity not only represent the cosmos but also in fact is that Universe. No mathematics will ever measure the thickness, because as $\Pi$ is standing still and it cannot have a width at all. The moment a width appears which one can measure or calculate, the line will become part of the factor forming the divided and not the divide. The instant when space connects, the spin direction will produce the partisanship of space and spin. Any form of space including even in the most minute will produce a favouring of direction. Through such motion gravity comes about because the motion is gravity and that produces the time aspect $T^2$

thus thereby changing the direction by rotary motion. The minimum is $\Pi^0$ going onto $\Pi$. It is the spot forming dots.

The moment there is an area there is a measurable rotating brought about and no longer a non-interfering divide that is there where $\Pi^0$ forms $\Pi$.. Such a line holds space in a position that runs far beyond the boundaries and limits of the three-dimensional. Another factor of such a line would be that the radius (let us substitute the radius r with the using of Kepler's **k**), **k** would be immeasurably small. The factor **k** cannot be zero because infinitely close to that first **k** is the start of the third dimension where time plays the part as the fourth quarter. The presence of **k** is undeniable and recognisable yet it is not visible. The fact that **k** is there albeit stripped of any influence disqualifies it from being zero and therefore not being there. With **k** already beyond any measurable space, leaving $\mathbf{a}^3$ as a factor of one and not being able to pin any volume measure to that one **k** will have to be to the power of 0 being $\mathbf{k}^0$. It is exponentially zero but zero it is not. In Kepler's formula $\mathbf{a}^3 = \mathbf{T}^2\,\mathbf{k}$ the area $\mathbf{a}^3$ would be one because of the dimensional non-existing of measured sides in any direction. If $\mathbf{k}^0 = 1$ and $\mathbf{a}^3 = 1$ the only alternative $\mathbf{T}^2$ could possibly have is also one. The factor of $\mathbf{T}^2$ identifies the time in the formula and when the formula indicates time as one. The time component must therefore be eternal. Only time in eternity does not change. In all human calculations an end as well as a beginning has to enrol the whole concept being calculated and when that is not possible the calculation as well as the way in which it is calculated becomes suspicious.

I know somewhere there is vast numbers of highly intellectual mathematician that will disagree in all sincerity about my next statement but nevertheless I have to state it. Mathematics has to use a beginning and an end to bring about coherency. That secures fixed values and implying numbers in relation to determine size differences or quantities. However since that may be and however clever it would make the mathematician appear, it is an irony and it focus on stupidity. Such quantities and numbers are only deceptions and have no place in the cosmos since there is no size differentiations or numerical number quantities to calculate when applying the cosmic standard rules. Gravity bends light is a statement very synonymous with Newtonian cosmology. To determine the gravity such determining formula requires a diameter. To find a diameter of a star one has to measure such a diameter by means of looking at the light the gravity bends the light so no true measure can be reached by using the bended light to measure the gravity, which is bending the light in the first place. One cannot use an unreliable measure to measure that, which is enforcing the unreliability of the measure. In that case what Kepler introduced makes most sense of all. Such was the case with Kepler's findings where each provides separate applicable relevancies and Newton's dispute thereof. Newton saw no reason except for his denunciation of what he thought was Kepler's inferior mathematical skill in Kepler's choice not to use the $\Pi$ while Kepler never thought of $\Pi$ because the cosmos never gave $\Pi$ in the formula whereby the calculations were done. But behind the cosmos not including $\pi$ naturally is a Universe filled with reasons.

Newton observed a need for the more commonly used mathematical formula applying when the calculation of the sphere comes about is $a^3 = 4/3\,\Pi r^3$ where it places one third dimensional but lesser factor in direct relation to another third dimensional relation and all that in relation to the form that is applying. By using the normal mathematical equation $a^3 = 4/3\,\Pi r^3$ it is definitely not what Kepler said when he produced his formula. The square Kepler allocated to time and by the same margin allocated the cube to space in the same formula. This means of mathematics does not find representation in the normally used mathematical expression of $a^3 = 4/3\,\Pi r^3$. In mathematics we find a deliberate lack of the time factor when using the equation $a^3 = 4/3 \ \Pi r^3$. This must prove the differences in what Kepler found to express and what Newton thought Kepler tried to express.

However when using Kepler's manner of measure in calculations concerning the cosmos there is no criss-cross matching of dimensional accumulating. If I am reading the situation correctly Newton saw Kepler's mathematical skill somewhat below the dimension of Newton's genius and that spurred Newton on to bring changes to Kepler's formula but in doing so many things went missing through incorrect translations and miss interpretations of the use of such a genius. Kepler was not referring to a mathematical space on a flat sheet of paper, which was what Newton saw. Kepler produces a value linking space to gravity being time of space in motion in as much as calculating space in the geodesic forming a motion and measuring time in the geodesic of space in motion and bringing about factors in formulas through figures that geodesic space informed Kepler about. It is the motion or time part Newton did not notice.

If Kepler was mistaken then the Cosmos was wrong about the cosmos and if Newton was incorrect it was about mathematics that Newton then was wrong about. Kepler places time in the square directly in relation to space in the cube in association where time shows two distinct qualities. But thirdly this relation applies to a second space flowing in contrast and securing a three dimensional cross point that gives every aspect in individuality. That is the part Newton missed. The one factor is time in the circle rotating while the other is in the linear or the straight line implicating the position that the other would have. But in all cases the straight line responsible for the factor **k** will lead naturally into a circle $T^2$ and the circle forming $T^2$ will grow into forming a larger **k** which will become a larger $T^2$. The cycle shows no end and that disqualifies mathematical calculations which demand a specific start and a specific end. Any starting point demands motion because the space that requires identifying can only be identified if that space is in motion. This proves that time or gravity (with time and gravity being the same thing) is the method that the Universe uses to control the dimensional growth of the Universe. It is $k = a^3/T^2$.

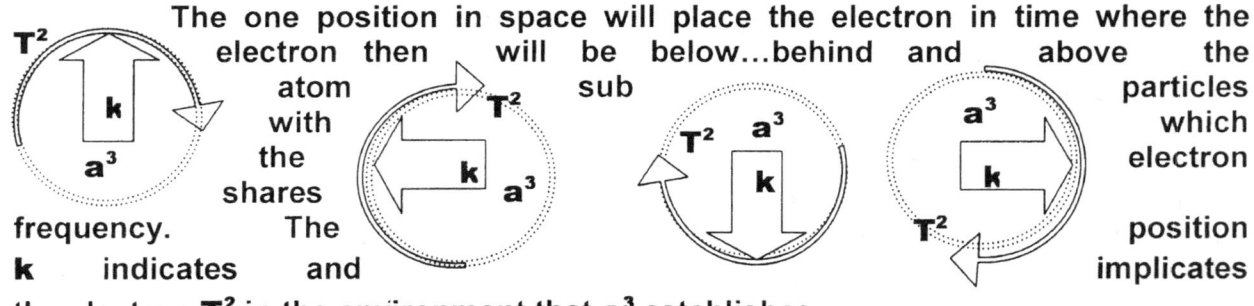

**The one position in space will place the electron in time where the electron then will be below...behind and above the atom with sub particles which the shares electron frequency. The position k indicates and implicates the electron $T^2$ in the environment that $a^3$ establishes.**

The space we find between the planets is representing the space we find between the Sun and Oord cloud. It is the same space and what we find between the planets we will find way beyond the planets all the space links to the Sun centre $k^0$. That space we will find outside the solar system in the regions between the solar system and the next star because it is the same space and still is controlled by the same centre. That space is filled with the same filling and it comprises the same substance. To see Kepler's finding only applicable to the Planets is so Newtonian as thinking of gravity the ability of a towing rope that can be used for pulling all innocent material with an invisible rope. Space is everywhere the same and holds the same value because space started out being the same substance everywhere. That is what Kepler introduced. Tycho Brahe and Johannes Kepler were the firs cosmologists.

Official astronomy defines Kepler as follows being The German mathematician and astronomer KEPLER, JOHANNES (1571-1630)

German mathematical and astronomer became Tycho Brahe's assistant in Prague in 1600 A. D. where he undertook to complete the tables of planetary motion Tycho had begun. Kepler first calculated the orbit of Mars. He spent much time trying to reconcile Tycho' s accurate observations of the planet with a circular orbit, but concluded (in Astronomia nova, published in 1609) that Mars moved instead in an elliptical orbit. Thus, he established the first of his laws of planetary motion. A theory that the Sun controlled the planets by a magnetic force led him to the second and third of his laws, which were published as part of his treatise on theoretical astronomy, Epitome astronomiae Coernicanae (1618-21). The Rudolphine Tables (named after Tycho's patron, the Holy Roman Emperor Rudolph II) of planetary motion appeared in 1627 and were still in use in the 18$^{th}$ century. Kepler also wrote De Stella nova, on the supernova of 1604 and Diptirce on optics and the theory of the telescope. The overall view followed in this book **Matter's Time in Space** places the true significance of his work in true contents. In KEPLER'S EQUATION is the equation that relates the eccentric anomaly of a body in an elliptical orbit to its mean anomaly. The equation is $E - e \sin E = M$., where $E$ is the eccentric anomaly, $M$ the mean anomaly, and $e$ the eccentricity of the orbit. It is important as one of the mathematical relations enabling the position of a planet about the Sun, or a satellite about is planet, to be calculated from the orbital elements for any time. However this only relates to the solar system, and KEPLER'S LAWS only apply in the contents of the solar system.

The three laws governing the orbital motions of the planets, discovered by J. Kepler is as follows: The first law states that the orbit of a planet is an ellipse with the Sun at one focus of the ellipse. The second law states that the radius vector joining planet to Sun sweeps out equal areas in equal times which as it says refers to time and not the circle. The third law states that the square of the orbital period of each planet in years is proportional to the cube of the semi major axis of the planet's orbit. The first law gives the shape of the planet's orbit; the second describes how the planet must continuously vary its speed as it follows its orbit, moving fastest at perihelion and slowest at aphelion. The third law gives the relationship between the planets' average distances from the Sun and their periods of revolution. Instead of placing, the true value to Kepler's laws I. Newton placed his own interpretation to Kepler's laws, and in doing this, he wilfully destroyed the principle working of the Creation.

Through Newton's tunnel vision, he applied his own miss interpretations to the correct presumptions of Kepler. Newton reduced the implication that Kepler findings hold by introducing to the law of gravitation. He then went about and changed it to three laws of motion. I. Newton generalized Kepler's first law, verified the second law, and showed that the third law should be amended to the form; $4 \pi^2 a^3 / T^2 = G (m + m_p)$. In this, the value of T and a are the period of revolution and semi major axis of the orbit of a planet of mass $m_p$ about the Sun of mass m, and G is the gravitational constant. The major aim of this book is to correct these misgivings of Newton. I shall return to the statement about $4 \pi^2 a^3 / T^2 = G (m + m_p)$ In all instances of measuring the distance the orbit travels around the Sun as the space displaces or space covered by travelling in the time it is covered and dividing such a ratio one find the distance of the orbiting object from the Sun the in relation to the other factors form one or very close to one.

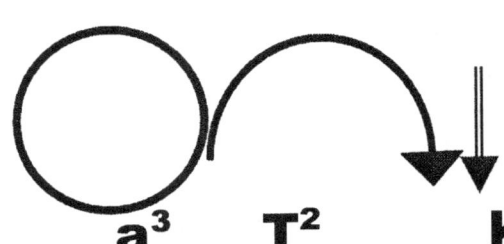

$a^3$     $T^2$     k

What Kepler saw was more of a dimensional nature than the practical mathematic symbols and values. On the one hand was a value to the third dimension, which equalled two-dimensional values one the second dimension, and one to the first dimension.

| Planet | Period years | T | T² | Distance | Space a³ | Ratio |
|--------|--------------|---|----|----------|----------|-------|
| Mercury | 0.241 | | 0.058 | 0.39 | 0.059 | 0.983 |
| Venus | 0.615 | | 0.378 | 0.728 | 0.381 | 0.992 |
| Earth | 1.000 | | 1.000 | 1.000 | 1.000 | 1.000 |
| Mars | 1.881 | | 3.54 | 1.524 | 3.54 | 1.000 |
| Jupiter | 11.86 | | 140.66 | 5.20 | 140.6 | 1.000 |
| Saturn | 29.46 | | 867.9 | 9.54 | 868.25 | 0.999 |
| Uranus | 84.008 | | 7069 | 19.19 | 7067 | 1.000 |
| Neptune | 164.8 | | 27159 | 30.07 | 27189 | 0.999 |
| Pluto | 248.4 | | 61703 | 39.46 | 61443 | 1.004 |

**At the first glance, Kepler's formula seems to be numbers and positions applying between the sun and specific but different planets in the At the first glance, Kepler's formula seems to be numbers and positions applying between the sun and specific but different planets in the solar system.**

In that it is just about numbers but about that it is much, much more than just numbers. The numbers paint a picture and tell a story. It is the only time ever that the numbers first was produced and concluded the formula applying arrived at. In all other case numbers are the result of a conclusion brought about by the use of a specific formula. Kepler produced a formula from the numbers and not numbers from a formula. The numbers brought about answers fro which we fail to ask the correct questions. By seeing a mathematical circle one miss the total picture and the story the numbers tell. The figures explain dimensions working in conjunction and together they combine dimension where the picture behind the story becomes a colour spectacle in comparison the mathematical grey that a mathematical circle produces. . In the case of the solar system it is relevancies carried from the Sun and the Sun is the governing singularity representative for the entire solar system. This is about relevancies applying throughout the Universe and not just locally. This balance is much, much more than what the figures say. It underlines and it explains gravity as a life form indicative of the cosmic life in the cosmos and is very different to what we consider our life to be. In the argument Kepler made he had hide much more facts into one formula than what I think even he realised.

Well, it is much more than that the Accepted Policy Protectors Of Science ever came to realise. He officially formulated space-time, he officially coined not the name but the origins of the Universe being the Big Bang and he was the first to put the speed of light in relation to cosmic development...and all of that with his rather simple formula. He said the space $a^3$ not the circle (a) or the circumference $a^2$ but in the circle $a^3$... where such a circle represent a factor in the third dimension. The formula he compiled was not rather but very specific about the area being a third dimension area and to prove it beyond doubt he placed it in the relevancy of the formula in a ratio of presenting the third dimension in space. He said $a^3$ is equal to $T^2$ **k.** That specifies space as motion. Newton and Newtonians came afterwards and played with mathematical toys as to challenge their mental capabilities but brought little new ideas to the table. Newton introduced a $4\Pi^2$ to indicate the presume circle on the one hand and on the other hand he brought this lot equal to $\{G\ (m + m_p)\}$ which he then presumed to be the general Universal gravity constant (G) and the sum total of the two structure mass. Newton saw a ring circling around a centre having $4\Pi^2$ to indicate such a ring outside a centre and he positioned $\{G\ (m + m_p)\}$ where the two mass factors combine the gravity effort in the general grand gravity constant in space.

I have had so much resistance in the past from all Academics but that is not what I see what Kepler see. With what I saw what Kepler's saw I shall trace that back even as far as to the centre of the start of creation. In their eagerness to calculate the Mathematicians calculated a formula to measure the circumference $a^2$ of a circle being $\Pi r^2$. I have seen an Astro physics examination question paper where they use $4\Pi r^3\ /\ 3$ as the formula to calculate the Sun and other stars volumetric space! Not one mathematician was for one second in doubt about the manner they may interpret the radius with light and gravity bending each other and all that!! They formulated the measuring procedure of the circle being in the third dimension that will show how big the volumetric space is of a sphere at $a^3$ being measured with the procedure being $4\Pi r^3\ /\ 3$. Then some Mathematician and an Englishmen of Importance in academic standings came onto the idea of gravity. Being a mathematician the Englishman placed the Universe at the feet of mathematicians. He saw circles where Kepler saw three dimensions interlocked in an ever intergrading relevancy where one feeds on the others and feeds the other factors forming circles where the one circle grew from a straight line forming a circle which comes down to the interaction of dimensions feeding and being fed by one another.

Newton saw gravity as pulling and shoving of material where Kepler saw gravity as space in motion forming the interaction of dimensions. Kepler knew time had to be somewhere as something and then covered it by pronouncing the circle cycle as space in motion that is responsible for the time aspect. What then is it that Kepler saw as he formulated $a^3 = T^2$ **k.** At the normal flow of time it takes the electron a certain time to spin around the atom. That is in short space-time. The atom uses space $a^3$ and the atom is a certain length **k** that forces the distance the electron has to travel in one cycle period $T^2$. The atom $a^3$ space connects the electrons **k** to go about circling around the space $a^3$ at the time of gravity $T^2$. The relevance **k** produce to support $a^3$ is to point $T^2$ to two positions the electron will be in the duration of one specific time. The electron travel will be cyclic and periodic in relation to the space the atom holds. The space stands related to the gravity with which the Earth reduces space and with the space and speed with which the atom travels through space.

Einstein fathered the perception that light travels as fast as time flows but I disagree with that idea. Nevertheless the perception is there that the speed of light is as fast as travelling of any sorts can reach. We accept the electron's travelling speed imitates the speed of light as much as it is permitted by gravity to do so. By this imitation the electron come as close to time as it can ever come (that is in accordance tot he general persuasion of science). The electron rotates around the atom nucleus indicating an atomic border of some sorts. From the centre one can draw a line pin pointing the position of the electron during the duration of a time period. In this time will indicate movement of the electron through space. The time indicated must be $T^2$ should space be in the third dimension $a^3$ and singularity will connect through **k.**

By linking space $a^3$ to singularity, which produces time $T^2$ it will have to indicate the influence of singularity through the single dimension connecting of **k**. In the relation only **k** will be representing the single dimension factor since that places the Universe in space and time. Should one place the time factor as a cosmic relevance, but presenting time as **t** in a single dimension role the two dimensional time have to disappear with the three dimensional space that also will disappear.

If not for any other reason then simply because the moment space holds one position a double time unit, eternity sets in and space by dimension disappear. That is what Kepler said when he said $a^3 = T^2k$ and that is what Newton missed when he corrected Kepler's $a^3 = T^2k$. The moment motions ends space and time falls into the single dimension. In that space and time must disappear. We all can accomplish that task by taking a photograph and print the image on paper and call it still photography. Then the paper will hold time in the square while the paper is in the cube all indicating individual and complimentary **k** pointing to individual and complimentary singularity producing space-time. The image returned to the single dimension while the photographic paper serves time in space. That makes that the electron cannot stand still.

By taking the line to **k** back to where the line or **k** cannot reduce further $k^0 = 1$ establishes such a value where **k** then finds a position in the single dimension. But in that case $a^3$ also is equal to one and so is $T^2$. In fact **k** still has to produce a line and we find that **k** represents $a^3$ to the full as well as $T^2$. The point where **k** forms the least or the most slightly distance the area $a^3$ establishes a value outside the single dimension because $T^2$ adds a value. The fact that $T^2$ comes in as a factor in the presence of the first sign of **k** appearing, **it** indicates the start of motion taking $a^3$ from one location to another specific location. It indicates the travel of the planet during a month or a day or an hour. Trying to freeze time will place the electron in two positions because the electron then is from ...(a point visible in time) to ... (another point visible in time) where it is seen as two individual points in that moment one wish to freeze time. It is not the points of electron visibility that forms time but the space occupies by time being between the two points showing a position of the electron. The electron cannot freeze, because if it does, there is a nuclear bomb about to begin.

**By taking the line to k back to where the line or k cannot reduce further $k^0 = 1$ establishes such a value where k then finds a position in the single dimension. In that case, $a^3$ also is equal to one and so is $T^2$. In fact, k still has to produce a line and we find that k represents $a^3$ to the full as well as $T^2$. The point where k forms the most slightly distance the area $a^3$ establishes a value outside the single dimension because $T^2$ adds a value. The one position in space will place the electron in time where the electron then will be below...behind and above the atom sub particles with which the electron shares frequency. The position k indicates, implicates the electron $T^2$ in the environment $a^3$ establishes.**

It does not indicate a circle except at the end when completing one cycle. $T^2$ is the distance in time $a^3$ that will take **k** from indicating one point to indicating another point. The formula points to a referring of the very time space was indicated by position location and time. The astonishing part is not as much the way Kepler formulated his formula to cover the movement and the position of the electron in relation with the rest of the atom, but the brilliant way the mathematicians neglected to see the fact. Kepler saw a three dimensional $a^3$ something in a specific position in time $T^2$ relating to a specific density $k$ of the atom. With space in a cube as it cannot ever be otherwise the time too has to be in a square because placing time in the single dimension of $t$ the time then becomes part of a single dimension such as one may find in a photograph picture. One can justly use the same formula to implicate the electron taking time to complete the distance between two points indicating the area from the centre of the atom. Allow me to establish by crude illustration time used to space produced in relation space using time within the atom, which is in motion other than circling the proton and thereby creating gravity through motion. When we view the atom we see an electron spinning as the electron forms the atomic boundary

The simplistic concept science holds on the Universe and cosmic travel is rather less to very little thought through. Mainstream science is of the opinion that the future space travellers are very much comparable to the adventures Columbus endured. It is about seeing new places, meeting new faces and an all around adventure. They are quite to the book of the opinion that travelling around the Earth by ocean as Columbus did and travelling through space is very much a similar journey. Columbus set out to travel across a sea but Columbus and his crew was at the same time remaining on the Earth. The Earth forced time and gravity onto the travellers. On his way he came across some islands where he was able to allow his crew some shore leave.

There was life on the islands breathing the same air he required to sustain life. To leave the island he only had to set sale and be off if the winds were blowing. This was in the time of Columbus that was today's *news-braking-space-travel living-on–the-edge* where the most developed minds participated on achieving the most inspiring mind provoking futuristic travel thought up by an exploring human mind. But let us face it, breathtaking it may be, it is far from a journey to the moon. Travelling through space on an Intergalactic hopping one need to go much faster than sailing speeds will allow. Let us study the following concept in more thought through detail on our part. When we glance at an object travelling from the Earth to Mars we see the Sun rotate during the motion the object travels by. The Sun might turn a few degrees but those degrease is immense when standing in relation to the motion of spin we find in the material atoms forming the travelling objects. We must remember direction has changed because our travelling now includes cosmic laws and not the laws of King and country.

The cosmos is not to the outside where we glance at the sky because we are the outside or the final limit of the cosmos we form. We are where our Universe ends because we are in the centre of our Universe. What we see that we think of, as the Universe is an array of light travelling through space. To our outside or away from us it is the light forming space-time that we have contact with and not the space we think we see. The space we see is light telling us to form a concept about what time tells us and not what space brought us. We see the history of space but the history of space is written in the ink of light on the paper of time and we are reading about the history of time but we use the ink of light to see the writings. It is telling us of space, while our minds know the space we think we see changed so much that we cannot see the space that changed. The space we are in contact with is to the inside. Space-time to us is towards singularity, towards the atoms forming our Universe of which we are the outer edge.

When we look at an object we see it in a specific colour. We think the colour we see the object in represents the colour the object has but in truth it is the last colour the object relates to because the object is rejecting the colour that we see. That we associate the object with by appearance is completely wrong because the object rejects the colour we see and by it is represented by all the colours it associates with which is not one colour we see. The object we see associate and cling onto all the colours there are that falls outside the spectrum we see. That in spite we still connect the object to the one colour the object rejects. The object rejects the colour because from all the colours there is that specific colour the object is not. The very same argument concerns outer space. We see the Sun as a star shining bright but with all the light it deflects the Sun has to be as dark hell. On the other hand with the night sky as dark as it is it is keeping all the light to itself because it is not passing out any light to us. It is keeping the light it has under control and that makes the darkness we see being the light it is holding back.

The growth we find in the Universe is from the point where singularity charges space-time and space-time extends from the centre to the outside. The atom cannot disappear, as it cannot vanish. The space we think we see by the light coming to us has already disappeared and has already vanished through the interfering of time. After all what we think we see is energy by heat that expanded. That what we think we see to the outside is the "outside" we believe we see forming outer space. It is the light that came about after the Overheating sparked a Universe out of complacency forming the light to be.

**THE COSMOS IS NOT OUTSIDE; IT IS INSIDE, INSIDE EVERY ATOM. THE ATOM CANNOT DISAPPEAR, AS IT CANNOT VANISH.**

**To us everything but us is all just energy. Yet, it can be relocated back to singularity. Travelling through space has this same relation because where I am I am the border of space-time. My relation with singularity applies only because my singularity relate to the singularity in outer space. My singularity will prove its worth in space-time by the relation in time through space it has with singularity points other than those under my control. My singularity, which is in my control is in relation with other points of singularity not under my control and by that I am in fact under control of those my singularity stands relevant to. If the photon and therefore light was travelling at the speed of time, (t= C) the photon has to cross the entirety of space at the very instant it enters space. After all, when light is time, time cannot restrict light, yet it does to the value of C. Time can restrain light therefore light cannot be time.**

The cosmos ends where you are located and the reset...well that is just space-time. To us everything but us is all just energy. Yet it can be relocated back to singularity. Travelling through space has this same relation because where I am I am the border of space-time. My relation with singularity applies only because my singularity relate to the singularity in outer space. My singularity will prove its worth in space-time by the relation in time through space it has with singularity points other than those under my control. My singularity, which is in my control is in relation with other points of singularity not under my control and by that I am in fact under control of those my singularity stands relevant to. If the photon and therefore light was travelling at the speed of time, ( t = C) the photon has to cross the entirety of space at the very instant it enters space. After all when light is time, time cannot restrict light, yet it does to the value of C. Time can restrain light therefore light cannot be time. Time is motion because while in motion set by the Earth motion the point I would hold would place my general or my governing singularity direct in relation to that of the Earth gravity.

The relevancy would apply as long as I hold a motion relation of equilibrium in relation to the motion standard set by the Earth governing singularity. The motion sets the trend of gravity applying and sets the deviation in motion equilibrium or the maintaining of such equilibrium. That is where the factor **k** produces its input to relevancy applying. If motion equilibrium is sustained **k** is duplicated by precise measure. The motion in double value holds on space with the other part valued at $\Pi^0$.

However when gravity does not set equal motion reproduction the relation will change where the Earth demand one value and the independent space in motion will have to contribute a motion higher and on top of that of what the Earth produces. When there is motion of equality between the Earth and a satellite orbiting in gravity equilibrium to that of the Earth there is **k** applying as well as a related $\mathbf{k}^0$, which proves to duplicates an extending of **k**. In order to be either $\mathbf{k_1}$ or $\mathbf{k_4}$ depends on the motion difference there is in the mutual gravity applying motion.

**When there is motion of equality between the Earth and a satellite orbiting in gravity motion forming equilibrium to that of the Earth, there is k applying as well as a related $\mathbf{k}^0$, which proves to duplicate an extending of k. In order to be either $\mathbf{k_1}$ or $\mathbf{k_4}$ depends on the motion difference there is in the mutual gravity applying motion.**

$$a^3_4 = T^2_4\, k_4$$
$$a^3_3 = T^2_3\, k_3$$
$$a^3_2 = T^2_2\, k_2$$
$$a^3_1 = T^2_1\, k_1$$

In the atom **k** forms a relation of space $a^3$ to distance **k** connected to the double proton time factor $T^2$. In this case, **k** represents the atom spinning and the atom represents the Universe.

Due to factors from the Big Bang and the fact that from that all singularity connects the same, the more the k that connects the atom to the sun centre will increase by momentum, the more will **k** within the atom reduce. Such reducing will affect the space $\mathbf{a}^3$ that will affect the time $T^2$ period. The period of time will increase because the time duration will reduce.

The reducing of time and the reducing of **k** results in the reducing of $a^3$. That is because there is more duplicating of atomic space by motion of the atomic unit as a whole.

To move the already rotating object from $\mathbf{k_1}$ to $\mathbf{k_4}$ requires that **k** duplicates by extending on top of the required rotating speed and that extending must be linear in relation to the circular motion gravity sets the equilibrium conditions of duplication to. It is when the duplicating produces an extending factor as a compliment to what already are in motion that the relations that is applying change somewhat. By increasing the motion to a forward motion, many compromises must be accounted for since many relevancies change drastic. The higher the forward velocity is, the more drastic the changes will be.

However this are many relevancies that require compromises to remain in equilibrium because the Universe insists on equilibrium above all else. To allow the one factor to expand another factor has to reduce. That is what the whole Big Bang event is all about. It is about expanding producing compromises. By moving fore ward there is a time differentiation coming about. It is a factor of time to space $T^2 = a^3/k$ and as **k** increase the unit space-time it has to reduce the object in motion space-time. It is all a question of rotating timing just as one would find in gears rotating in synchronised motion. The reducing will be in space but as I said there is only space inwards and light to the outside. If one wishes to apply motion reaching the speed of light, the space towards the inside must reduce to the size of a photon and the space to the outside must incorporate the entire global Universe we so desperately but incorrectly whish to incorporate into our personal standards. Our Universe is the total compliment of our atoms.

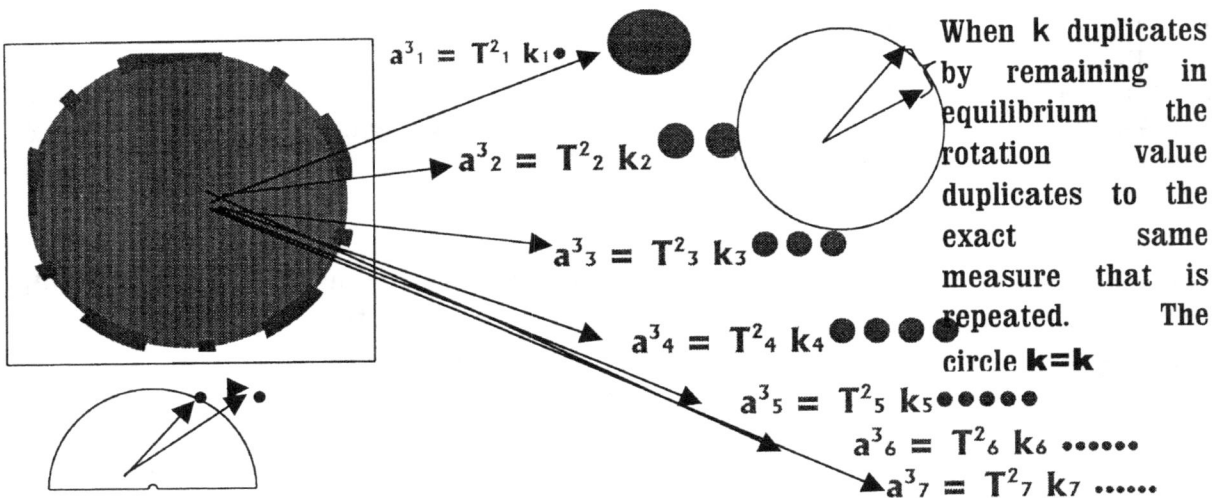

$$a^3{}_1 = T^2{}_1 \; k_1 \bullet$$

$$a^3{}_2 = T^2{}_2 \; k_2 \bullet\bullet$$

$$a^3{}_3 = T^2{}_3 \; k_3 \bullet\bullet\bullet$$

$$a^3{}_4 = T^2{}_4 \; k_4 \bullet\bullet\bullet\bullet$$

$$a^3{}_5 = T^2{}_5 \; k_5 \bullet\bullet\bullet\bullet\bullet$$

$$a^3{}_6 = T^2{}_6 \; k_6 \bullet\bullet\bullet\bullet\bullet\bullet$$

$$a^3{}_7 = T^2{}_7 \; k_7 \bullet\bullet\bullet\bullet\bullet\bullet\bullet$$

When k duplicates by remaining in equilibrium the rotation value duplicates to the exact same measure that is repeated. The circle **k=k**

**The duplicating requires a repositioning of the aligning of k from a certain position to a more fore ward position in relation to and in that k will also have to extend a value when moving from $k_1$ to $k_2$**

**If the proton starts to move at the pace the electron does which is apparently, C the neutron would have to compromise space-time to allow the proton to go faster. The electron is directly related to the proton by the neutron and therefore the motion the proton has will directly influence the motion of the electron. The electron will never allow the proton to catch the electron because that would destroy all space the neutron holds and the neutron forms all that is the atom. When the proton moves with the electron in front, the space has to reduce. When the electron is in rotation behind the proton with the proton moving at the speed that electron normally move the heat will compensate for space-time relevancies returning to what was about when the Big Bang became a fact. However now we change our perception in allowing the atom to move while everything about the atom is moving as before and this brings about the second relevancy of independent motion of the electron in relation to motion around the proton.**

## Route the electron follows

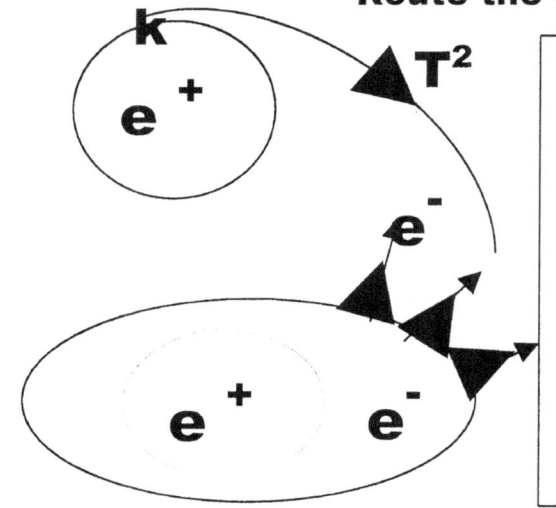

There is **k** that forms the distance between the proton and the electron while the electron is spinning **T²** around the proton **k⁰**. While all this action is going on, we think of the atom as being very still and satisfied with being a small part in a lump of metal we call iron. It could be any element but I use iron just as an example this time. The lump of iron is as motionless on earth as anything can be while being.

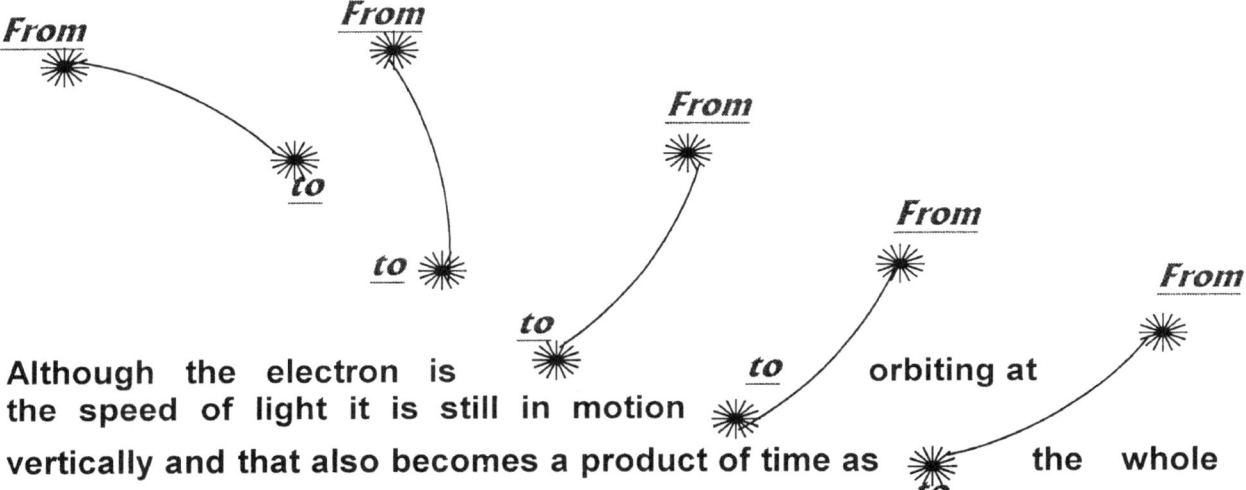

Although the electron is orbiting at the speed of light it is still in motion vertically and that also becomes a product of time as the whole

structure is repositioning the relevancies, it had a moment

before to what it will have the next moment. Such motion will again have an influence on the relation in the position the electron forms with the rest of the Universe while the lump of metal is now travelling as a spacecraft destined to other galactica. It is if we use the logic those intellectuals calling them Academics show and those Super-Educated that advocate how we may travel to far away galactica while we go on skipping the nearby galactica that is only two to twenty million light years away. Since the electron is duplicating by motion the motion links the electron to a time constant. The time constant is linked to the speed of light but time as such, is part of the speed of light. The faster we take the electron to go straight in the motion man produces, the less time there will for the electron be to circle around the atom. If we make **k** bigger in relation to increased motion, the smaller will **T²** produce a usable space.

This is because moving the atom faster in a straight line will reduce the electron time to complete the circle and that reduces the space the atom occupies. The relevancy linking everything is motion. Even the electron and the atom is a result of the manner in which motion brings about gravity by studying the Coanda principle and applying that principle to the working of the atom set up we find the manner or method of the forming of the atomic bonding as it clearly indicates the working of gravity by motion.

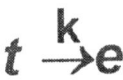

**The atom working becomes real to our thinking if we place the space the atom claims as $a^3$ and the motion of the electron as $T^2$ while the distance of space the atom has as k. By motion of the electron does the atom form a bonded unit precisely in the manner as the Coanda effect must have. The liquid neutron takes the electron spinning around a spherical proton, which personifies the solid, but roundness and that serves all the requirements we are looking for to establish the Coanda principle.**

The Coanda effect is not even considered by science to form part of gravity and yet it is what sustains gravity.

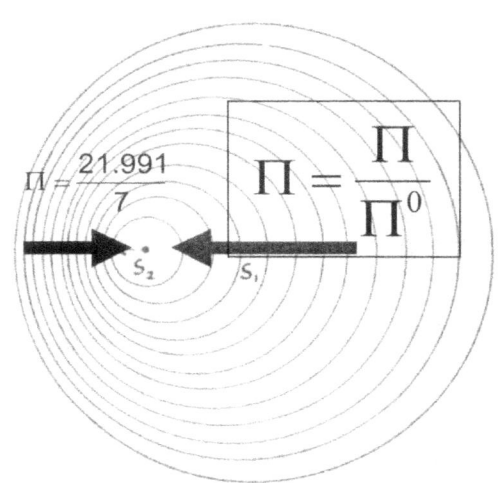

$$\Pi = \frac{21.991}{7} \qquad \Pi = \frac{\Pi}{\Pi^0}$$

**Coming toward the earth from space the object travels straight but the spin of the earth by 7° re-directs the object to follow an inclining line of 21.991 /7. At such a point the object moves towards the earth but the object cannot have weight yet, since the object does not connect directly with the earth. When the object touches the earth it relates to $\Pi\Pi^0$ and at such a point the object receives a value of mass by which it becomes a unit within the earth. By connecting to singularity in terms of $\Pi\Pi^0$ or in terms of $\Pi=21.991$ /7 puts the object in liquid or in solid.**

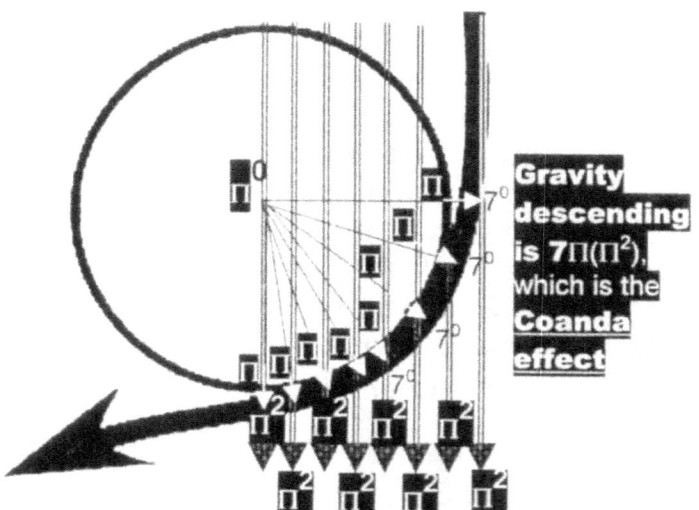

Gravity descending is $7\Pi(\Pi^2)$, which is the Coanda effect

The movement of the circle $\Pi^2$ extends the line forming $\Pi^0\Pi$, which is thought to be the radius. The faster that the rotation speeds the longer would the 7° be that forms $\Pi^2$ and since we saw with the top that it is $\Pi^2$ that produces singularity at $\Pi^0\Pi$ then therefore the longer would the extension be that forms the line $\Pi^0\Pi$.

This ultimately gives the earth the atmospheric layers that form the atmosphere around the earth. It is because of the influence the Coanda effect establishes that we have weather and cyclones.

This is how you do it I have shown how to locate singularity.

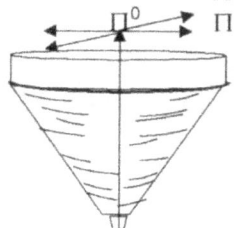

When a top or any circular object starts to spin there are two factors working in combination where $\Pi^0$ finds relevancy with $\Pi$.

Any changers occurring in $\Pi$ will lead to a an unequal triangle providing two different values to r and will alternate the link between r and $\Pi^2$ bringing about different form ($\Pi$) and time ($\Pi^2$). When singularity forming the lines of the triangle is not in equilibrium the triangle will destroy the matching of half circle.

**In every sector the directional flow will provide a distinct meeting of $\Pi$ linking r to $\Pi^2$ and this allow the time component in the rotation.**

**$k^0$ to $k$ at no speed**          **$k^0$ to $k$ at half the speed of light**

As $k^0$ now applies a relevant motion $k$ in distance, has to comply with motion $k^0$ produced by changing relevancies at a specific rate other than just circling about the atom and the compromising of $k$ in motion had to endure.

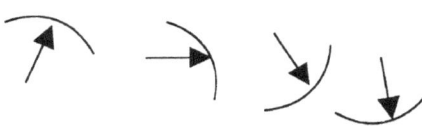

Indicating a very much- reduced $k^0$ to $k$ at travelling half the speed of light speed while sustaining the atom.

The motion provides a circle in which the atom claims space. The slower the motion is in which the atom is operating the bigger the circle is that the atom claims space. The motion and the claimed space form a unit where the two share a common unit but each one of the two steaks a different claim to the unit. The space is not the same space as although both share the same unit. The one enlarging of the one will produce the reducing of the other. The one will have an inverse affect on the position of the other in the same unit the two shares. The faster the circle will have to move and the encircling will have to reduce. As the motion enlarge the total circle the motion will reduce the claimed circle but since the Universe keeps the circle intact the relevancy of change will fall to the atoms claim on space.

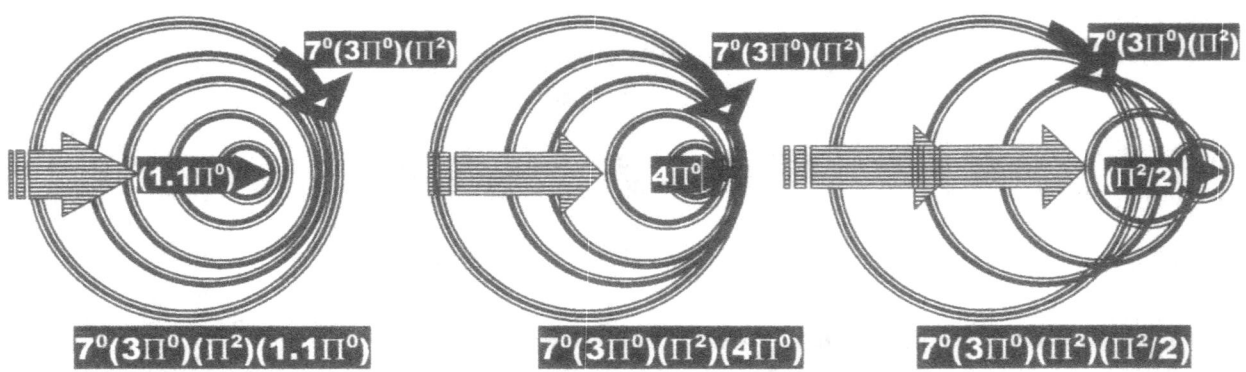

The Newtonian way of thinking is by their focus they are concentrating on the motion provided by the electron that is rotating the proton cluster. There their focus on all motion stops. Yes that is true but while this is going on the electron is doing so at the speed of light. It goes at the speed of light to run around the proton core. Should the proton core move in any direction this would have to reduce the electron circle because the motion coming from the proton has to be reducing the time it takes to have the electron circling the proton. This is not even taking into account when there is a rise in the atmospheric density through which the electron then has to struggle but on that matter I shall return at another event.

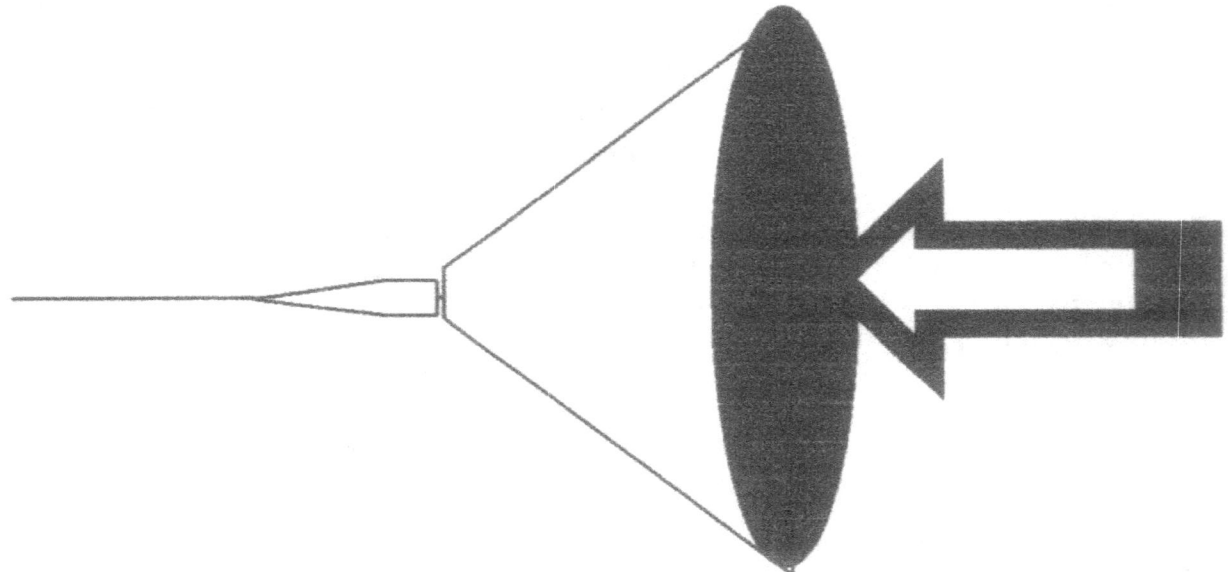

When the proton moves forward the aligning the electron has with the proton changes. This means the aligning has to change with the compromise, which will and can only come from the electron reducing the circling of the proton. By reducing the time about which the electron has to complete the circle the electron has to discharge heat stored inside the walls confining the atom bordering what is in the atom used as material to the outside of the atom on the outside of the electron in the form of heat for that matter as time. Remember the gain of material coming from the atom has been going on since the Big Bang and ever since the compromise of time is stored in the way of heat inside the walls of the atom. By the time of the Big Bang gravity by contraction was the ultimate and therefore motion was non-existing. Since then motion became a commodity of valuable currency and heat gained by material was the compromised to that heat from outer space. If the proton shifting the electron circle has to reduce when the singularity within the proton distributes heat by allowing motion, which pushes the heat back towards what it was in the direction of the Big Bang. Only a photon can travel at the speed of light and nothing bigger has that possibility.

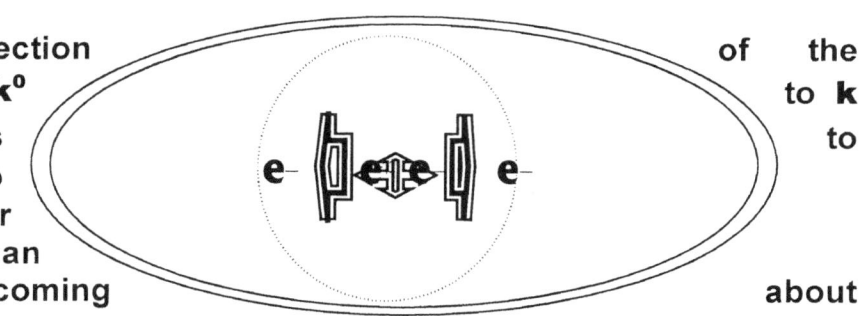

In the opposing direction moving direction of $k^0$ the component **k** has reduce to compensate for motion other than cosmic motion coming and thereby reduce the distance between singularity and the electron. It means that $k^0$ to **k** has to reduce in distance between the atom centre and the new atom borders that form but in the compensating other factors has

of the to **k** to

about

The principle, which I described in the motion of the spacecraft travelling through time, and space in space-time also apply in the same manner in the Coanda effect. It is the same motion producing gravity by applying a space relevancy that changes in the same duration as when not in motion but through motion, the space changes to, a new motion that comes into play. In the motion producing a relocated centre, a new centre will play its part in the compensating motion that will force upon the relevancies a new space-time dispensation.

to compromise for such changes brought onto the reducing the line formed a $k^0$ to **k**. With the line $k^0$ to k forming the indicator that indicates the space between the atomic centre and the electron forming the atomic border. That means there is second motion coming from the object. While the object is travelling through space, which arrange its atoms in the same time duration as that it would have used while being stationary on Earth, that pushes the electron circle out of the normal sequence that electron had while being in a steady state of motion on Earth and under the control of Earth time.

The space is not the same space as although both share the same unit. The one enlarging of the one will produce the reducing of the other. The one will have an inverse affect on the position of the other in the same unit the two shares. The faster the circle will have to move and the encircling will have to reduce. As the motion enlarge the total circle the motion will reduce the claimed circle but since the Universe keeps the circle intact the relevancy of change will fall to the atoms claim on space.

That means the second motion coming from the object travelling through space which arrange its atoms in the same time duration as that it would have used while being stationary on Earth pushes the electron circle out of the normal sequence that electron had while being in a steady state of motion on Earth and under the control of Earth time.

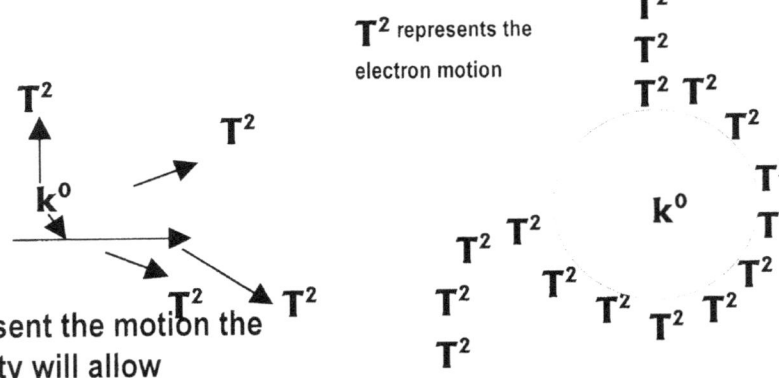

$T^2$ represents the electron motion

$k^0$ represent the motion the singularity will allow

By applying motion the flow of which represents as much as bring about a new space-time as the factors has to compensate for the motion and the motion in itself bring about a new controlling $k^0$ in the position elected by all atoms forming part of the relevant motion. The motion introduce a securing of the position that is elected by all the atoms and is forming the new part that will serve as a controlling singularity by substance producing the motion or the substance of the solid securing the position of the newly elected controlling singularity. By applying motion the Earth gravity is interrupted by the space producing motion that through motion take charge of the space and turns the vertical flow of Earth space into a circular motion although only for a short while.

This is how electricity is charged but the motion then involves the reducing to the

the liquid around the round solic rein acts singularity the motior

The principle, which I described in the motion of the spacecraft travelling through time, and space in space-time also apply in the same manner in the Coanda effect. It is the same motion producing gravity by applying a space relevancy that changes in the same duration as when not in motion but through motion the space changes to, a new motion that comes into play. In the motion producing a relocated centre a new centre will play its part in the compensating motion that will force upon the relevancies a new space-time dispensation.

flowing electrons to compensate as much as producing speed of light Other elements and their characteristics are also involved and in the charging of electricity the Coanda principle personifies the gravity that comes about in the condensing of space-time around singularity.

It is not only the relevance applying from the centre to the electron but also the electron holding an allegiance with the centre in the precise manner as planets do in relation with the Sun. It is because of the dual relevancy that the electron stubbornly clings to the newly elected centre before being overpowered by the Earth providing the controlling centre. It is all about relevancies attaching centres that formed when creation came about.

The time that Newton froze on paper in his establishing of a single t that is representative of time is effective in remembering the viewer of an event but that cannot be the event that is part of the present any longer. If we reduce the moment to a snapshot the picture we focus on can only be what we see for a very short instant and then forms that

All Newton's changing was possibly done with good intensions but even that I doubt. The end result however was in some cases far from good, as it does not do such great credit to Newtonian insight into cosmic affairs. Only Kepler and only Kepler unaided without the intimidation and interfering of Newton can explain the Coanda effect. I grant the fact that the Coanda effect was discovered before Newton saw himself fit to change Kepler, but only Kepler can explain the Coanda gravity effect when Kepler is without the attentions of Newton.

which was how the event occurred during the time from where the camera shutter opened $T_1$ to where the camera shutter closed $T_2$ and the time frame $T^2$ was then during the open period of the camera shutter. But as soon as the shutter shuts, time moved on and another $T^2$ formed leaving the image taken as time never to repeat again. Afterward the image we see as the picture represented t and when looking at the picture, the looking of the picture became an event during a specific $T^2$ that went from where one is taking the first look to where one is looking away from the paper carrying the first dimensional image of an event gone by and that is at that stage a representation of t in another milieu of $a^3 = T^2 k$. The t in the single is when mathematically presented as only t indicating a mathematical single flat dimensional view of time and is then correctly applied because it represents a reminder of a four dimensional event $a^3 = T^2 k$ that went single dimensional because the moment in the fourth dimension was then frozen in a single dimension on paper with the paper being part of space-time while the fourth dimension $a^3 = T^2 k$ soldiered on and time will always be representing $T^2$ as Kepler stated. Time is in the square, and that is allocated to space having a cube. Kepler said gravity is $a^3 = T^2 k$ at a time even before gravity got a name. But reducing the dimension of time to a single t one will find the ability to mathematically design the paper on which the photo image will be printed in time $T^2$ using space $a^3$ in the third dimension to apply the ink in the third dimension. Printed on the paper is an image that is

In $4\pi^2 a^3 / T^2 = G(m+m_p)$
$a^3 = T^2 k$
$a^3 / k = T^2$ but
$k / a^3 = 1 / T^2$
$k = a^3 / T^2 =$ singularity
$a^3 / T^2 = G(m+m_p)/4\pi^2$
and $a^3 / T^2 = k$
then $k = G(m+m_p)/4\pi^2$

But I showed that $k = a^3 / T^2$ and Newton's claim is that $a^3 / T^2 = G(m + m_p) / 4\pi^2$

not part of space-time while the ink used is space-time and the paper is space-time. The ingredient of ink on paper all hold different values since the image has a value we as humans grant the image to carry such a value. The image, the ink and the paper all hold different relations but the image only relate to thought in our mind. The image has no k indicating only references with and to what forms part of a realistic different singularity coming from the Earth centre and connecting to individual atom groups, forming individual as well as group space-time.

In the formula MC$^2$ I have been protesting that the speed of light cannot double or go square and by that I seriously had doubts about the formula. Then fortunately I managed to correct the error not of the formula but of the interpretation of the man that produced the formula. All I can say is that even Einstein had no idea what to his formula presented to the world. Being a person to the likes of Einstein, one must at all times presume what he says has merit on the grounds that he says what he says. I was surprised at his error but afterwards I saw Einstein for what hew was. He was a brilliant mathematician but he sometimes was a very poor translator and very often accuracy went array in his interpretation and translation from the written mathematics to the spoken languages used by all people. What he saw had no bearing on the speed of light by C going square. Einstein saw the conditions applying when the atom was introduced when the atom came into use as an era related principal.

His C$^2$ was not directly about harvesting the speed of light in what light represent. The Einstein formula is depicting the Coanda gravity and it works on the fact that where the neutron places the electron, the space-time displacement goes to gravity at C which is the speed of displacement of space-time the electron is charged with. In the past the no one even thought of placing Coanda principle in line with gravity. By following Kepler's formula we find that the Coanda affect IS GRAVITY. Gravity is moreover the Coanda principle than it is any other variety of forces or concept of contractions. The Coanda principle is gravity. It is the how all gravity is charged and distributed. That forms the basic principle of Cosmology.

The Coanda affect is proof of Kepler's statement. It proves that motion (**T$^2$**) establishes a centre at the distance of **k** and such motion claims the space (**a$^3$**) from a centre within that motion. The motion creates a centre and a centre establishes a gravity field. The motion of the liquid water or in the case of the liquid neutron proves to be the time aspect because it claims the space in the time it is running. The faster the motion is the stronger is the gravity that the motion generates in the space it claims by the gravity it generates. Even Einstein's square of light is not the square of light but refers to gravity being equal to the speed of light. At the time of the Big Bang outer space was one big electron giving the k relevance the value of C. With the Coanda effect charging gravity that means the gravity goes $\Pi^2 = C^2$ and the space goes $\Pi^3 = C^3$

Space is created from one position to another position and the duration it takes to complete the distance is time. This the Coanda effect proves as water flows past a round object and the contact the flowing water makes diverts from the normal route the Earth gravity will enforce. Gravity is the very same but it is the recalling of the space by creating motion in the space. By duplicating space through motion in relation to singularity the flowing water diverts from the normal route. By recalling the space it is also reducing the space because it is counter acting the time expansion provided. That then is clarifying the reason why gravity will always on the limit be stronger than light. At a point it slows the time component down to being on the limit and of the time light takes to move. Gravity is motion that is going way past that point and the motion gravity produces is past where the limit of light is. Gravity as motion, can be much stronger than light which too is just motion. At a point gravity slows the time component down to such that the space reduces faster in that time than what light can produce motion. This must be time because the Black Hole contracts light back into the star. That is also precisely what Kepler introduced as.... This is why the Coanda effect applies. $a^3 = T^2 k$ then being $k^3 = k^2 k$ and this is showing that the space $k^3$ is equal = to the motion $k^2 k$ of the space $k^3$ seen form one specific point.

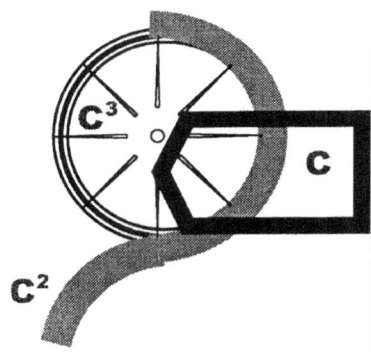

The entire Universe broke into 3D by the margin of C and in C the atom was

The $C^2$ value Einstein came up with has no bearing on the speed of light, but is equal to the gravity displacement the neutron has when entering the atom. It shows the stopping of the flow of space –time and the atoms overheating in the process when the two atoms touch and stop one another's space-time flow bringing about cooling. The $C^2$ is the gravity at the point of the atom presented because of the neutron centralising the gravity $C^2$ at the rate of the electron $C^2$. What Einstein saw was gravity at the point where gravity equals the speed of light, but it has nothing to do with the actual flow of light going square. It is the flow of space-time. Going nuclear takes the atom back to what happened at the start of the Big Bang.

captured in space in timer and in gravity. **The atom is in accordance with the Coanda principle moving at c with the electron excluding all else from the space $a^3 = C^3$ in relation to the centre C which the motion of the speed of light confirms at c keeping the gravity that applied at $C^2$. Space is created from one position to another position and the duration it takes to complete the distance is time. By duplicating space through motion in relation to singularity the flowing water diverts from the normal route. By recalling the space it is also reducing the space because it is counter acting the time expansion provides. That then is clarifying the reason why gravity will always on the limit be stronger than light. At a point it slows the time component down to such extend the space reduces faster in that time than what light can produce motion. This is why the Coanda effect applies**

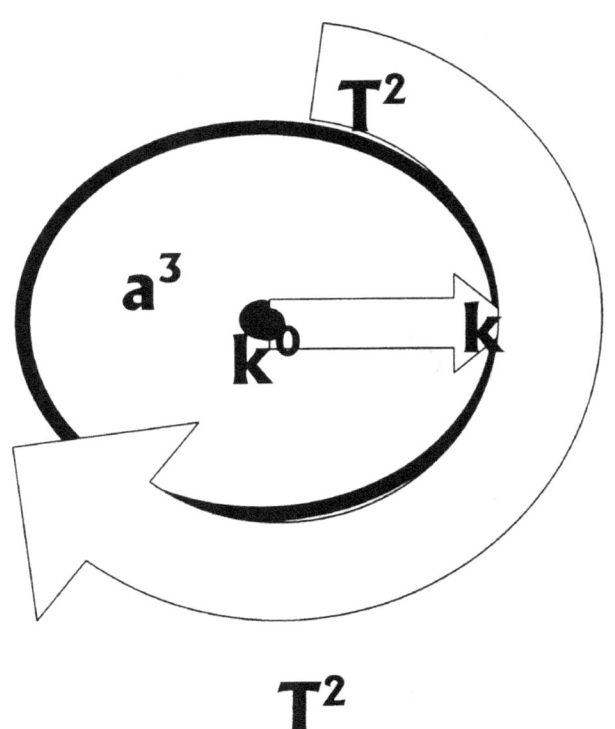

The whole process of gravity comes to the surface when Kepler's version of gravity is used. The neutron providing the spinning motion manifests the motion **T²**. Since the neutron produces the space for the proton to have, the proton and the neutron establish a symbiotic relation of equality. The motion of the proton uses space to dismiss while the motion of the neutron uses space to duplicate the space in which the neutron is as well as the space in which the proton is dismissing the space in which the neutron is. To secure a balance or a favouring of any one of the positions, the electron adds or removes it's favouring of placing a dynamic to the atom's form.

**Having that, we find that all stars go through a developing that ranges from being all about liquid heat to becoming liquid material as the neutron is and ending as space solidness by dismissing all that the neutron may provide in duplicating space-time. In this we trace the beginnings before the Big Bang**

In the past no one even thought of placing the Coanda principle in line with gravity. By following the Kepler's formula we find that the Coanda affect, in fact is gravity. Gravity is moreover the Coanda principle than gravity is any other variety of forces or concept of contractions. The Coanda principle is gravity. It is the how all gravity is charged and distributed. That forms the basic principle of Cosmology

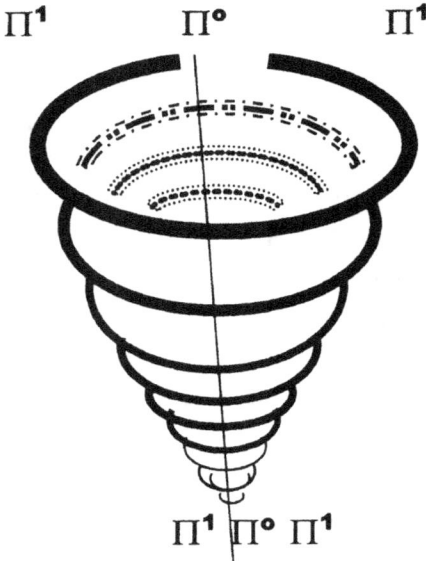

In this manner we find various natural phenomena acting in accordance with the Coanda gravity principle. It is always motion that is forming circles around a defined centre where the centre holds calmness. In my saying this I am referring to hurricanes, tornadoes and other natural occurring disasters.

The wind circles from the outside and by motion around a specific centre the motion is empowering the centre that gives the centre that ability to form a devastating lift. The action we find strongly reminds of what the motion and action the spin of the top produce as the behaviour that the spinning top presents. One can again see the spinning top being the example in the occurring of the winds in question.

As to why the disasters occur and why the wind reach such limits, that I debate in another book named the <u>Seven Days Of Creation</u>.

The motion of the liquid which the neutron is proves to be the time ($T^2$) aspect because as it increases it claims the space ($a^3$) in the at a distance **k** of time ($T^2$) that the running has increased. The faster the motion is the stronger is the gravity that the motion generates in the space it claims by the gravity it generates. In the Coanda affect we can read how the atom became the Universe. The proton is the substance performing as the solid on both sides of the Universe, The neutron being a liquid is what establish the gravity, which helps the singularity secure space-time by heat compensating for overheating. With the neutron forming a liquid and the proton providing the spin, the neutron established the space that forms the atom to the inside of the electron.

In the past no one even thought of placing the Coanda principle in line with gravity. By following the Kepler's formula we find that the Coanda affect, in fact is gravity. Gravity is moreover the Coanda principle than gravity is any other variety of forces or concept of contractions. The Coanda principle is gravity. It is the how all gravity is charged and distributed. That forms the basic principle of Cosmology

The Coanda affect is proof of the functioning of gravity inside the atom. It proves that motion ($T^2$) of the neutron establishes a centre in line where the compliment of material forming the atom will secure a controlling singularity that is governing the entire atom. That forms the centre of the Universe. Singularity then finds a position at the distance of (**k**) and such motion claims the space ($a^3$), which is the atom by construction from a centre within that motion ($T^2$). The motion ($T^2$) creates a centre at the line of (**k**) and a centre of the space ($a^3$) the motion ($T^2$) establishes a gravity field all along the lines and at the distance of (**k**) in the space ($a^3$) that the motion ($T^2$) created.

## Kepler said

$$a^3 = T^2 k$$

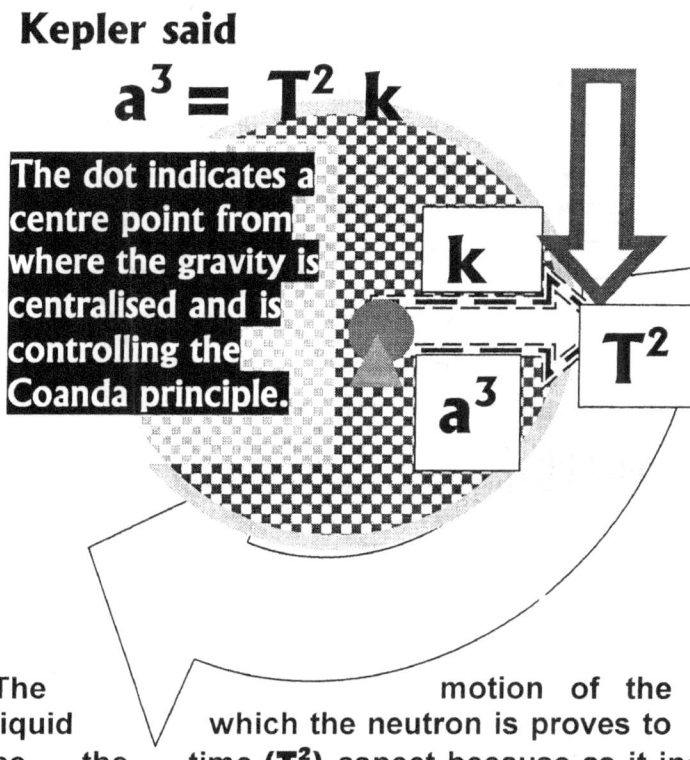

The dot indicates a centre point from where the gravity is centralised and is controlling the Coanda principle.

The motion of the liquid which the neutron is proves to be the time ($T^2$) aspect because as it increases it claims the space ($a^3$) in the at a distance **k** of time ($T^2$) that the running has increased. The faster the motion is the stronger is the gravity that the motion generates in the space it claims by the gravity it generates. In the Coanda affect we can read how the atom became the Universe. The proton is the substance performing as the solid on both sides of the Universe, The neutron being a liquid is what establish the gravity, which helps the singularity secure space-time by heat compensating for overheating. With the neutron forming a liquid and the proton providing the spin, the neutron established the space that forms the atom to the inside of the electron.

With the facts well established in the Coanda principle, much more of Newtonian visions come into question, one being his all-to-famous view about gravity and on the establishing of gravity. Gravity is precisely what Kepler first said it was. The Coanda effect is the best prove of that and from the Coanda effect comes a sound explaining about the sound barrier. Gravity is created by motion of space within space forming motion that is served by points. Without gravity space would not be there at all. $k = a^3/T^2$. The proof is there in every atom. Every atom manifests the gravity being all over and that gravity can be everywhere. The establishing of a space providing gravity by motion proves the validity of the Coanda effect in the producing of gravity by the atom reforming space by motion using the same period of time to apply the motion.

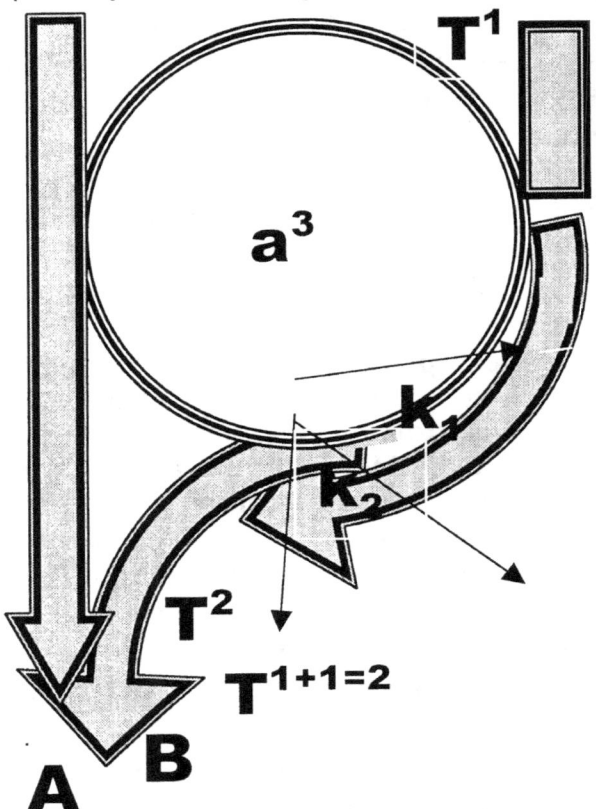

The Coanda effect is proof of gravity coming about through space forming motion. In the case where water diverts the normal directional flow the space that translates to the motion is deflecting singularity with the flowing water charging the motion. In the centre of the object having the round form, singularity is duplicated and by transferring $\Pi$ to form $\Pi^2$ and the motion of the water creates a line of gravity that pushes the flowing water to follow the direction that the newly gravity applies to the water. This again proves Kepler's statement of $k = a^3/T^2$ that specifically states that space (in this case the object transferring singularity to a new position within the round object) and with the motion of the water redirects the gravity flow of the water to new space in new time. Only Kepler can explain the phenomenon but only when Kepler stands alone, correctly interpreted and divorced from Newton's opinion about Kepler's statements.

Normally water will run down to the centre of any gravity point, as A shows. By allowing the flowing water to come into contact, an object of a specific form the flow will divert (B) from the normal line and follow the contour of the object presented. For that to take place there is one condition that has to come about.

The Coanda effect is the very evidence that exists about the proton or then singularity maintaining a direct link with the electron, which then forms contracting gravity as a result of motion. $k^{-1} = T^2/a^3$ Take the case of a wheel running along a wet tar road where the water becomes the neutron, the water on the road becomes the electron and the wheel holding singularity within the stub axle becomes the proton. In the case of the water the water becomes the neutron or the factor proving motion. The motion coming from the centre is provided as the car move forcing more motion of water around the wheel and this process is producing more retaining of the space that the centre confirms by conforming the motion of the water onto the centre by space-time in gravity. When the centre does move, as is the case of a wheel on a car, the centre then confirms the water or the liquid onto the wheel in much stronger terms, than would be the case when the water is flowing by individual initiative. Such motion of the wheel grants the water a state of solidity that enables the car to run on an inch of water. The wheel holds singularity and therefore the wheel becomes the proton while the water becomes the neutron, which provides the motion. Remember motion is gravity and while the electron and the proton both are restrictors of heat flow and thereby shows the indicating of mass, the neutron show pure gravity or motion and shows no restricting of motion because the neutron exemplifies gravity or motion. The motion of the water provides the gravity that secure the flowing water onto the wheel and when the wheel is moving as a result of an independent initiative that also provides motion, such motion reflects onto the liquid or water which then enables the flowing water to become a solid part of the wheel where the car then can run on an inch of liquid water, which under normal conditions would prove quite impossible. Motion is gravity and the moving of the venture does affect the concentration of the electron state.

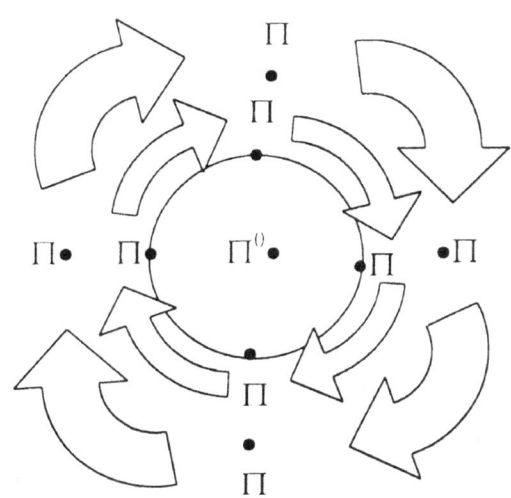

**The atom creates a unit wherein the electron establishes space and from the electron to the centre we call the atom but when the centre of the atom brings about a relation where such a centre has to comply with changes in space holding time such motion redefines the space the atom claims. This is the way that motion creates space independent from space within space. The following movement of a continuing stream of atom forming a circle relations with atoms in a solid structure forms what Kepler introduced as space-time $a^3 = T^2k$. The position k will establish, as a point will come into affect as soon as motion or time $T^2$ sets any space $a^3$ apart from the rest of space. That rule applies to all motion on Earth. That rule applies to all motion in the entire Universe and that rule is the Universe. That is what happens to an aircraft and that is why an aircraft will go through the sound barrier. The aircraft is setting a new filled space $(3\Pi^2)$ at the position of $7^0$ diverting from singularity and the independence will allow a motion of $\Pi^0$, or $\Pi$ to $5\Pi$ within the boundaries of Earth's gravity, but as soon as the limit of $5\Pi$ is shattered so is the sound barrier and a new limit sets in place being the Roche factor by half $= \Pi^2 / 2$. The complete formula comes to $7(3\Pi^2) = 207$.km per hour, which is the maximum velocity any object in free fall, will hit the Earth.**

The Coanda effect is most probably the strongest suggestion that my theory is correct. I have a theory where I show the singularity coming about be reproducing a circle motion that is resulting in the creating of the circle centre where the circle enhances or motivates the centre singularity to produce gravity point that secure the hold of singularity in the form of applying gravity through motion. There has to be motion either by space or by material and the flowing water on this occasion is the liquid and the liquid falls into the solid round space that holds the gravity part of the cosmos. The flowing water is producing the motion and such motion represents the space where the water is flowing creating motion, which is creating gravity We are very aware of the Coanda effect where the water takes another route other than the normal way water will fall when using the gravity the Earth supply. Adhering to the Coanda effect the water flowing will invest in a new centre holding the flow of water related to the space the flowing water will establish thus proving that the motion **k** in the time duration $T^2$ create a new space unit $a^3$. The natural thinking is acknowledging Newton's mass pulling mass and taking the shortest route to flow but the Coanda effect denounces this very prominently. As the water flow around the bowl it will divert in direction by producing a longer circle to flow and eventually loose the grip it had secured on the round bowl. The restraining of the circle forming stands in relation to the velocity of the liquid or the solid that supplies the motion. By clinging onto the surface of the bowl the water tries to flow upwards and stay realigned to the centre it created by motion producing such a centre. It conforms Kepler $k^0 = a^3 / T^2 k$ in defiance of Newton's $F = G (M.m) / r^2$.

$a^3 = T^2 k$ then $k^3 = k^2 k$ and this is showing that the space $k^3$ is equal $=$ to the motion $k^2 k$ of the space $k^3$ seen form one specific point.

In the shown formula introduced by Kepler years before Newton changed it, the formula shows that a space, which is clearly defined by a property of being individual, has the ability of motion and that is what Isaac Newton saw and named as gravity. He saw the falling and initially named that falling gravity. However Newton was of the opinion that a force was in charge of the pulling the one object closer to another object. He put the emphasis on the material having the status of pulling each other. To my mind that is incorrect because I see the flowing of space that may or may not contain material and the space containing material or not is flowing towards singularity. I announced years ago there is no such thing as gravity after which most Academics thought me to be a loony and even my friends would have me institutionalised behind my back. I then changed my song somewhat not because I do now believe there is gravity because the gravity Newton referred and which Newtonians accept, that there is not. There is no contraction between particles pulling them closer.

The circle forms in defiance of the natural flow that the Earth gravity will enforce that is according to Newton. The water secures a longer route around the round bowl as the water clings to the bowl by running around the edge of the bowl and even tries to run up the bowl circle. When motion is introduced to the bowl, as a car tyre would do the clinging becomes much exaggerated because a wheel of a car can produce 25 mm layer of water clinging onto the wheel at high speeds. This is even where the clinging water supports the mass and the motion of the entire car. The Coanda gravity keeps the water in place where the circle of water surrounding the entire tyre can withstand the force of the car's entire mass, as well as the mass that comes about as a compliment to the motion that is exaggerating the initial mass of the car. The situation renders high speed travelling in wet conditions to be very much unsafe and it is well known to all drivers of motorcars. But in essence this is confirming the Coanda principle as the Coanda principle confirms Kepler statement because as the wheel has more motion $T^2$ through maintaining a higher velocity it will secure more liquid **k** onto the solid tyre structure $a^3$ and thereby establishes more space $a^3$. However creating a circular singularity is one aspect of the Coanda gravity alteration but there is more establishing of such a singularity. By increasing both more $a^3$ in supplying wing contact that creates more space duplication as well as increasing motion $T^2$ by producing a turning momentum. This is very apparent in the motion of a propeller through which singularity is established as it then forces the singularity to recognise new linear singularity relevance by producing a neutral **k.** We call this flying**.**

Let us again review the location where I traced singularity. Looking at a motionless top which children spin in amusing their playful minds, the top is a solid structure with every aspect of the top being the same. The antigravity comes from the motion we deliberately invest in the top as we bring about a spinning motion to the top from where the top then finds liberation by motion performing as artificial antigravity, which are strong enough to provide a release from the Earth's confinement. But it can only be possible if motion applies in a spin and no other way. Throwing a ball horizontally brings about different influences coming from singularity, which I explain, in another book **a Cosmic birth... Dismissing Nothing** ISBN0-620 -31609 .

The ball has to spin even when travelling in a straight-line to secure direction control. Even by shooting a round ball through a vintage cannon will not establish direction control by force. The producing a spin secures a point that is creating the dynamics of singularity and that singularity in charge will strive to create independence for the space $a^3$ that is **(=)** now spinning $T^2k$. The motion must come from the way of spinning and the spinning must establish a centre. Then only can the top move about an individual axis and that axis grant the top independence and freedom from the confining gravity the Earth apply to secure the tops position as part of the Earth. So we find motion establishing a centre and from that centre the motion in the top secures the space that is the top releasing the space that is the top from the space the Earth grant the top when the top does not apply motion This spinning top is using the principles we find in the Coanda effect. This way the Coanda effect is the generating of electricity, is using the principle in the crank and connecting rod action of fossil fuelled engines and is every motion we conceive. That all are the Coanda effect in living proof and in acknowledging Kepler's formula of $a^3 =$ $T^2k$. It shows the top having control over space claimed from the Earth by applying motion and thereby defying gravity confinement. The motion establishes a centre $k^0$ from which there is space claimed $a^3$ by rotation motion $T^2$ relating to the solidness of the ground **k**. It is once again $k^0 = a^3 /$ $T^2k$. That is the circular effect and then there is the linear effect.

**However, the Coanda effect also is more than only that.**

$$a^3_1 = T^2_1 k_1$$

$$k^0 \blacktriangleright$$

$$a^3_2 = T^2_2 k_2$$

By moving linear, the wing becomes the **k** in Kepler's formula and the earth is $T^2$. That is the reason why the aircraft can fly. By creating an uneven $k^0 = a^3$ on top of the wing compared to the $k^0 = T^2 k$ below the wing can lift or descend. It is a balance going unbalanced.

**This all confirms that translating Kepler's mathematical expression $a^3 =$ $T^2 k$ correctly to the verbal statement in English Kepler said that there is a space $a^3$ which is equal = to the motion $T^2$ thereof between two specific points which holds a relation to a centre being k a straight line.**

Normally water will run down a straight line to the centre of any gravity point, towards the Earth centre. By allowing the flowing water to come into contact an object of a specific round form the flow will divert from the normal line and follow the contour of the object presented. For that to take place there is one condition that has to come about. The form has to be circular to influence the movement and therefore there has to be movement.

**Space is created from one position to another position and the duration it takes to complete the distance is time. This the Coanda effect proves as water flows past a round object and the contact the flowing water makes diverts from the normal route the Earth gravity will enforce. Gravity is the very same but it is the recalling of the space by creating motion in the space. By duplicating space through motion in relation to singularity the flowing water diverts from the normal route. By recalling the space it is also reducing the space because it is counter acting the time expansion provide. That then is clarifying the reason why gravity will always on the limit be stronger than light. At a point it slows the time component down to being on the limit and of the time light takes to move. Gravity is motion that is going way past that point and the motion gravity produce is past where the limit of light is. Gravity as motion can be much stronger than light which too is just motion. At a point gravity slows the time component down to such extend the space reduces faster in that time than what light can produce motion. This must be time because the Black Hole contracts light back into the star. That is also precisely what Kepler introduced as.... This is why the Coanda effect applies. $a^3 = T^2 k$ then being $k^3 = k^2 k$ and this is showing that the space $k^3$ is equal = to the motion $k^2 k$ of the space $k^3$ seen form one specific point.**

**$a^3 = T^2 k$ then $k^3 = k^2 k$ and this is showing that the space $k^3$ is equal = to the motion $k^2 k$ of the space $k^3$ seen form one specific point.**

**Space is created from one position to another position and the duration it takes to complete the distance is time. This the Coanda effect proves as water flows past a round object and the contact the flowing water makes diverts from the normal route the Earth gravity will enforce. Gravity is the very same but it is the recalling of the space by creating motion in the space. By duplicating space through motion in relation to singularity the flowing water diverts from the normal route.**

The circle forms in defiance of the natural flow that the Earth gravity will enforce that is according to Newton. The water secures a longer route as it went along the edge of the bowl around the roundness of the bowl as the water clings to the bowl by running around the edge of the bowl and even tries to run up the bowl circle. When motion is introduced to the bowl, as a car tyre would do the clinging becomes much exaggerated because a wheel of a car can produce 25 mm layer of water clinging onto the wheel at high speeds.

This is even where the clinging water supports the mass and the motion of the entire car. The Coanda gravity keeps the water in place where the circle of water surrounding the entire tyre can withstand the force of the car's entire mass, as well as the mass that comes about as a compliment to the motion that is exaggerating the initial mass of the car. The situation renders high speed travelling in wet conditions to be very much unsafe and it is well known to all drivers of motorcars. In essence this is confirming the Coanda principle as the Coanda principle confirms Kepler statement because as the wheel has more motion $T^2$ through maintaining a higher velocity it will secure more liquid **k** onto the solid tyre structure $a^3$ and thereby establishes more space $a^3$. In this light the Coanda effect proves there is no Newtonian gravity. Since the effort of getting Newtonian mindset off the addiction of Newtonian gravity has gone beyond the worth there of I changed my tune to comply with the harmony Newtonian abstinence insist on. Nevertheless I might have bend somewhat, but I refuse to admit I am wrong. Newtonian gravity, which is related to magical forces, which conspires to pull at the will of mass, that there is not. Kepler nailed it by saying space is there in the motion of the space in the time it takes to move the space and the essence in that is manifested by the "inexplicable" Coanda effect.

---

**During the Big Bang, the relevancy of motion was the speed of light C, the gravity applying was $C^2$ and space was confined to $C^3$.**

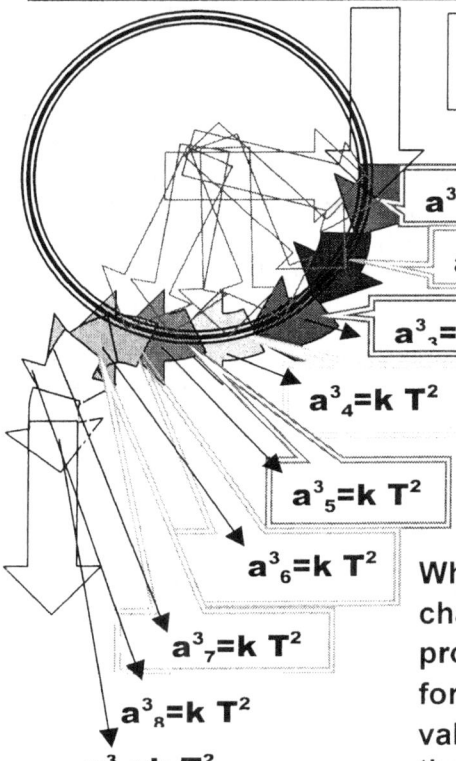

**There has never been a better and clearer explanation about the Coanda effect.**

$a^3_1 = k\ T^2$

$a^3_2 = k\ T^2$

$a^3_3 = k\ T^2$

$a^3_4 = k\ T^2$

$a^3_5 = k\ T^2$

$a^3_6 = k\ T^2$

$a^3_7 = k\ T^2$

$a^3_8 = k\ T^2$

$a^3_9 = k\ T^2$

Every position the water establishes by motion in relation to **k** from the centre $k^0$, such motion introduces new space and the space is a continuing of direction changing in relation to the centre prevailing.

Every time an $a^3$ comes about claiming a new space spot for the flowing water the motion of the water determines the spot in space created.

When the water fills a new spot $a^3$ it is the directional change of singularity running from $k^0$ to **k** that produces the two points of **k** between $T_1$ and $T_2$ forming $T^2$ creating small instances of new relative values where $a^3 = k\ T^2$ is equal to the motion within the boundaries set by singularity. It proves Kepler correct. It proves that space is the duplication thereof through motion.

---

By recalling the space it is also reducing the space because it is counter acting the time expansion provide. That then is clarifying the reason why gravity will always on the limit be stronger than light. At a point it slows the time component down to on the limit be stronger than light. At a point it slows the time component down to such extend the space reduces faster in that time than what light can produce motion. This is why the Coanda effect applies.

In the past no one even thought of placing the Coanda principle in line with gravity. By following the Kepler's formula we find that the Coanda affect IS GRAVITY. Gravity is moreover the Coanda principle than it is any other variety of forces or concept of contractions. The Coanda principle is gravity. It is the how all gravity is charged and distributed. That forms the basic principle of Cosmology. The Coanda affect is proof of Kepler's statement. It proves that motion $(T^2)$ establishes a centre at the distance of **k** and such motion claims the space $(a^3)$ from a centre within that motion. The motion creates a centre and a centre establishes a gravity field. The motion of the liquid water proves to be the time aspect because it claims the space in the time it is running. The faster the motion is the stronger is the gravity that the motion generates in the space it claims by the gravity it generates.

For these phenomena to occur some independent singularity spot has to prove their independence by overshadowing the Earth's gravity two conditions has to apply. A round object **k** has to be in place allowing the establishing of a space $a^3$ within the space, and then movement $T^2$ has to establish borders to such a space. That proves Kepler correct. And that proves gravity can come about at any given point by establishing singularity independent from the space it holds, the movement will generate heat which will energise gravity setting borders and confining that singularity in $a^3$ by and confirming the singularity energised with **k** producing the boundary of motion $T^2$. That was what Kepler said before Newton's interfering...he said that space comes about by the motion of time from singularity $a^3 = T^2 k$.

As the example of the atom in motion through space showed motion changes relevancies. But in that it changes the relevancy of space-time affected by the altering of the motion and not applying relevancy of singularity. The Coanda example shows that motion establish singularity dominating space-time controlling motion and the direction thereof. That means the motion establish the space-time and the position as much as the direction of the flow of space-time. Motion creates space-time as much as space-time is supplying motion to form space. The only way to enable that to become a reality is that motion creates space as much as space follows the direction of motion. That is what Kepler said when Kepler said the space is equal to the motion thereof $a^3 = T^2 k$. There is no solid $a^3$ Universe but all interrupted by positional changes that recreate the space in the and according to the new direction singularity will create as singularity allow space to flow by motion by fragmenting space into time sectors.

By any object applying motion such motion is reducing the time the object is occupying the space it holds as the occupier of a position in that space. By motion the object occupies more space and by occupying the space it holds the motion creates the reducing of the space in favour of the time coming about and duplicating the object reduction of such an occupation of space in virtual size. **The** faster the object will duplicate such occupation by motion the more the space will reduce in favour of the time remaining the same. That means the duration of the time will produce smaller space but longer occupied space.

Some pages ago I gave a specific example using an atom in motion in space-time by moving through space-time. Within the realms of the Earth the space tends to remain constant while the time duration changes accordingly. Motion is space duplicating and comes from the point where singularity contributes to space-time. Singularity is at the point where Mainstream Science now view that that point holds nothing. The growing in size of the space occupied is a result of relevancies coming about. Since **k**, moving towards $k_2$ is growing by the increase of $k_2$ the position $a^3$ holds in relation to $k_2$ and $k_1$ will result in a larger $k_2$ running from $k^0$ to $k_2$ through the previous point of $k_1$. Therefore $k^0 - k_1 - k_2$ is overall larger in ratio by measure of $k^1 - k_1$ was, therefore in ratio the space unit grew as $a^3$ remained the same. But that is the virtual position. In reality the space $k^0 - k_1 - k_2$ is a unit, and that unit grows by the Hubble constant and by no other means.

Now the same principle comes into effect, which takes place in the sound barrier and which takes place as the spacecraft enters the Earth's atmosphere. But the Earth holds its atmospheric space as a controlled unit. Outer space holds its measure by time. Therefore the object extending the motion has to comply with changes coming about. We know that the unit forming the craft, holds the unit by space to size taken from the accumulated compliment of all the atoms taking part to form the unit. The two units in relevance are also in dual because the object in motion holds relevance to a larger space than it did before. But seen from the role the space has, outer space holds the same measure by the same means. Outer space does not compromise because as far as outer space is concerned, outer space remained the same. In addition the space the object unit claims by atom occupation is much inferior to the gravity that outer space insist upon, therefore to adapt to the new relevancy, the spacecraft atom unit has to adopt the new relation.

The spacecraft atom unit has to reduce the claim it has on the space of the totality in order to compromise for the bigger overall space in time in relation to a larger space in time by the craft repositioning its time it has with a larger velocity. Therefore the compromise is that the lesser applying gravity will reduce the space it claims and such reduced space compromise will translate to heat in liquid taking up the discrimination.

The very same scenario occurs when the craft enters the Earth and since the Earth's gravity is totally dominating the craft's gravity, the craft's gravity has to adhere to the Earth's gravity and the craft has to compromise in size. In the case of the Earth atmosphere versus the craft's, the extending of the $k$ factor comes about as the extending of the $T^2$ factor since the compactness of the Earth's atmosphere translates to more space per measure compared to the craft's atomic space claimed. But all this might seem to us to be in the outer whereas in truth it is in the inner Universe of the Earth.

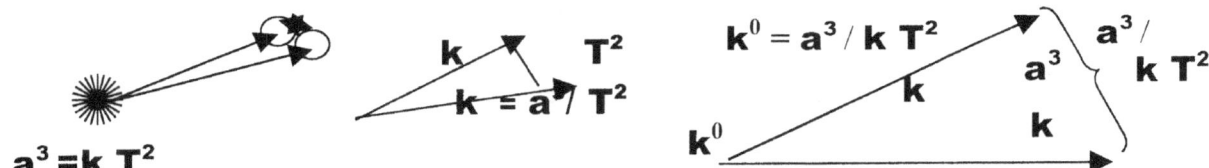

$a^3 = k T^2$

$k$   $T^2$

$k = a^3 / T^2$

$k^0 = a^3 / k T^2$

$a^3 / k T^2$

$a^3$

$k$

$k^0$

$k$

**Motion is space duplicating and comes from the point where singularity contributes space by motion. Singularity is the point Mainstream Science now view that holds nothing as a value. The only definite place one will locate zero is in between the starting point of the lines going in opposing direction in the position the lines hold before there was the least of directions applied, but that is only because there is no such a position, not because any line is coming from there. The two lines are still one holding the opportunity of parting as an option but have not yet parted and therefore are on the very precise same spot. The line coming from there is already there because it already has the choice of going in any and all opposing directions and when it starts running it will place filled space in that location because the space was already filled with a line starting and not with a line not there at all. When reversing a line we might find a better idea of what is in place and where it is in place.**

From the centre of the atom the electron will align in a position that it will maintain in relation to such a centre. As the relevance of the increase in motion outside the atom reduce the space occupied and by holding synchrony to singularity in the centre the electron will use the same duration to cover less distance since some of the distance is increased on the outside of the atom. That means the electron will take the same time to complete the rotating circle but since there is less space to cover the motion would be relatively slower because the motion is shorter. The duration of space duplicating or space interrupting at that speed favours the interrupting part more than the space part in relation to the situation where the relative motion of the particles were stationary.

The six sides in the six-sided Universe is a confirming of the past presently running into the future. Motion is the driving of space through time by the duplicating of space in time. The future is the present becoming the past by duplication of motion of space using time to manifest space. The past becomes the picture as motion confirms material by allowing time retarding to compact heat into a space we find as being space –time.

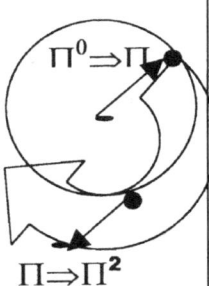

There are two ways of looking at this issue. The one is looking at it from the centre that is keeping the rotating object honest or there is the rotating object forming space in relation to the centre and placing the centre in the centre. It will always be one taking prominence to the other and where Kepler introduced the formula it is indicating motion producing gravity which is gravity that is keeping form outside the sphere. Gravity is motion but the motion we see is much different from the gravity we experience while we know it has to be the same with only relevancies changing.

**From singularity, there is motion as three markers, which is moving to fill the position the three other markers where the space will be and are pointing. That is space filled by motion. The space is a direct result of the motion coming about as the outside of the sphere has six sides establishing the space. Then what about the sphere? We can only see the differences there are between calculating the cube and calculating the dimension of the sphere when using the manner that Kepler calculated. By turning Kepler to a more precise stance it shows that the calculating of that space is just as much different from the manner that Newton used to calculate the sphere. There is distinctly no comparing in the calculation methods used in either one of the methods in measuring the sphere versus measuring the cube and that should show that the answers Kepler arrived at do not indicate the normal mathematical calculation methods normally used.**

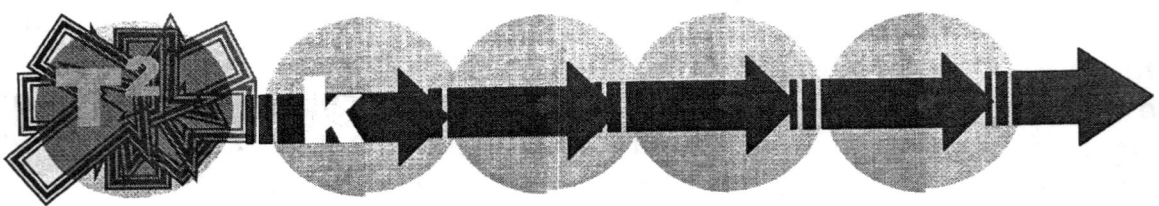

By going faster the duration of lost space is filled with liquid heat because the motion that reduces the **k** factor also takes the moving object back in time going closer to the conditions that applied at the Big Bang. Time slows down as space occupied decreases with singularity dynamics coming closer to the position singularity holds. The time does not in affect slow down as much as the duration of the time it takes to produce the interrupting of space. The duplication will be more, which is taking longer in relation to the time in the space that is viable. One has to break time down to $((1836)^3)^2$ multiplied by the speed of light, which is C in order to establish the frequency.

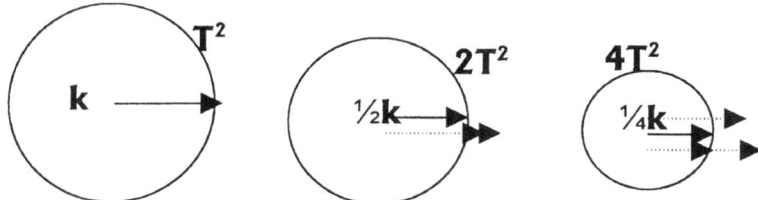

By producing motion to the object that relates to a change in the space-time relevancy changing as the motion applying alters the relevance in time producing space. As motion or gravity doubles the atomic border distance of the electron position would half whereby the space being occupied will be the quarter of that which it was before. The linear motion will establish a centre singularity $k^0$ just as the circular motion does but since $k^0$ comes about from $T^2$ time changing the relation and where **k** must be responding **k** will no longer set a **k** confirming the relation of the border but **k** will reduce the relation to compromise for the changes that alters the $k^0$ position at every alternating point there is of establishing space once again. Where the object in motion has an elected $k^0$ that holds a relevance to $k_1$ the faster motion will produce more duplicating and that duplication will stand relevant to more space outside the atomic accumulated space. In that manner $k^0$ of the elected atom unit that is going faster must reduce its claim on space occupied because the space not occupied is more, but if the relevance is placed on the other side the space remains the same and therefore $k^0$ holding the claimed space has to reduce. The motion coming about is also gravity increasing because as I shall show later, gravity is purely motion and gravity is not a force of the magical kind. By altering the space as the space reduces it allows the electron to remain in harmony as the space duplicate according to gravity.

By going faster the duration of lost space is filled with liquid heat because the motion that reduces the k factor also take the moving object back in time going closer to the conditions that applied at the Big Bang. Time slows down as space occupied decreases with singularity dynamics coming closer to the position singularity holds. The time does not in affect slow down as much as the duration of time it takes to produce the interrupting of space and the duplication will be more, which is taking longer in relation to the time in the space that is viable. But one has to break time down to $((1836)^3)^2$ multiplied by the speed of light, which is C in order to establish the frequency.

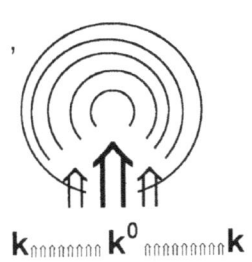

$k_{\cap\cap\cap\cap\cap\cap\cap}\ k^0\ {\cap\cap\cap\cap\cap\cap\cap}k$

**With singularity placed in infinity within the centre of every rotating object, therefore every atom is in singularity. The atom holding singularity is relation to its surroundings including other atoms form space-time diverting from the point holding singularity as far as rotation goes because every object holds three relative positions in as far as where it was, where it is and where it will be in relation to singularity providing time. I shall elaborate this a little.**

When the line came from singularity and expanded into space forming space and by using Kepler's formula we can see two dynamics coming into place. There was singularity remaining at value holding six positions in relation. This is a how Kepler saw the involvement of space-time in relation to singularity. However, there is another involvement. When we observe the formula Kepler introduced we must see where the formula holds value and by doing that we find three factors that is equal in relation to forming a relation.

In the one formula Kepler introduced the relevancy suggests that space is producing motion to becoming space filled by motion. $a^3 = T^2 k$. That is an indicator used by material forming a circle outside the sphere to accommodate the motion of the sphere. Clearly such an indicator would not use the direct value of singularity because it is not in direct contact with singularity. In this relation I showed where the factors are the same although the symbols used suggest otherwise and that made me realise to search for a symbol that will produce equality. There are three sides in space and the three sides are moving in the direction forming the next three points the previous three points will form. What may seem to us being as realistic as man can be, we see a solid structured Universe boxed into a nice cube. Instead, we have a hugely flickering Universe, which is only relevant when viewing space-time from singularity.

Kepler stated that planets orbit in relation to

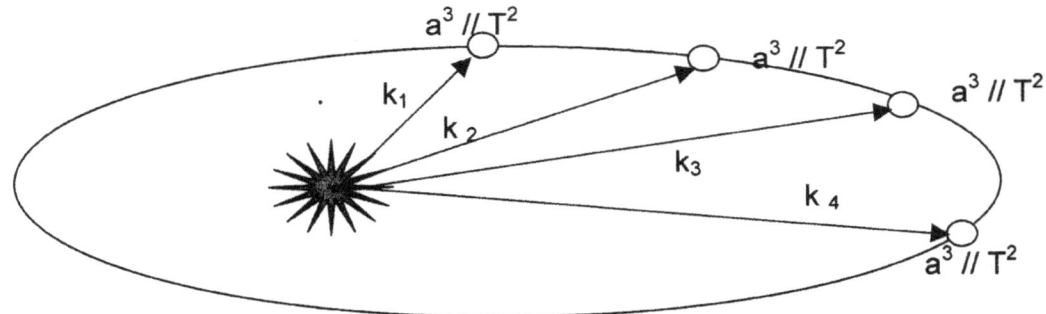

The dual action that all cosmic motion presents is always present whenever motion comes about and to be in the cosmos is to move about in the cosmos since all motion is about confirming space in time to singularity. It is providing singularity to express space in time by confirming singularity and in that one may never remove the rotation there is from the line forming or the other way around. That is another conformation, which the confirmation found that Kepler's formula provides. Space $a^3$ is always equal to both in compliment of $T^2$ being time in rotation as well as eternally linked to space in line **k.**

In the one formula Kepler introduced the relevancy suggests that space is producing motion to becoming space filled by motion.$a^3 = T^2$ **k.** That is an indicator used by material forming a circle outside the sphere to accommodate the motion of the sphere. Clearly such an indicator would not use the direct value of singularity because it is not in direct contact with singularity. In this relation I showed where the factors are the same although the symbols used suggest otherwise and that made me realise to search for a symbol that will produce equality. In space on the one hand there are three sides in space and the three sides are moving in the direction forming the next three points the previous three points will form. What may seem to us being as realistic as man can be, we see a solid structured Universe boxed into a nice cube. But instead we have a hugely flickering Universe, which is only relevant when viewing space-time from singularity.

$[k^0 = a^3 / T^2\ k] =$ 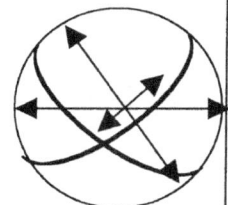 $\boxed{7}$ Edges come from solid structures within the space having the six sides in relation to the sphere and from the material in the sphere, there are six points securing one another by $90^0$ and $180^0$. In the centre of the sphere where all the pointers connect as well as cross, there is one point all points refer to when locating every point holding a relative position. That then must be one. This still would not have made much sense if it were not for reapplying the Kepler formula but this time to suit the connecting in the sphere.

This still would not have made much sense if it were not for reapplying the Kepler formula but this time one must use the Kepler formula to suit the connecting sphere. Since a centre from a centre in which the centre is part of a sphere defiantly controls the planets in the sphere, which we named to be the sun. In the cube we find the one side provided the other side space but the sphere goes beyond that. The solid structure of the sphere not only relates but also places the one in connection with the other and this is directly placing the relation to a centre where such a centre must be one. That suggested to me that centre that mathematically according to Kepler there must be a centre that will provide a connecting. $k^0 = a^3 / T^2$ **k.** From there I divided the line **k / 2** that brought me to singularity and from singularity I could value singularity and find the connection that the cosmos uses to bring about cosmology in the form of $\Pi^0 = \Pi^3 / \Pi^2\ \Pi.$

From singularity there are motion as three markers, which is moving to fill the position the three other markers where the space will be and are pointing at. That is space filled by motion. The space is a direct result of the motion coming about as the outside the sphere have six sides establishing the space. Then what about the sphere? We can only see the differences there are between calculating the cube and calculating the dimension of the sphere when using the manner that Kepler calculated. By turning Kepler to a more precise stance it shows that the calculating of that space is just as much different from the manner that Newton used to calculate the sphere. There are distinctly no comparing in the calculation methods used in either one of the methods in measuring the sphere versus measuring the cube and that should show that the answers Kepler arrived at does not indicate the normal mathematical calculation methods normally used.

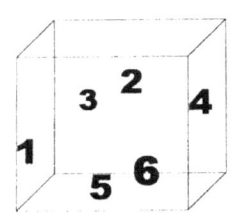

Outside the sphere, we find six sides where three sides oppose three sides. The six sides place a specific dimension but they do not define specifically impenetrable edges. The sides are vague and undefined but from the sides one can introduce borders that will come about by connecting angles on the outside using $90^0$ as connecting angles.

There is front 1 to back 2. There are left 3 to right 4. There are bottom 5 to right 6

The practise clearly points to another method because Kepler did arrive at and where Kepler found answers and in fact the formula came about from the answers he arrived at. His work did not need the changes that Newton introduced to find the calculations that came to form the formula. The calculations produced the formula and not the other way around. This is bringing the crux about cosmology to the foreground. It proves that the mathematics applying to cosmology is not standard maths used where mathematicians are in the designing of a high-rise building. In the cube there are six measurable sides.

The layout of the sphere demands a totally different perspective since the sphere on the inner parameters are about specific borders in the stringiest control from the centre and the length of the edge calculated from the centre is in measure precise to five other opposing borders. From such a precisely located centre there are no margins of flexing the radius which will alter the points forming the edge or border of the sphere other than adhering to precise cosmic principles concerning gravity. Gravity is the strongest where space is the least and that is in the centre of an evenly space sphere where such a centre produces the gravity in precise measure to accommodate every possible point in six opposing sides at any position on the edge of the sphere.

But as one can see I also realised gravity is relations of motion applying in two factors. There is no separation of the two of the factors acting as one but both have different application and values in the unit. It was what gravity was because this action prevented expanding whereas centre commanded the borders and assisted in the collapse of such borders. This is the result of singularity having three parts acting as one but giving three distinctions in application. Gravity is as much part of dismissing space as it is about making contact with space in time. Since the connection comes about as a circle, the connecting points will relate to $\Pi$ as the value. Due to the spinning nature of such a point with all surrounding the point will be alternating direction favouring change every instant of a time frame used and in that the value to such a point can only be $\Pi$ because of its constant changing. Using r would specifically oppose another r from every angle because the use of r will bring about a static relation to the previous and following instant and therefore it will cancel the constant spin flow. By reducing the line to its maximum possibility one end with $\Pi$ being the minimum but that $\Pi$ is actually $\Pi^0$ which can also be $k^0$ or $a^0$ or $T^0$, which all indicate positions in singularity. Only when forming a value past singularity does independent identification come about. When the atom formed that atom applied a relevancy of ten positions where seven positions are included in the atom spinning and three positions are part the exterior of the atom spinning but all the positions relate to singularity but as space flight taught us such relevancies can change when an object is within the space boundaries of a larger structure or roaming free in outer space. Within the boundaries of the atmosphere where the sphere border touches the space borders the space borders hold six positions and the sphere hold seven points. But at the precise place where the points make contact with the sides one side fall away in favour of the point it connects too leaving five sides relating to seven and where one of the six sides takes control in removing one of the cubical sides by replacing that side with a sphere point position the object then becomes directly controlled by singularity positioned in the centre of the sphere. The object seems then to fall from space and enter the atmosphere becoming a shooting star. What the Coanda effect proves above anything else is that gravity in control of space-time comes about from a centre and such a centre can be created by motion applying to a liquid in relation to a solid. That means there is undisputedly a flow of space-time towards a centre and the centre has to diminish the space-time reaching such a centre to create the flow and therefore the control from such a centre. That's the one pivot of gravity.

In the cube we find that the one side is providing the other side, which is the side directly opposing each other with space by duplicating space through motion of space but the sphere goes beyond that. The solid structure of the sphere does not only relate by opposing sides mounting opposing borders, but there are no precise centre such as the case is with the sphere being in precise contact with the controlling centre. By having precise cross referencing on each other makes the sphere superior as there are then also placed in relation a precise centre where six point on the outside are relating by one in connecting with the other. The opposing sides run a connection in a precise duplicating of the one onto the other as the cube does but goes beyond that and this is directly placing the relation to a centre where such a centre must be one centre to every possible crossing line running from any given point to the directly opposing point through such a centre position. That suggested to me that there is a controlling centre in all spheres that can mathematically, according to Kepler, form. There must be a centre that will provide a connecting and by reducing such a connecting line from both sides will be where I then will be able to find $k^0$ or singularity as suggested by the formula translated from Kepler $k^0 = a^3 / T^2 k$. From there I divided the line $k / 2$ that brought me to singularity and from singularity I could value singularity and find with that the connection that the cosmos use to bring about cosmology in the form of $\Pi^0 = \Pi^3 / \Pi^2 \Pi$. The way I concluded the value of singularity as $\Pi^0$ going to form $\Pi$ I have already explained.

But as one can see I also realised gravity is relations of motion applying in two factors. There is no separation of the two factors acting as one but both have different application and values in the unit. It was what gravity was because this action prevented expanding although gravity is about expanding as much as contracting. This is the result is that singularity has three parts acting as one but giving three distinctions in application in two opposing variations where both points in motion is controlled by the centre. Gravity is as much part of dismissing space as it is about making contact with space in time by duplicating space in using time. Since the connection comes about as a circle, the connecting points will relate to $\Pi$ as the value. Due to the spinning nature of such a point with all surrounding the point will be alternating direction favouring change every second and in that the value to such a point can only be $\Pi$ because of its constant changing although such changing is remaining the same. Using r would specifically oppose another r from every angle because the use of r will bring about a static relation to the previous and following instant and therefore it will cancel the constant spin flow. By reducing the line to its minimum possibility where all ends on one end it will be with $\Pi$ being the minimum but that $\Pi$ is actually $\Pi^0$ which can also be $k^0$ or $a^0$ or $T^0$, which all indicate positions in singularity. Only when forming a value past singularity does independent identification of distinct value differences come about. When the atom formed that atom applied a relevancy of ten positions where seven positions are included in the atom spinning and three positions are part the exterior of the atom spinning but all the positions relate to singularity.

As space flight taught us such relevancies can change when an object is within the space boundaries of a larger structure or on the other hand roaming free in outer space. Humans are taller in space than they are on Earth. Within the boundaries of the atmosphere where the sphere border touches the space borders the space borders hold six positions and the sphere hold seven points. But at the precise place where the points of the cube and the points of the sphere make contact with the one another's sides the domination of the seven contact point, we find in the sphere, including the controlling centre will dominate the six sided cube and will bring about that one side of the cube will fall away when making contact with the sphere in favour of the point it connects to. By one side removed, this is leaving five sides in the cube in relation to the seven sides in the sphere where the relating to seven eventually stand in for one as the Lagrangian system spawns from this. Where one of the points in the seven-sided sphere takes on the six sides, the seven will take control in removing one of the cubical sides by replacing that side with a sphere point. This is when the sphere centre takes control of the position.

The object then becomes directly controlled by singularity positioned in the centre of the sphere. The object seems then to fall from space towards a centre point and enter the atmosphere becoming a shooting star. What the Coanda effect proves above anything else is that gravity in control of space-time comes about from a centre and such a centre can be created by motion applying to a liquid in relation to a solid. That means there is undisputedly a flow of space-time towards a centre and such a flow contributes to space being liquid. The centre has to diminish the space-time when space-time is reaching such a centre to create the flow that produce the motion contributing to gravity being in place and therefore the control must come from such a centre where space is dismissed by the lack of possible motion. That's the one pivot of gravity.

Since the Coanda effect shows gravity is the control of space-time by motion flowing towards a centre, which then allows to control and manage the space-time within the realm of such a centre that then also prove as it explains the one part of gravity. The part it explains is that space reduces space by increasing time towards a centre and that is established by motion and the lack of space establishes a lack of motion in that very centre. Because there is no space in such a centre there can be no motion of such a space in the centre and since there is no motion of space in such a centre there can be no duplication of space in such a centre, which then brings about the killing of all space before and after the space can form.

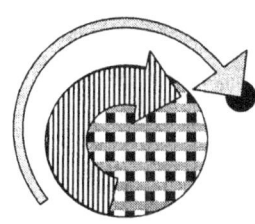

When the object is released from the atmosphere of the dominating space, this very same gravity $k^0 = k\ T^2/\ a^3$ ratio will still be enforced since it is not the law of the Earth prevailing but it is the law of the Universe applying. Outside the atmospheric borders the Earth no longer have the means to remove one of the cube sides that form the lesser object space and where the cube reinforces position by keeping the rotating object in position floating above the Earth.

Newton did not see it that way. Newton did not see relevancies. Newton ignored the vital role motion plays in gravity. Newton placed one relevancy between the objects as to serve both objects. But there is so much more to the relevancy because the motion applies from both ends in regard to both ends.

A larger object will allow more time to duplicate the bigger space it has in the same time duration than would a smaller space find a need to duplicate a smaller space in the same time duration. But in the cosmos there are no big or small and only relevancies produce links with and too singularity. By having less space while having to produce the same time the illustration of the atom once again enters the equation we think of. By having s smaller space to duplicate than the larger partner a shorter **k** is needed when the same time $T^2$ is enforced. But as Galileo's swinging pendulum proves the space diminish to allow the time $T^2$ component that the Earth enforces on the smaller object to be adhered to by the smaller component. Since the time does not match the duplication required to sustain the **k** factor and the time factor cannot change the space of the smaller object has to reduce in relation to the diminishing space of the Earth forming the atmosphere. Therefore the smaller space seeks to find a position where the **k** component will match the time component because the space is offered as compensation in any case.

This is the very same principle applying when there is a lesser object such as a planet in relation to the Sun floating in orbit with the Sun. The minor object then received an individual $4\Pi^2\ a^3/\ T^2 = G\ (M_s + M_p)$ What I am trying to say is by using Newton's symbols which he added to fill the picture he saw helped nothing. What he did was to repeat what Kepler said by duplicating what Kepler said. He only extended what Kepler said. The value of **k** is what he suggested as $G\ (M_s + M_p)$ and $4\Pi^2$ is representing **k** by implicating gravity to the full time value of **4.** Later I shall show that the Earth or any free orbiting structure in relation to the Sun is $4\Pi^2$ and in that sense the relevancy of all stars committed to fusion has a relevancy value of 7/10 multiplied by $4(\Pi^2 + \Pi^2)$.

If we think of the statement as Newton put it, it comes down to the manner how the Earth as a planet connects directly to the Sun with having the Earth only as a relevant mediator and not as a controller. It is because $\Pi$ is the basic basis which the Universe apply as a value. Because $\pi$ is already the principle the use of $\pi$ was included as the base value and that is the main reason why the cosmos did not care to involve the use of $\Pi$ when it produced the formula which was unveiled to Kepler.

The only value that must apply is that the distance from the centre holding the factor **k** holds space and time valued. The Earth then sets the speed of motion or the gravity factor demanded by the Earth of the orbiting structure orbiting the Earth. After the glitz and glamour that accompanies a launch of a rocket into space, the celebrated factor of such a launch is placing the object as a Sun controlled factor that after the launch became independent. Such independence would have to rely on the gravity that the orbiting structure produce to validate the sustaining of such a relevancy in continuing the independence as the object puts itself in the role as another satellite with semi solar factor to encircle the Earth as a satellite. To sustain such motion the object must be in a position to charge the required amount of heat and offer the heat as payment to the centre of the Earth and then set off the orbiting object.

Remember never, for one second, to forget that the launch and the orbit are manmade and directly resulting from the intervening of the will of life and are therefore as far as the cosmos go totally artificial. Where such orbiting took place in relation to cosmic law such an orbiting object had to secure just the rite amount of heat to establish the velocity it must have in order to bring about the gravity it need to stay relevant with the motion of the Earth. In order to maintain the freedom received by the orbiting structure that structure must apply a time in motion in relation to the centre where this orbit speed must match the distance from the centre in harmony with the Earth.

If the object is going to slow it will retreat by falling towards the Earth centre and when going to fast it will apply more distance taking the object further away from the centre. It is the motion that the object has in relation to what is coming from the motion of the Earth dictated by the centre in the Earth. As it should be clear to every one, one find the gravity that is then applying to be a relevancy as speed or motion is a relevancy where space in motion relates to a specific duration of time from point to point $a^3 = T^2k$.

With the term micro gravity such a term is placed in me to bring the idea across that the cosmic rules does not apply any longer in outer space or if it then applies, it applies in a very diminished sense. It is documented that an astronaut can pick up something like an object that will show an Earthly mass of about four tons of mass when on the Earth when he is working with that object in outer space. The condition to this lifting is that the person then must have contact with a device that is in control of a much more dominant singularity. This then makes that again there is a precondition attached.

Such a precondition is also a relevancy of some sort bonding is one of the contradicting ideas. This time a much larger object must secure the position of the astronaut such as the spacecraft when he does this lifting of the four-ton object. The larger object then produces a controlling point serving as the elected singularity governing to all the space-time that is forming the structure unit. This is the result only when the objects are not being captured in the space claimed by the Earth but is still captured by the motion the Earth brings along.

From the centre in control of the Earth this motion will apply to space duplication in relation to the Sun without the Earth applying boundary control over such motion in that specific space. The only control coming from the Earth centre is the speeds forming the gravity or gravitational position the orbiting object has to secure. I guess one may say the objects are in the region represented by the ominous G (using Newton's formula) as in the Gravitational constant that is everything but a constant or has any Universal equal applying. The bodies at that position relates to three independent markers in singularity where every marker is duplicating space in relation to what the Sun demands and what the Earth provide but where every object still responds to the individual dismissing of space according to individual singularity protection. It is all about relevancies applying attachments of which mass is but only one implication of such a relevancy

The Pulsar is much more Universe because it holds much less of our Universe and is much closer to the true Universe being the singularity it claims, standing in relation to the singularity I claim. It is the singularity within that Universe and not the motion providing space to be within the creation.

The space the pulsar has relates directly to the time it takes to replace the space but not only that the space-time inside the pulsar is relatively in opposition to what we find space-time to be. The motion creates our space much slower because we create much more space and much less heat. The time it is taking the pulsar to create space is in our view instantaneously because there is comparatively very little space to create but there is enormous heat involved and very little time to recreate the space. Motion duplicates space and the more space being created by motion applied the more solid the space will seem. The Universe is not what we are looking at but the Universe is what we cannot see. The Universe is where singularity comes together or where singularity parted to form what we see. But what we see is the extending or reality and not reality. The pulsar is much more universe because it holds much less of our universe and is much closer to the true Universe being the singularity it claims standing in relation to the singularity I claim. But it is the singularity within that is the Universe and not the motion providing spec to be within the creation. Because the Pulsar is flickering so fast we have the fortune to be slow enough to be solid by space and motion. The space becomes the motion.

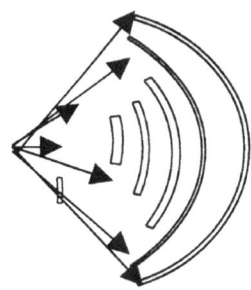

Although space and time is the same the one is the opposite of the other. $k^0 = a^3 / T^2 k$. Space is the same but the inverse of the tome there for less space constitutes to more time. Looking at the pulsar we see the a little space to duplicate at the inside where the Pulsar is much closer to singularity. From the inside looking at us the space will seem eternal and the time it takes to duplicate the motion will have an eternal relevancy just because we see the pulsar to act instantaneously

If we observed the Pulsar as equal to our time the durations would have matched but since the space is obviously. With the space reduced but we see the time as the reduced factor means we have an inverse view on the practicality of space –time towards the inside of a star. Our perception must always dictate our observation and not the other way around. That is what brings about the relevancies we have to acknowledge.

Because the Pulsar is flickering so fast, we have the fortune to be slow enough to be solid by space and motion. The space becomes the motion. The heat that is directly released from a pulsar will be able to destroy large regions of the part of the Universe we occupy by cutting the space-time into cosmic threads. Still, we also must remember not to regard the pulsar much alien from us, because within every atom we compose of and that forms us as well as our Universe, hide all the aspects that govern the Pulsar. The pulsar only used a huge number of more protons to get it unified in the singularity it holds than we have at our disposal. As time (not space) changes, bringing about different space, we are heading in the direction the pulsar froze space-time. It is only a matter of having other aspects in our atoms, which in our Universe takes prominence by importance compared to the Pulsar's Universe.

It is the measure the singularity sustaining the drive of the Pulsar has gained by destroying space-time and duplicating heat through light reflection as it incorporates all singularity within the confinement of the star as a unit forming eventually only one atom the size of many galactica combined. With that much singularity focused into one unit, the relevancies of the Pulsar have to change considerably from the relevancies governing our little yet-to-be developed planet.

In all units forming in space, there are six sides in relevance with three sides in attendance held by singularity in relation to relevancy of space-time and three sides coming about through motion or repositioning singularity space-time. Before singularity overheated that brought space-time about and placed space-time as a form of heat in a position to heat we have to presume that all space-time that formed in the form it did, was there already present but it was only still in the theory part of Creation.

The development thereof came about as gravity came about and formed relevancies by applying the motion of the three sides. The motion established relations by three in three with three in forming a balance to indicate relevancy with singularity. It is most likely that it is ten points each in independent singularity where each is forming a relation. The ten plus one point nine plus ten is in relation to seven which forms pi, but that means there are always ten points in seven to pi. Only when such relevancies exist is where the relevancy establishes a partnership or connections. Gravity can only come about as a relation of motion differences between points occupied by singularity in maintenance. The motion of space-time inside and outside between atomic particles establish gravity by motion of space-time flowing.

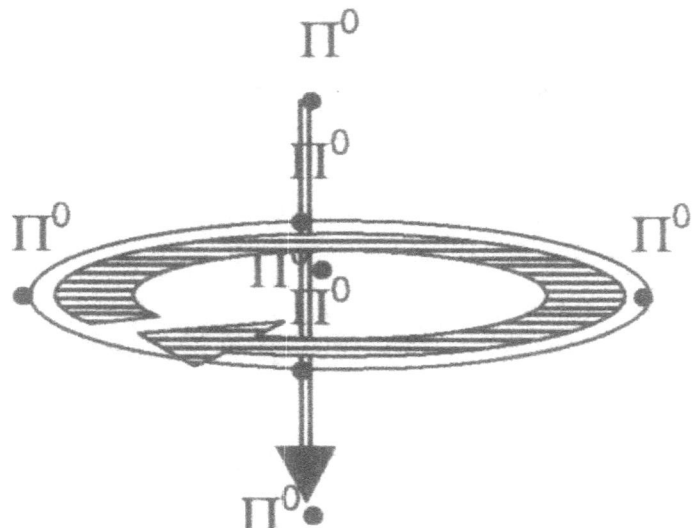

Through the rotation came boundaries, which we now consider to be particles in time. It is the time (from $T_1$ to $T_2$) that it takes **k** to swing into a different relevance to $\mathbf{a}^3$ space. It established movement in the area holding the least space. The relevancy started when **k** claimed space by motion from singularity. That is what Kepler claims. What it does say is how Kepler incidentally forgot to improvise for the claim of a circle but accidentally forgot about including $4\Pi^2$ on the one side and G (m + mp) on the other side. Kepler placed the growth **k** directly relating to the area $\mathbf{a}^3$ separating as the spin $\mathbf{T}^2$ provides the space. Kepler showed with his formula what gravity is. Gravity is the least space (singularity) $\mathbf{k}^0$ claiming space $\mathbf{a}^3$ through spin $\mathbf{T}^2$ throughout the entire distance of **k**. Kepler announced space-time, the Hubble constant, the Big Bang and other later cosmic developments. He said space comes by the motion thereof.

Only when such relevancies exist is where the relevancy establishes a partnership or connection. Gravity can only come about as a relation of motion differences between points occupied by singularity in maintenance. The motion of space-time inside and outside between atomic particles establishes gravity by motion of space-time flowing. Then a moment arrived where **k** developed from $k^0$ to **k**. This brought about a revolution of cosmic proportions and this was the only time that such an expression is not exaggerated. Matter divided from singularity as matter claimed space. The growth of **k** had to produce $a^3$ which is an interpretation of the space **k** will bring in place. The most tiny and slightest of growth established **k** coming from $k^0$ to **k**. We can never bring about any concept by which to understand this. By establishing **k** that very second $a^3$ also came about. Through Kepler we can see how singularity achieved space to come about. It came through spin.

**Everything is space-time by confirming space in establishing time**

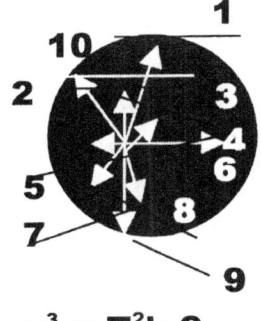

$a^3 = T^2 k$ 9

When the Universe was in the beginning with the entire cosmos still in a single dimension there were no limits as we know limits to form in the Universe we use and no borders indicating limits because after all it is the single dimension where there is only one dimension holding so much diversity. The dots referred to in this case have no space but were as close as singularity is when singularity has no sides but only shapes and the lot were the same, the very same one with a time delay parting them. The borders were part of development because we can witness the legacy of such borders in the present day holding the 3D in place.

Remember that space was at a premium like Creation afterwards never can repeat again when the first moment arrived for the very first time. In contrast matter had to expand as heat surged in search of space to Create cooling. Because of the overheating, space was a desire but the heat had no means to expand to find a form of release from the eternal grip of singularity being space less. Kepler said that $a^3 = T^2 k$. Space broke away from singularity by applying spin. Gravity is spin. When **k** extended it secured $a^3$ being space but through the spin of $T^2$ the space separated between particular points holding singularity, to establish individual singularity. It secured individual space from space by rotating space.

Our need that we have to form the understanding of this notion that space produce time to duplicate space by dimension is all inclusively to underline what space-time is as much as it is critical and demands our full attention in forming our understanding concepts we have about the Universe and Creation at large. My believing Kepler then tells me that the space we have is set in a solid state by the motion that the atom brings about. But way down in the atom where singularity takes charge every action is the vibrations of pulsating protons, flickering to set motion to the immovable singularity. The atom finds the way to draw "flat" and re-establish space-time again. By taking full control the singularity that is elected by the compliment of atoms begins to govern as it unifies all individual atoms into one structural compliment. A star such as the Pulsar is an atom that by now is in a position where the size of one star is fast becoming. That is the direction the Universe is heading one atom. That reflects what happens in the final stages of the Universe before singularity once again steps in and takes control as it does in the Black hole. At first singularity spawned and in the end singularity will grow to capture and remove all spawned uncontrolled space-time to get the Universe under direct control of singularity once more. The pulsar expands and we see that as light streaming from the star.

Through the rotation came boundaries, which we now consider to be particles in time. It is the time (from $T_1$ to $T_2$) that it takes **k** to swing into a different relevance to $a^3$ space. It established movement in the area holding the least space. The relevancy started when **k** claimed space by motion from singularity. That is what Kepler claims. What it does say is how Kepler incidentally forgot to improvise for the claim of a circle but accidentally forgot about including $4\Pi^2$ on the one side and G (m + mp) on the other side. Kepler placed the growth **k** directly relating to the area $a^3$ separating as the spin $T^2$ provides the space. Kepler showed with his formula what gravity is. Gravity is the least space (singularity) $k^0$ claiming space $a^3$ through spin $T^2$ throughout the entire distance of k. Kepler announced space-time, the Hubble constant, the Big Bang and other later cosmic developments. He said space comes by the motion thereof.

Our need that we have to form the understanding of this notion that space produce time to duplicate space by dimension is all inclusively to underline what space-time is as much as it is critical and demands our full attention in forming our understanding concepts we have about the Universe and Creation at large. My believing Kepler then tells me that the space we have is set in a solid state by the motion that the atom brings about. But way down in the atom where singularity takes charge every action is the vibrations of pulsating protons, flickering to set motion to the immovable singularity. The atom finds the way to draw "flat" and re-establish space-time again.

By taking full control the singularity that is elected by the compliment of atoms begins to govern as it unifies all individual atoms into one structural compliment. A star such as the Pulsar is an atom that by now is in a position where the size of one star is fast becoming. That is the direction the Universe is heading one atom. That reflects what happens in the final stages of the Universe before singularity once again steps in and takes control as it does in the Black hole. At first singularity spawned and in the end singularity will grow to capture and remove all spawned uncontrolled space-time to get the Universe under direct control of singularity once more. The pulsar expands and we see that as light streaming from the star.

This provides space by motion expanding the heat and increasing the space. Then motion produces contraction and the star starts a spin that removes whatever heat/space there was taking it by spin back to singularity. We might observe a broken stream of light as flickering and it may occur let's say twenty times per second. Inside the star one such cycle of flickering will last say one thousand years. I am very conservative in an attempt not to shock the reader witless but in other books I use much more realistic periods in time duration. The time on the outside where we are located and the time duration which the pulsar froze when the pulsar developed individual space-time from that of the Universe has a differentiation in period duration of billions of years.

As space $a^3$ expands, time $T^2$ has to change since **k** enlarges. Being in the star reduces the time so much that if the pulsating cycle is one cycle every thousand years then by the same standard will be that the time on the outside where we are located in is having the duration of twenty thousand years per one pulsar cycle. Remember there are possibly forty flickers per our one second. It takes a thousand years duration for the cycle to end and the full duration from the start of a flash to the end of a flash or the start of a dark faze to the end of a dark faze is twenty thousand years in one faze. The star was shining for twenty thousand years or it was not shining for twenty thousand years making one cycle duration becomes forty thousand years.

The time going on in the star is Π by$10^3$ years, which is a normal relevancy in a star cycle of that development. Then the other side of that Universe take charge and all the factors repeat. But since we find ourselves part of a six-sided 3D Universe our set of time frames per space Unit that is largely out of synchronise with such a massive star the star is in harmony with our inner atoms. Such a star may embody our entire galactica many times over into one growing singularity as far as material occupation positions a star. There for time on our end is pretty slow to the pulsar as the space frames takes a much quicker duration to complete. We see the flash at a pace of forty flashes per second but to realise the other side we have to place ourselves in the position of the pulsar. We have to place our minds in the time from applying in the star.

If the pulsar takes forty seconds of our time to complete one period or faze and that phase takes twenty two thousand years in the time frame of the star to complete we must present a pretty motionless picture to whatever is looking at us from the inside of the star. At such a rate, it is no wonder we find us in a solid Universe, which we know that that Universe, has to flicker but the flickering is so fast and uninterrupted, we find it to produce a solid state of affairs. When the Sun broke free from the Milky Way, it secured individual space-time. That also happened when the Earth secured an individual atmosphere. But the capturing of the space by singularity also accomplished the freezing of the time duration that applied at the time when the Earth or the Sun or any star captured space-time.

Space increases as time in duration decreases **k** = $a^3/T^2$ and by becoming individual it holds space and time as part of space $a^3$ time $T^2$**k**. Even if the star had the ability to breakdown all space-time and reposition all space time at a rate of forty times in our second then in one second of the time relating to such a star we would still go very slow. The well-developed star would then find one thousand years worth of time in every second we have. To argue that the star and us uses the same time component in duration is quite thoughtless and laughable. It is pretty Newtonian to think of space-time in that way. One cannot alter space and not time while in the same breath science promotes the idea of space-time. What happens to space forces the reverse to happen to time. It is precisely what Kepler said when he said **k** = $a^3/T^2$.

There has to be some difference to time when space changes, since it takes the light coming from such a star many millions of years to reach our part of the Universe and there was some Hubble development since then, when the light was sent on its journey to now where we are. We see it to be one second but the inside of the star see us as moving much more likely only once every million years because our moving about might take us a twenty four hours to complete our cycle.

Compare that in relation it takes the pulsar to complete twenty-one and a bit times multiplied by many thousand years in one second and multiply that with thousands of years in each cycle. When looking at us being on the other side or the outside we travel so slow that in the view of such a pulsating star from on the inside of such a star we, on the outside of such a star have gone solid. Our time duration seems never ending from the position the developed star has because our second is a matter of thousands of years compared to what is coming from the inside of that pulsating star. Those on the inside of such a star see us standing without motion for one million years and then in that million of their years we do the travelling that we do in one second of our time.

To them our Universe froze solid and that is what happened. That means by using a ratio of one million years to one second of our time we freeze the moment in our time to a standstill for the duration of one million years. If we did not freeze space-time we would not have had the fortune of being solid in the 3D that is producing the six dimensional Universe captured in space-time we enjoy. We are motionless because we are taking such a long time to apply motion. Our motion remains relative to the motion of the pulsar and what we see in the star, the star see the very opposite in our behaviour.

The space the pulsar has relates directly to the time it takes to replace the space but not only that the space-time inside the pulsar is relatively in opposition to what we find space-time to be. When saying this we must not stray too widely away from the flickering because that flickering is also inside every atom that forms our Universe. The motion creates our space on the outside much slower because we (our atoms) create much of the space-time we enjoy but is using the same heat to manufacture the space-time we use as frozen material. But to get the heat as concentrated as it is in the pulsar it takes a million years to concentrate the space to the amount of heat used to produce motion in such a dense environment.

The pulsar froze a small part of the Big Bang space-time as the rest of space-time being outside the star developed away from what was space-time when the pulsar star came of age. When we look at such stars we see a Universe whose stars froze in space-time to our benefit. It is the converting of such an amount of space from the amount of heat that makes up the duration differences because the slow part of the star is in motion at the speed of light and the really fast part of the star is way beyond the speed of light. If space as well as time did not change since the Big Bang to where we are, the only scientific explanation about the star since the Big Bang up to now would be to think of it as magic.

If we observed the Pulsar as equal to our space-time, and used matching space and we used matching time duration as well then the time in durations between us and the Pulsar would have matched we could have had the attitude that all things between the Pulsar and us are on par. Since we can see that the applying gravity producing space-time in the Pulsar exceeds the entire space-time the Milky Way generate but the space claimed is much smaller that the Sun claims there has to be many discrepancies and none lesser that time and space equalities.

The Scientific way of thinking being practised by the Super-Educated wants to take our Universe and the Pulsar Universe into the same arena. Doing that ignores the space aspect since the space is obviously different to what we have. With the space reduced by that much, we see the time as the reduced factor which means we have an inverse view on the practicality of space –time towards the inside of a star. Our perception must always dictate our observation and not the other way around. That is what brings about the relevancies we have to acknowledge.

While the pulsar is casting away its three dimensional space-time it is exchanging it for singularity. Three-dimensional space-time formed by singularity is a state where time delay becomes material clogging space into density. It is time or heat being delayed to the point where time is heat running into its own space. Time being singularity that is overheating and in delaying the time as the new heat forms space it is stretching the space it forms in the time supplying the space it forms. This delay of time finding new relevancies by which more singularity manifest becomes space forming as a result of space-time where the space part becomes solid material and the time part that which are movable in time. In that way space forms time and time grants material space within which it can be forming. In essence what the third dimension constitute of is space being duplicating space by motion. As much as space is filling empty space, which I shall prove is time, space is emptying space, which I shall also prove to be time.

If the Sun held a relevancy of 10 relating to seven with one or two or three of the planets...well yes that might be coincidental, but when it shows such a relation with all of them where all of them includes planetary fragments being between the Earth and Mars making such a coincidental claim on ten structures perfectly distributed then any claim on coincidental is more than ducking the truth. To honestly be honest about the finding of scientific truth and discarding such evidence then as coincidental is being unfaithful to the search for truth. Coincidental can be one planet holding such relation and on a stretch maybe two but when all of the planets shows this precise tendency without skipping one or even giving the fragmented planet material a miss that truthfulness then shows it holds the ten to seven ratio in extensive regard and becomes some aspect that is shouting to be investigated since the law of average outlaws coincidental ness and then exceeds the being coincidently by miles.

While considering you as level minded person, who all scientist that is truthful should be, and being an Academic pillar they have to be level minded under every circumstances life may introduce them to. They then should judgemental to the truth in facts consideration without any form of self-righteousness due to their standings in life. Never ever should they think about lesser beings that are in reality so far down the social ladder they are hardly aware of such little people? They must always consider everyone in terms of other persons to be a participant equal to an opinion than may matter and distance them from any form of bias. While being all this in the same breath they must be big enough to consider the smallest opinion and never blow away such clear evidence because the candidate is much unlikely to have an opinion of importance. Their judging should be honestly considered as gross clear dishonesty.

Then if one takes the importance of the cosmos who is the actual presenter of facts being so clearly presented by the cosmos requires investigation by those claiming to be non-bias. To go and dismiss this certainty as coincidental because it does not fit into a Newtonian Universe is stretching the truth to beyond the accepting norm. Science should always remain an open field that avoids the path of being a one way thinking street while science should not try to focus on one person even if he is Isaac Newton. They should rather focus on an image of such a created spot or dot where all came from, because there is no image of any person that will uphold cosmic laws. The cosmos shows evidence and in such evidence we must seek the truth. We must research and investigate such obvious evidence where we have to go beyond the obvious until we find singularity hiding outside the Universe we see, feel and experience. Such a singularity divide may be beyond our viewing capabilities but it is still a point although that point is beyond the visibility being only where we can reach mentally in reading singularity by using intellect. Singularity is elected wherever singularity is needed. The spinning top that uses the Coanda principle is the indication of that.

By rotational motion, the top creates such a line and by generating the line, the line charges gravity. The gravity is what drives the top as the top and as long as the top spins

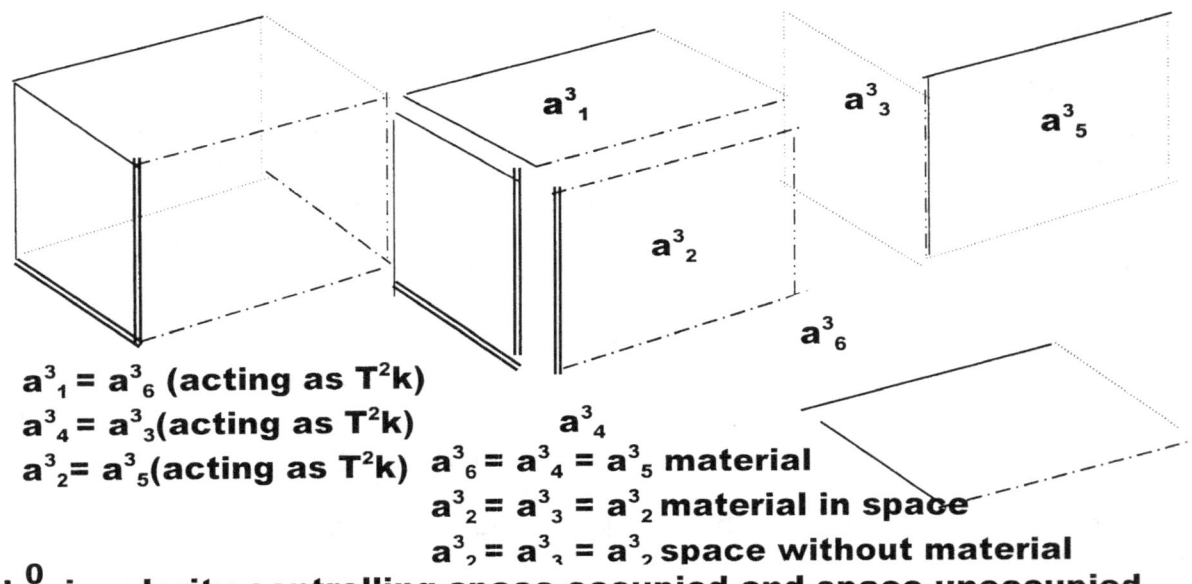

Singularity $\Pi^0$

There is an influence generated by the spin of the top that keeps the top upright while the top is spinning. The line is generated but the line is far from magic. The line is where the centre of the Universe is. The Universe is the particles that fill the top. The particles in motion generate motion by electing a centre from the centre of every particle in the top. Such an elected centre becomes the centre of the universe as far as the top relates to a Universe.

Then if one takes the importance of the cosmos who is the actual presenter of facts being so clearly presented by the cosmos requires investigation by those claiming to be non-bias. To go and dismiss this certainty as coincidental because it does not fit into a Newtonian Universe is stretching the truth to beyond the accepting norm. Science should always remain an open field that avoids the path of being a one way thinking street while science should not try to focus on one person even if he is Isaac Newton.

$$a^3_1 = a^3_6 \text{ (acting as } T^2k)$$
$$a^3_4 = a^3_3 \text{(acting as } T^2k)$$
$$a^3_2 = a^3_5 \text{(acting as } T^2k) \quad a^3_6 = a^3_4 = a^3_5 \text{ material}$$
$$a^3_2 = a^3_3 = a^3_2 \text{ material in space}$$
$$a^3_2 = a^3_3 = a^3_2 \text{ space without material}$$

$k^0$ **singularity controlling space occupied and space unoccupied**

Space developed sectors through time applying differences as singularity changes the universe from **k** to **k** through $a^3$ and $T^2$. The factor **k** positions the centre of the universe and then sets rules applying in that Universe as far as setting space in time by applying space-time. Space = $a^3$ and time $T^2$ is coming about from singularity $k^0$ pointed by **k**.

## Space-time $a^3 = T^2 k$

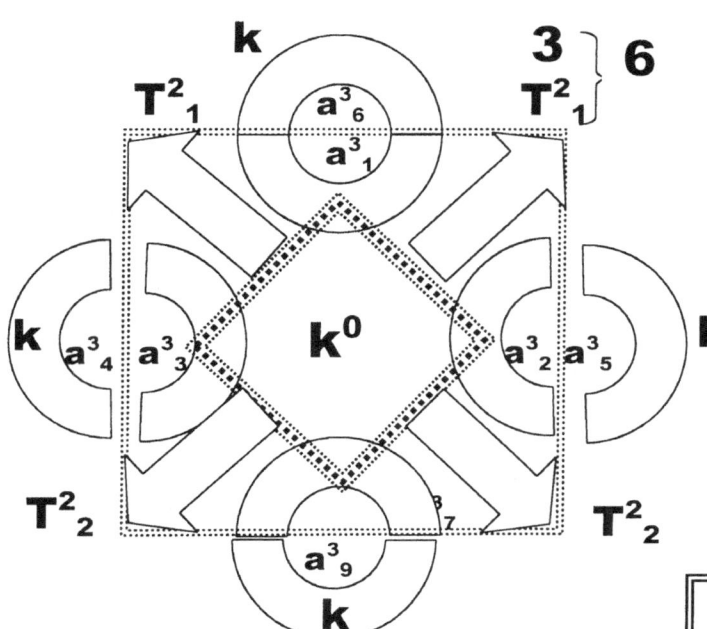

From singularity twelve points rises in forming a relevancy and puts a relevancy on space and on what we find to be time. Since two of the positions in singularity never concern us, where we in space and being part of 3 D, it is only worthwhile concerning us with ten positions. The moment space cracked the motion confirming the crack was also the cause of the crack.

According to Newtonian statements, it is said that the Titus-Bode law is no actual law. It is said that the law has no theoretical basis, but it shows how orbital "resonance" can lead to "commensurability". By following the guidance of Kepler the Titus-Bode law is most certainly a principle law affecting every aspect of gravity. It stands miles above the Newtonian denouncing of such a law as being coincidental. How the Newtonians would have the lack of cosmic vision to dilute such a law to the level of being coincidental goes beyond my understanding of the Newtonian mentality.

One cannot remove motion from the forming of space because the forming of space is the motion in which the space is then formed. With the motion space duplicates what singularity establish but singularity cannot be discarded as a factor. Therefore singularity is the referring factor to the other six sides forming space in motion.

When the top is spinning it will spin most of the duration in the time it use to spin standing upright as if it is a tin soldier. By applying an effort that brings a high velocity with impact the top will start to spin at a high velocity thrusting from side to side. The effort will seem as if there is a fictitious wall that the top wants to climb. The top will try to lift as it bends over to one side and then changes sides bending over too the other side. One does not need any imagination to see the top is trying an utmost effort to lift itself from the ground on which it is spinning.

The line dividing that the top create by motion is a product in the cosmos which is the centre part of any particle and that run through every particle, no matter how large or small it is or even if it might be beyond our vision. Such a small line might be so small it is not even noticeable to the cosmos, it might be in the 3D but nevertheless it is large enough to part the cosmos into sectors. It splits the biggest there is into particles and we are not even able to notice the precise location of such a split. It divides a star as massive as a Pulsar into bright and dark periods, yet it is so small it still remains invisible. When observing singularity from our stance, in truth there is no top or bottom that we living in 3D can see because there is no large or small that we can see. Standards measures borders and sizes are all man's creations because man is part of the Creation. We shall have to use a general conception brought about by intelligence when we observe the cosmos. One's intellect tells one about such a spot, but that is all because that spot is on the other side of the Universe (quite literally).

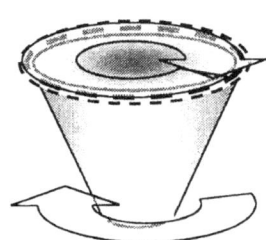

**On the surface, at first glance the top is an ordinary piece of dead wood that is machined into a sloping shape. The top is normally fitted with a sharp needlepoint at the bottom and the sharper the point is the better will the spin balance be.**

Our ability to investigate rather than play mathematical games is what should lead us to finding the pre Big Bang era. By referring to the playing of games is the space whirl concept the critical density madness and all the other quests the mathematicians set their minds on. It is as if those with the super mathematical skills are of the opinion that only the Universe would have the potential of matching their greatness and by mathematics they can play the game of god. Only the Universe might offer the challenge to meet their mathematics. How the hell would the person find the Black Hole somewhere in outer space, which he then would line up with another Black Hole to elude time of all things. If mathematics says some action is plausible while logic and common sense says the opposite I think the opposite is the truth and then mathematics is the swindler. By motion that creates gravity applying from a specific established centre one can see how the Universe came about, and that the centre of the Universe is elected where ever motion duplicate space to the ratio of space-dismissing. The centre of the Universe is like the Universe; it is not a fixed feature but is created as motion centralise space to create gravity by time applying.

From the wheel we find $k = a^3/T^2$ which says the gravity line $k$ is confirmed to the areas of space $a^3$ by the motion of the liquid $T^2$. What is more simple than that...and yet those with the super mathematical abilities can find them designing space whirls without finding the common sense to explain the Coanda gravity relation. The spinning motion of the top brings about that space forms by centralising the gravity, where the top finds the ability to stay up straight. Because of what the motion produce, the Coanda affect allows the top centre to charge an individual gravity with such splendour it can become a moving object securing individual maintaining of the secured centre. From the one side there is space confirming the liquid as a gain to the space $k = a^3/T^2$ and balancing that is the liquid $k^{-1} = T^2/a^3$ where it shows the liquid bonds to the space by contracting to the space. What can be easier than proving the Coanda effect in mathematical principle by the using of the most basic and simplest of mathematics that the method I just showed...and yet those whom are the brave never dared! From the motion of the liquid relating to the area attaching the gravity $T^2 = a^3 / k$. The motion attaches to the space by activating gravity in relation to the centre of the space.

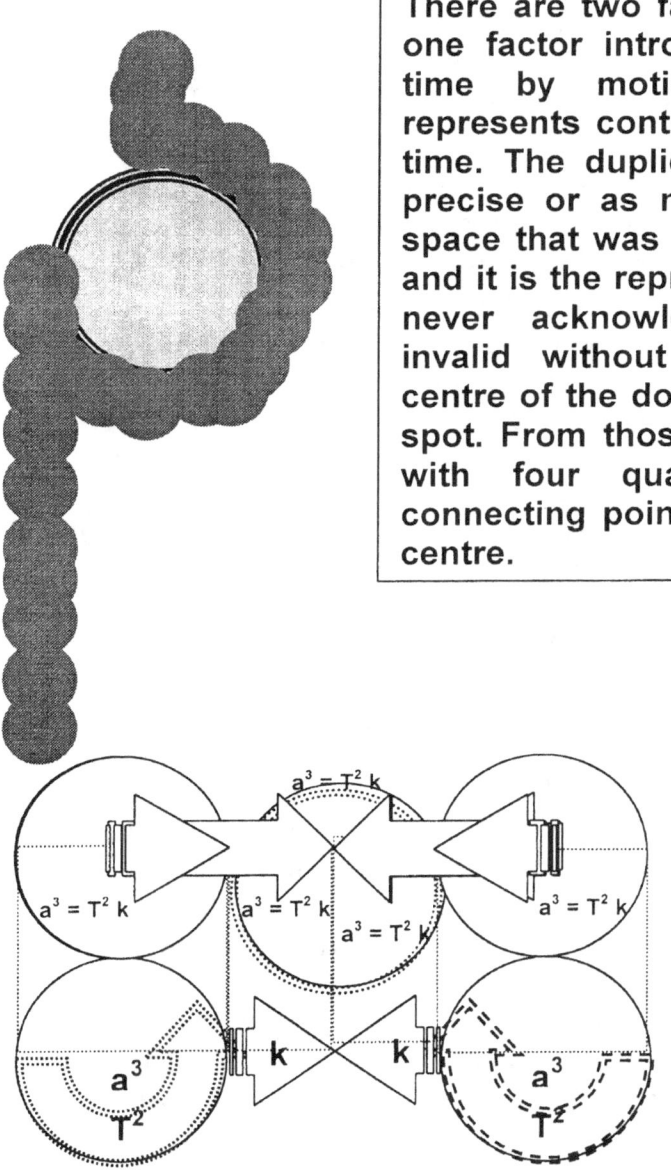

There are two factors present in gravity. The one factor introduces duplication of space-time by motion and the other factor represents contraction by reducing of space-time. The duplication is presenting space a precise or as near a precise replica of the space that was available the previous instant and it is the reproducing that science thus far never acknowledged. The contracting is invalid without the duplicating. From the centre of the dot, there is a top and a bottom spot. From those points, there is connection with four quarters. That produces six connecting points that are all aligning to the centre.

Motion is the time it takes to reform space from being three sided in half the Universe to a six sided six dimensional Universe extending from singularity by seven which includes singularity and claims another three parts in singularity under the motion of space. Such motion is time since the duplication of space in motion produces the time the motion takes to confirm the space then formed.

Space $a^3$ is both cyclic and linear in motion in $T^2 k$ by time

$$(a^3 = kT^2) = (a^3/kT^2 = 1) = (k^0 = a^3/kT^2)$$

Without giving the recognition that the Coanda principle is due, to science in general science will never rise from the slump of the Middle Ages and science may even fall back into obscurity. Even in the brilliance persons such as what Max Planck was, the way gravity is formed and the way the Coanda affect not only presents gravity but also becomes gravity. It is how gravity produced the atom. Some solids that was also protons became liquids which today we call the neutron

The higher the spin of the solid (or the liquid) is the more gravity there will be. The more gravity by motion there is the more space-time is secured in solidness. A spinning wheel secures liquid water onto a solid tire to the extent that a car runs on an inch of water at say one hundred and what... kilometres per hour. The gravity that forms around the spinning wheel secures more solid water that what ice can provide in density. One cannot drive a motorcar on one-inch thick ice but one can drive the motion car on one inch of water that is secured by the motion of the spinning wheel. Take this phenomena back to the atom and one would see how the atom formed before the Big Bang became in place. The liquid provides the gravity and being the only thing without mass, the neutron then is gravity. The neutron provides the space-time the dimensional flow between the electron and the proton and in this we find how the atom formed before the atom formed as the atom we now recognise.

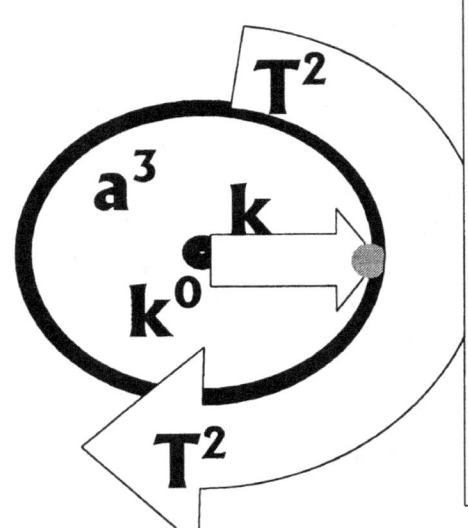

The whole process of gravity comes to the surface when Kepler's version of gravity is used. The neutron providing the spinning motion manifests the motion $T^2$. Since the neutron, produce the space for the proton to have, the proton and the neutron establish a symbiotic relation of equality. The motion of the proton uses space to dismiss while the motion of the neutron uses space to duplicate the space in which the neutron is as well as the space in which the proton is dismissing the space in which the neutron is

Having that we find that all stars go through a developing that ranges from being all about liquid heat to becoming liquid material as the neutron is and ending as space solidness by dismissing all that the neutron may provide in duplicating space-time. In this, we trace the beginnings before the Big Bang.

To secure a balance or a favouring of any one of the positions, the electron ads or removes it is favouring of placing a dynamic to the atoms form.

$k^0 = k\, T^2/\, a^3$ . Since any value to the power of zero is representative of singularity it is from singularity that the first line of material that did not dissolve. It therefore remained in direct contact with singularity. In the position, singularity holds gravity allows space to disappear thus presenting all material with the sphere as form. Singularity connects to seven points representing singularity in the sphere. Connecting to the seven is the space going into motion $a^3$ or the space coming from motion $k\, T^2$ but since that space is coming about through motion only three sides apply. The total is then the seven being in and being part of the ten, which include the none, and then it included three.

In this singularity idea there can be no sides and without sides there can be no drawing showing the explanation by means of illustrations. However, I do that to try and bring some understanding across from an unexplainable concept. Where I do just that, I ask for your forgiveness because being human means I have no capable means of performing an explanation. Yet I am forced to do just that and I have to allow the transfer of what can only be in my mind formed by the way of sketches, well knowing the implication that such act is not allowed. Let us again go to the beginning of creation.

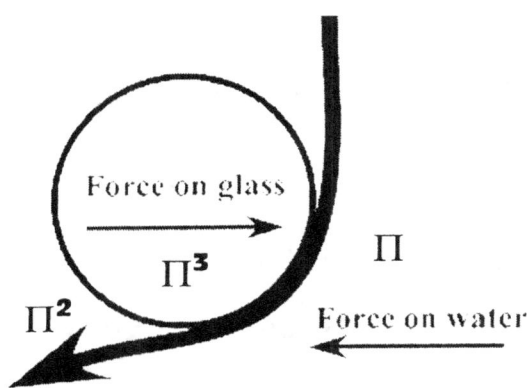

In the sketch one may take delight in the Newtonian application of forces forming. There is a spook or a force for the water and another spook or force for the glass and then Newtonians say they do not believe in spooks! More seriously though the Coanda principal indicate how motion forms the limits to space. The motion of the water attach to the surface of the sphere or circle which then forms what Newtonians describe as the curvature of space-time. It is only the form singularity applies than manifest as the roundness we see. In that is the proof about singularity taking on $\Pi$ as form. The water attach to the sphere as much as giving the sphere the gravity with which it does attach. In that manner we know that that was the way particles formed combinations just after the arriving of moment-Alfa (a term I introduced to describe the very first instant when heat parted from cold and $\Pi^0$ moved to $\Pi$.) Singularity brought the Universe but also singularity brought the divisions between the many Universes that followed the immeasurable many Universes that came after the flooding of Universes to follow the leaders. At this point mathematics renders it useless. Every slightest point in space became an opportunity of establishing a Universe with most different functions and ingredients there might form. This is apparent from the fact that it still takes place at the present moment by motion attaching new singularity through duplication and through duplication release previously attached singularity from serving the purpose of duplicating by motion.

When there is understanding about the process and one can read into what gravity is and what causes gravity, the reading becomes less complicated. It is a fact that science has this hype about mathematics but without understanding the basic if what applies where, mathematics become meaningless. I have been scorned on occasions for having this view but how many mathematicians can explain the Coanda principal. Mathematics is a language and mathematicians are translators of the mathematical coded language into a verbally used form of communications such as a language.

This translating of coded mathematical messages is about bringing across the correct interpretation of the coded message. When making incorrect deductions from incoherent translations it will sustain facts that produce a silly theory most people would find laughable. By starting a line from zero the starting with zero has removed the line and therefore the line cannot start…it is removed by zero. Zero cannot accumulate and zero cannot progress. Zero removes any value from the mathematical line because by zero continuing there is no progress registered as growth. Placing zero at any point removes any possible point from that location and therefore removes all future possibilities to development from that point. Zero removes and cannot construct. The Universe is about lines connecting objects because it is such lines that light has to follow in order to flow from point to point when connecting the objects that is connecting time. When reducing the length of the line to an infinite number ($k^0 = 1^0$) it leaves free, all possibilities that may grow from such a point including a line. Replace ($k^0 = 0$) and no possible growth can extend in any way imaginable. It is counter productive to place zero as the first starting point in the path of progress in any line. The Big Bang brought about size when form was already present. Think about it in this way…

The Coanda principle indicate that the gravity described in the previous page is generated by motion of liquid in relation to a solid anywhere motion can produce gravity. There is no mention of mass because mass is a derogative of the gravity which the motion creates. A centre is formed where the surrounding space-time forming the one group is relating a position from the "centre point". That forms one inclusive relevancy between points within the gravity field. The gravity field is holding "back" and "front" running through "the centre" where the other line is relating from "side" to " side" running through the "centre point". The fact of the line in the centre is that "it is there", but we cannot see it. Try as you may, no one will be able to calculate the very position that forms the lines, but as they change all particle characteristics, the lines are a reality as the spin of the matter is real. Being to small to hold atoms, the space holding such a centre line is no space at all and with that knowledge we may presume then therefore what ever the line constitutes of must become part of singularity, where singularity is a spot in the centre with two lines crossing the spot at an angle of $90^0$.  That is the basis of singularity, and since all the positions still relate too a centre of a circle, forming a part of a spinning circle, $\Pi$ must form the basic value. The second major reality that one has to recognise is that the only way singularity was broken was by motion. The only way motion can come about and break space less ness is by establishing heat which establishes expansion and the Universe became a possibility and later a reality by expansion. The heat swell into space and the space swelling is the motion that produces the gravity we find visible in the Coanda principle. The space at first was presumably filled with material because the expanding could only be material. The coanda principle alters time and establish with such alterations to space-time a new Universe with borders and all. By introducing motion it sets a new time standard by which the space created will apply a newly generated gravity.

One such a relevancy is the sphere.

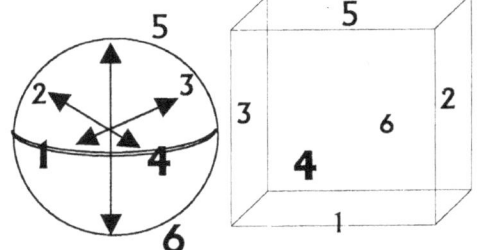

| | |
|---|---|
| The cube has six sides in three pairs relating to one another at all times | |
| Then connecting the six sides is a centre form where the control comes about that places these edges at specific related points and the points in return puts the centre at the precise centre | |

The sphere has six edges relating to one another at all times

Science presents a picture that portrait the Universe as being one big growing sphere. The picture of the one sphere moving into where the larger sphere holds space is simulating the cosmos, as the Universe is getting bigger. In those pictures they show spheres that is by measure depicting the progress in time of the Universe increasing in volumetric size. That now would then transformed a lot further by becoming the edge of the Universe. So what lies past where the Universe ends? There has to be such a point if there is an edge to the Universe because they claim they can see up to a point representing the edge of the Universe. At the time when the Universe was as big as a neutron, that what was as big as a neutron is still as big as it was because that same Universe is still today so big it cannot get bigger?  The Universe is not expanding, the Universe is shrinking into the oblivious as we are getting closer the centre of the Universe and therefore gravity is getting weaker.

When the water drop is in micro gravity floating it always forms a sphere. It will be the sphere when water is not pre-cast to have any specific form dictated by the Earth's gravity and therefore take on by cosmic pre-cast the sphere as form ...but why would the sphere form as the original form?

By merely blaming gravity pulling from the centre is rather avoiding the question with simplicity because the question arising from this answer is why would it then pull to a centre while the actual force should be a pulling to the centre of the Universe and where is the centre of the universe? Because the sphere is protected as it is protecting singularity by form $k^0 = 1^0$

The cube has six flat sides loosely connected at corners where the corners prove even weaker connecting points than the flat sides convey support to the structure as a whole.

At all times the sphere has six precisely controlled edges connected by a supporting centre that is in such a position the six plus one in the centre is in immediate support of any or all of the points at any time.

When touching one point the point reserves the strength to its disposal that is given by all seven points, which are backed by the entire structure. Try beat that for form strength and that is why a sphere is the ultimate form that provides structural strength

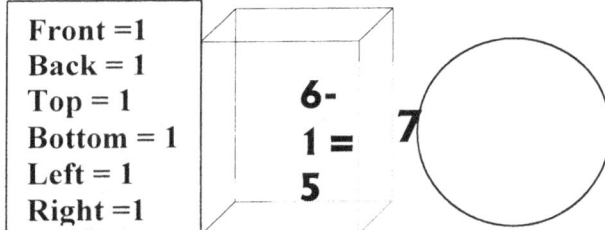

Front = 1
Back = 1
Top = 1
Bottom = 1
Left = 1
Right = 1

6 - 1 = 5

7

Where the sphere makes contact with the cube the sphere loses one dimension to the sphere. Because of the absolute domination the sphere has in form and in control coming from a centre the sphere removes one of the six sides the cube gas leaving the cube with five sides in relation to the seven the sphere has. That is another factor that gravity shows. This explanation also concerns the Lagrangian form.

In that manner we know that that was the way particles formed combinations just after the arriving of moment-Alfa. Singularity brought the Universe but also singularity brought the divisions between the many Universes that followed the immeasurable many Universes that came after the flooding of Universes to follow the leaders. At this point mathematics renders it useless. Every slightest point in space became an opportunity of establishing a Universe with most different functions and ingredients there might form. This is apparent from the fact that it still takes place at the present moment by motion attaching new singularity through duplication and through duplication release previously attached singularity from serving the purpose of duplicating by motion.

The 21.9991/ 7 might seem three-dimensional but it came to be on a line holding singularity. The line of time is eternal everlasting with $a^0 = T^0 K^0$. Then the event of heat was introduces and infinity interrupted eternity. That is when the 21.991/7 came about. A spot formed other spots formed on the line interrupting the line.

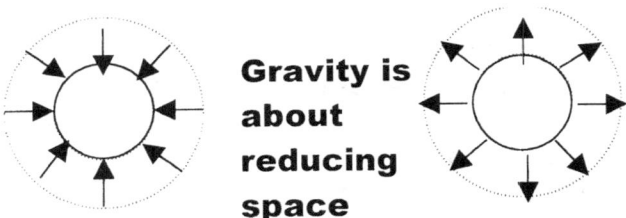

**Gravity is about reducing space**

**Expanding is all about heating. Heating takes up more space and gravity reduces space.**

This says it all and yet every person with a position of influence in science is missing all there is to see in Cosmology! Greatness in Cosmological terms is not in size, but the measure goes by intensity of density and lack of space. A smaller $(a^3)$ result in a larger $T^2$ where $(a^3)$ is the space the object holds and $T^2$ is the sizable gravity the cosmic object has. The suggestion confirms Kepler and disagrees with Newton. According to Newton, the ultra gigantic Red Giant Betelgeuse should be formidable when applying gravity whereas we all realise that the Black hole is the true undisputed giant! The red giant is sloshing around like the bowl of liquid heat-soup it really is while the gravity the Black hole unleashes gravity to the point where it devours even the smallest photons in the largest waves thereof that we can imagine. By taking the diameter, as the means to measure is clearly no solution in a method to calculate the gravity of any given star because it solves not one thing. They go on to even circumvent this failing. The Academics change their approach by applying the usual radius r forming the square in the use of the formula in those formulas, which Newtonians devised to measure gravity. However, instead of having the radius holding the square as is done in normal mathematics when calculating the gravity, those practising astro physics then gets really mathematical. Instead of squaring the radius they bring in the speed of light in such a place within the mathematical equation and put the C under the dividing line in the formula next to the radius. By them using the C to indicate the speed of light as C they place the C in the formula to bolster the radius value and that diminish the size the star has so many fold. By bringing in C they reduce the star because the radius then gets bigger by the multiplying of the speed of light. The star suddenly gets reduced by the factor that the speed of light produce and then they go on disrupting the truth even further by applying the square that should fall on the r in the normal calculations to the C that indicates the speed of light. The star then reduces by the square of the speed of light while the radius, which should carry the square value suddenly remains in the single. That is supposedly their way to put a measure there that somehow has the means to bolster the gravity. The speed of light is the worst or best form of antigravity depending on which way one looks at matters.

The line became an interruption of spots as $k = a^3 / T^2$ and the cosmos grew.

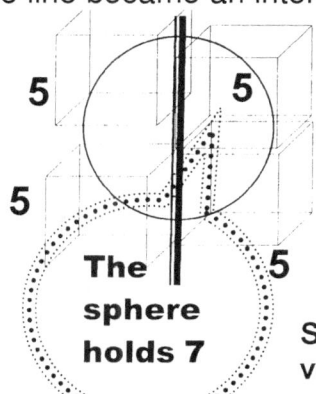

**5** **5**

**5**

**The sphere holds 7** **5**

Where the sphere and the cube meets there is this most robust form with seven supporting the one point that is in contact with the cube and this point will always completely dominate whatever the cube is resisting. That is not where it stops. The time factor that support the sphere claims four quadrants in total, which means the sphere holding seven relates to five times four totalling seven in all.

Singularity holds the eternal one or $\Theta^0$ or $\Pi^0$ or whatever value there is in the idea one whish to attach to the notion of the concept of the original.

Then one tenth less, which is one fraction of the square of space is the other side of the Universe and there singularity is a value of one minus one fraction of the square of space making the value one and one singularity measure (-.1-.01-.001- .0001- .00001) deducted from another one which that one being deducted from sits as singularity being part of the same singularity but on the other side of the world...and on the other side of the Universe. Remember there is no Universe towards the outside because there is no outside but image. The Universe runs towards the inside or away from the inside and every singularity is an individual Universe, only one Universe away from the next Universe. Since the Universe starts and ends in infinity and that end all definitive value big and small is merely human appreciation of what cannot be? It is a relevance of what came when and that is all. Everything past singularity is space created time driven temporarily substituted by the unreal. There is was and will never be one fixed solid Universe one can touch and smell, but the Universe is timely created space by motion of duplication in time delay. Once motion stops, time stands still and space falls into a black, Black Hole of eternal space less motionless reality where all the created concepts of space and time are contained in reality of eternity. That also is not religion but is physics. Time can only stand still in the Black Hole of empty space. There was the spot that became lots of dots. The dot had no borders therefore there was no separation and still we know there were more than one in a group of one. When $\Pi^0$ moved to form $\Pi$ the evidence of this move is very present in the cosmos at present and one can find such evidence all around us. The overall picture resulted in a ring or circle due to the release of from motion by all parts and all rings hold $\Pi$ to secure the form. The only form that existed then was $\Pi$ and therefore even today the borders use $\Pi$ to indicate positions. But in the single dimension such definitions were far from clear and the only distinctions came from securing singularity in preserving the position of singularity to apply gravity and thereby absorb all anti -gravity. But anti-gravity could not control expansion by counter acting contraction through gravity so the overheating continued forming non-existing borders. The borders appeared in some material that was infinitely solid just as Einstein predicted because this took place before light came about and therefore before the speed of light. That which we refer to at this point even pre-dates light and therefore light at that point was excluded as being part of the cosmos. The cosmos formed a partnership with one side overheating forming antigravity by expanding into space through the applying of the overheating. In the relevance which the Universe is all about there is another side and the other side formed gravity or contracting of the expanding space

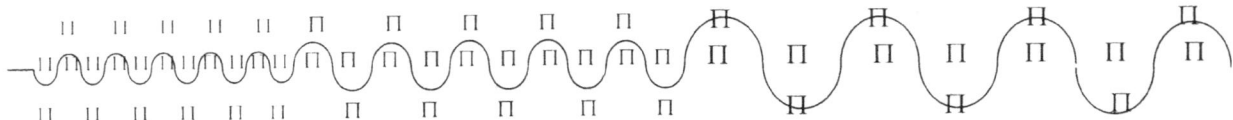

In this the sphere took shape in a shapeless Universe along the line of time.

The question we arrive at is how do we know the Universe started with Π? This we can see in the example where the Universe sets the top alive. There are one set of rules applying as cosmic law and to the cosmos it does not matter if it is atoms making the top spin or life giving the top a load of movement for a short while, the top adheres to the terms of the laws applying and gets its chance to break fee from a primary gravity. Looking at the top e see it is the four sides supplying movement that puts the axis in line. There for it is the four moving (2²) that forms the (3) and it is the three giving the time support to the space that time creates.

**earth**

Therefore it is (3 + 4 =7) that gives the turn while it is (3²) + (4²)= (25)^½ =5 that locates the new position. Every time an object turns it comes alive as it rotates through seven degrees. This seven degrees turning is responsible for time moving in which time creates non-existing space. In other books I explain this in much detail but not here due to space within the book. By establishing time (1) this timing in a ritualistic motion forms space (Π) and it is this process I that explain at this point.

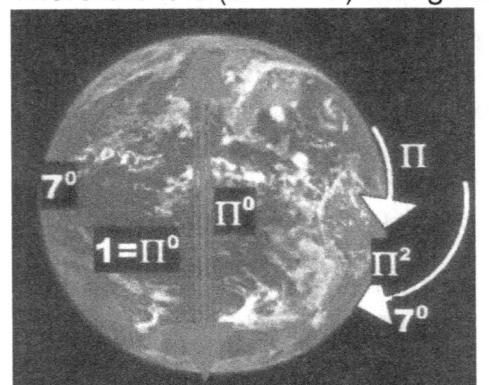

It is the manner in which the spot (1⁰) forms the dot (Π) and the dot (Π) is the start of space whereas the spot (1) is the start of time. This sets the two points responsible for the Universe one Universe apart.

$$1^0 \text{ going } 1^1 \quad 1^0 \blacktriangleright 1^1$$

This process of going from 1 to 1 eventually forms Π = 3.1416 / 1 but it is a long road getting there. It started with two dots forming and seen today it remains two dots forming time that then forms space or a Universe as

**becoming**

**the Dot**

This was the era of distinction, when separation brought an all-possible new Universe

what we see as Π = 3.1416 / 1.

● 1⁰ ● 1¹ Then the perfect era ended with infinity releasing eternity as much as eternity dismissing infinity. There was a past seen from the present and there was a future seen from the present. There were two dots moving from one position to the next position.

The Universe is filled with non-existing emptiness forming an overflowing void. Between that emptiness that can never start and that emptiness that can never end is a compacted void that fills up with the emptiness it forms the solidness we call a Universe. Time leaves non-existing space to form a patrician between instances that gives time distinction.

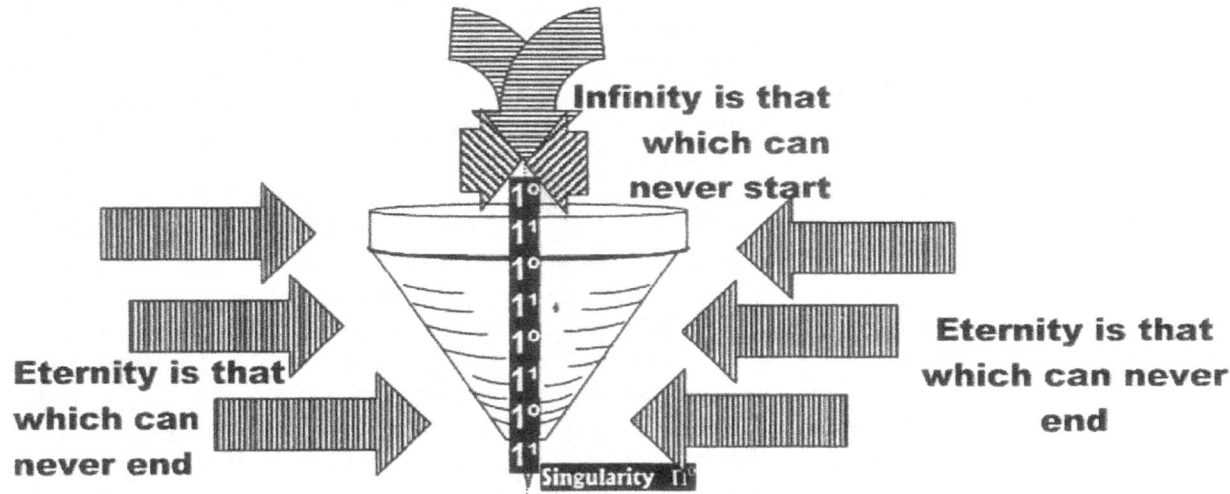

The Universe fills with the void of emptiness to the point of solidity we know as $\Pi$. There are two routes we can use to get to $\Pi$ where the one route is leading there is $(3 + 4)^2$ and the other is $(3)^2 + (4)^2 = \sqrt{(25)} = 5$. Both these eventually leaves a Universe.

$$\frac{\Pi r^2}{r^2} = \Pi$$

$$\frac{\Pi}{\Pi} = \Pi^0$$

The Universe is a circle $(7/10(\Pi^6/6))$ and everything in it is a circle forming a sphere. To get to the point where the Universe starts we have to reduce the circle to singularity. That means the circle is $\Pi r^2$ and that we reduce to $\Pi r^2 / r^2$ as to remove space $\Pi$. The Universe we are able to see is $r^2$ and the rest is form it holds as $\Pi$. Then to reach singularity we reduce the dot further by removing form and then we are left with removing the form $\Pi / \Pi = \Pi^0$. Every time an object in the Universe turns it does so by $7^0$. In that every turns it creates Universe at the object holds. This employing the Pythagoras. time any an point it does When turning occurs it object entire new the by goes

$$50 + 50 = \sqrt{100} = 10$$

displacement of 3 and 4.. There then is seven forming as material and ten forming as non-material space. That makes the worth of the Titius Bode law with 10 / 7 and 7 / 10 that influence as gravity in motion because gravity is movement and nothing else.

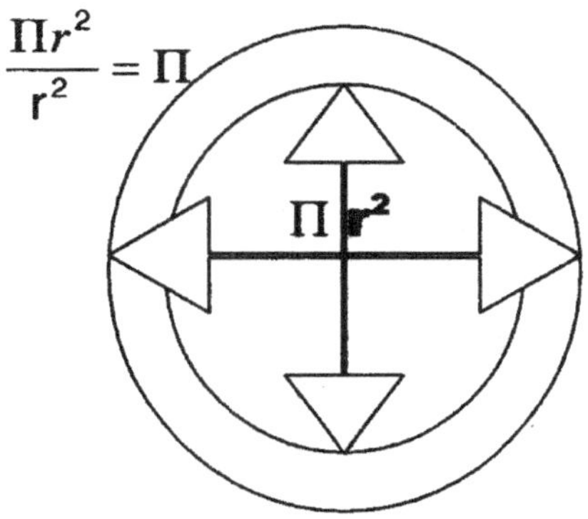

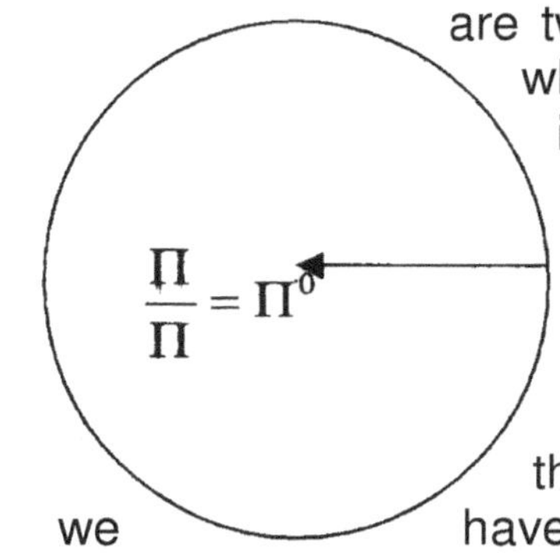

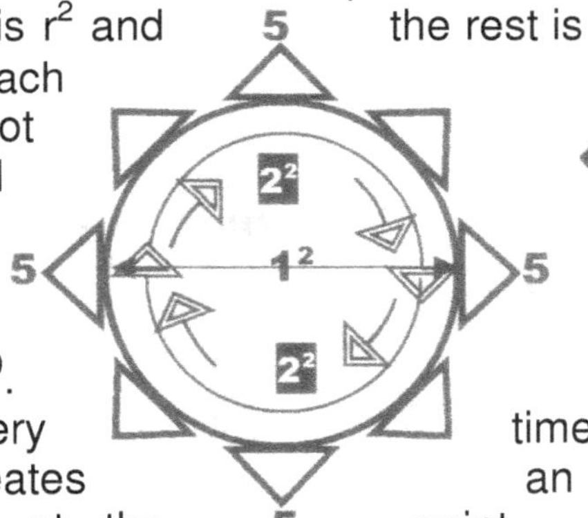

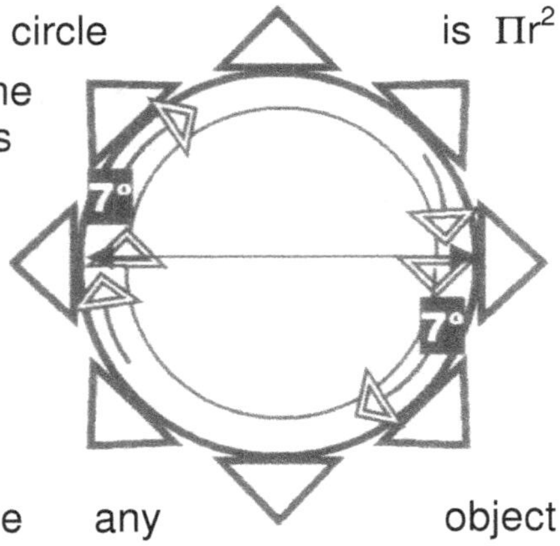

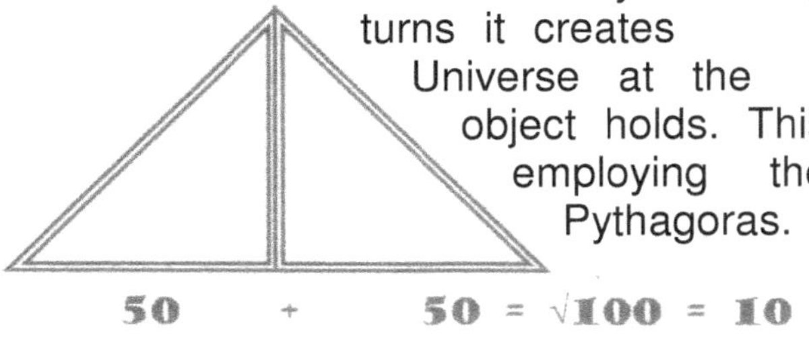

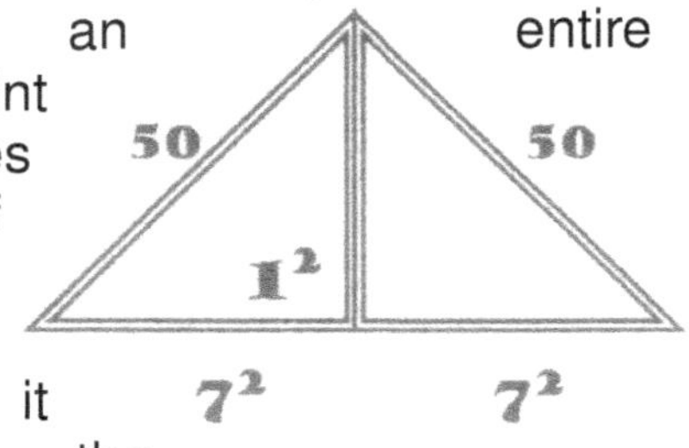

Because the seven forms by the same seven being on both sides there is always seven in relation to ten where the ten is on both sides of the divide. Determining the value of $\Pi$ goes as follows:

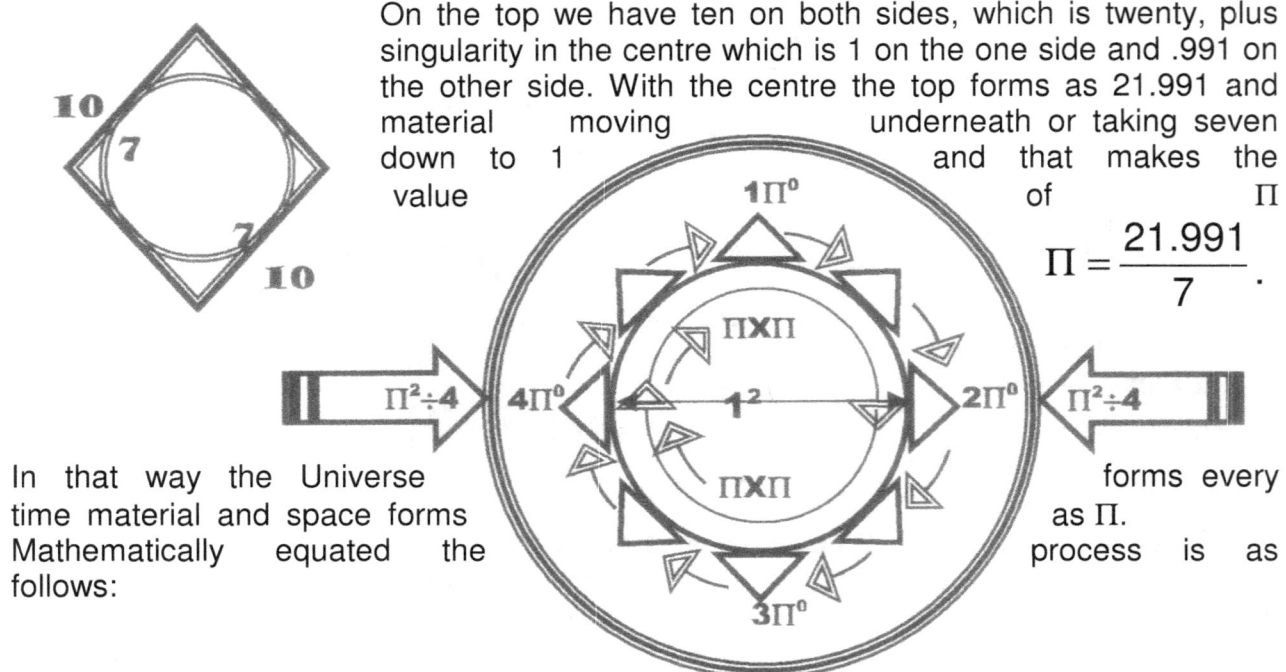

On the top we have ten on both sides, which is twenty, plus singularity in the centre which is 1 on the one side and .991 on the other side. With the centre the top forms as 21.991 and material moving underneath or taking seven down to 1 and that makes the value of $\Pi$

$$\Pi = \frac{21.991}{7}.$$

In that way the Universe forms every time material and space forms as $\Pi$. Mathematically equated the process is as follows:

**Matter in relation (part of) to the total dimension of space.**

(10 / 7) \ (7/ 10) = 2.04

1.4285 / 0.7 = 2.04 *Taking from both orbiting influences*

**SPACE DIVIDED INTO TIME**

(7/10) / (10/7) = 0.49

.7 / 1.4285 = 0.49 *Taking from both orbiting influences*

**SPACE MULTIPLIED WITH TIME**

7/10 / 7/10 = 1 and 10 / 7 X 7/10 =1 *Therefore not influencing change*

**THE PROCESS PARTED USING THE ROCHE PRINCIPLE**

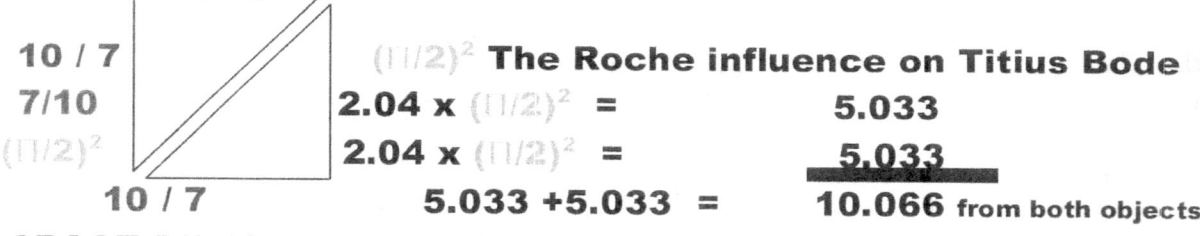

10 / 7

7/10

$(\Pi/2)^2$

10 / 7

$(\Pi/2)^2$ **The Roche influence on Titius Bode**

2.04 x $(\Pi/2)^2$ = 5.033

2.04 x $(\Pi/2)^2$ = 5.033

5.033 +5.033 = 10.066 *from both objects*

**SPACE DIVIDE INTO TIME**

7/10

10 / 7

7/10 / 10 / 7= 0.49

0.49

$\dfrac{10 / 7}{7/10 = .49}$ $\dfrac{10 / 7}{7/10 = .49}$

.49 + .49 = .98

.98 X 10.066 = 9.8 =$\Pi^2$

**TIME SPACE = $\Pi^2$ = 9.8696**

**TIME SPACE = $\Pi^2$ = 9.8696 = Space and time in a dimensional implication.**

Every time Π formed singularity it brought about motion as gravity $\Pi^2$, which still maintained the line. Gaps formed and became the atom.

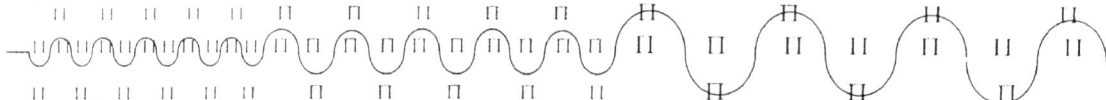

The line being singularity formed singularity going from the spot to a dot. The heat surged and expanded into new space as heat still does. The overheating claiming more space that formed Π. The motion was $\Pi^2$.

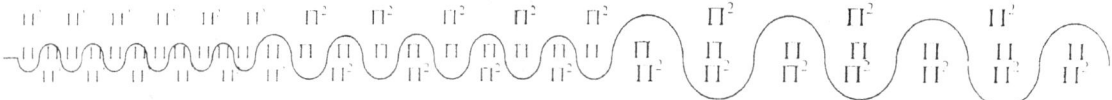

Every time Π formed singularity brought about motion by gravity in $\Pi^2$, which still maintained a line. The line is still there and can never become obsolete because the lime represents eternity. Gaps formed and defined the atom in space. The heat moving became the motion that became the gravity. The moving was between **$((7/10 + 7/10)X10)$ and $10/7 = \{(.7 + .7) = 1.4X10\}/1.42 = \Pi^2$.**

**$((7/10 + 7/10)$.** Material forming space became available provided by heat surging. Time became time the space in which material moves.

**(10)** The three in time-space or the space in which material can move.

**10 /7** heat cooling retreating as the contraction retarded the growth.

In this was a line, which I can't sketch since sketching relies three-dimensions, and at the time that was very flat today is three-dimensional.

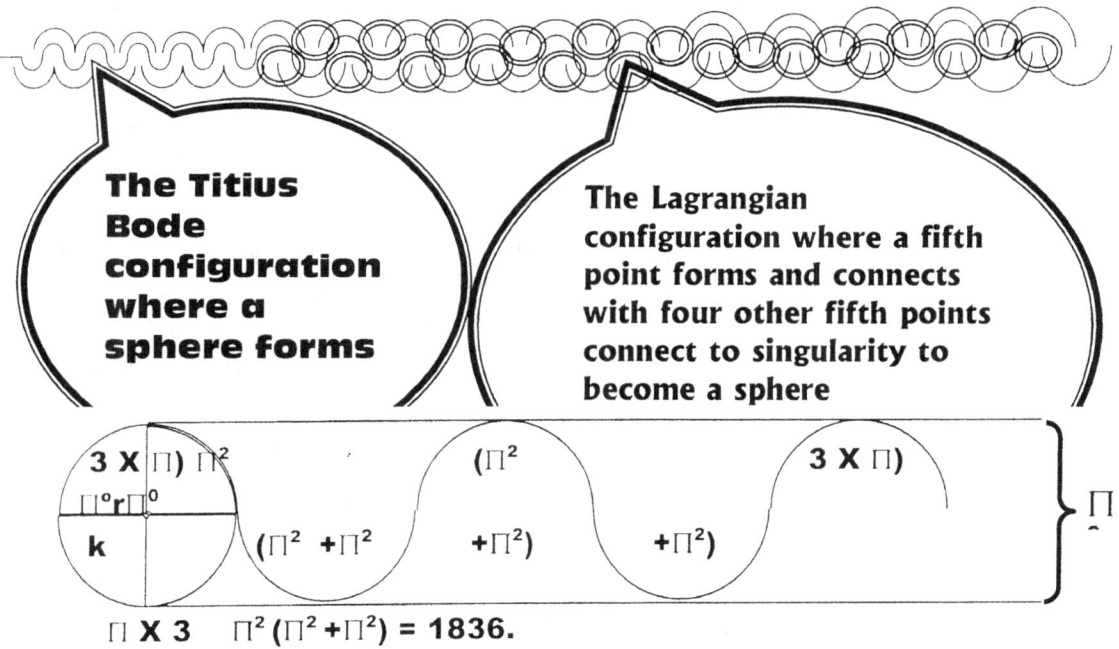

The Titius Bode configuration where a sphere forms

The Lagrangian configuration where a fifth point forms and connects with four other fifth points connect to singularity to become a sphere

$3 \times \Pi) \Pi^2$    $(\Pi^2)$    $3 \times \Pi)$

$\Pi^o r \Pi^0$

k    $(\Pi^2 + \Pi^2$    $+\Pi^2)$    $+\Pi^2)$    $\Pi$

$\Pi \times 3$    $\Pi^2(\Pi^2 + \Pi^2) = 1836.$

Finally with all the heat retarding the line formed a gap in time.

▶ $\Pi^0$

All the while time is just a spotted and dotted line running along time as space duplicated with heat surging and cooling as cold contracted much similar to the actions of stars in the process of pulsating known by what ever name one wish to use. The star takes time back so slow we can see the pulsation of gravity cycles. By cycling between material **$(.7 + .7) \times 10$** and space in time **10/7** the motion forms gravity $\Pi^2$.

Light is the strongest or the most intense antigravity there can be and to throw that into the Black hole by the square to hide the insufficiency of the methods applied to calculate gravity is once again another cover up to hide the Newtonian ideas, which are not functioning in cosmology. By producing $C^2$ in an attempt to bolster the gravity figure they supposedly are able to calculate in the gravity of a Black Hole and placing $C^2$ in conjunction with r symbolising the radius. This blatant further method of corrupting the corrupted is just the way not to improve the incorrectness, which their theory quite deliberately brings about as a measure to determine gravity. Inadvertently they did confirm that gravity is a speed ratio, but their confirmation of such a fact past them by without their noticing it. We can take the proof of my statement about the incorrectness of using the radius to measure gravity one step back and see how the method apply to the Neutron star or the Pulsar. Then what about the measuring of the gravity in the Neutron star by using the diameter and how do they explain the Neutron star, in principle. The Neutron star will either be stronger in gravity than the Black hole when also using the $C^2$ method, which is clearly not the case or it will be pathetically weak when not using the C method. If the Black hole has a diameter of 10 kilometres then the Neutron star has a diameter of twelve kilometres. By using $C^2$ in determining the gravity in the Neutron star the Neutron star suddenly have a larger gravity than the Black hole has or instead is much weaker in gravity than the moon is. It is either that it is stronger than the Black Hole or the neutron star is so weak it has less gravity than a comet.

This ridiculous scenario developing just shows how little mathematicians have any grasp about cosmology. Yet they and those are the persons denouncing my thinking and me. They (the mathematicians) should keep to building dams or skyscrapers where the use of mathematics is useful and is appreciated and stay out of cosmology. Nevertheless, I know that for this remark they will get back at me as they usually do. It is precisely because what we can learn from the incorrectness of mathematics that we can determine that mass is the result of gravity and gravity is not the result of mass. Gravity that forms restricted motion as gravity apply in and becomes mass due to of the gravitational restriction but mass certainly does not bring about gravity by some magical intervention as it then is applying as a pulling force. Gravity brings about mass but mass does not produce gravity as Science wish to advocate. For instance, there can be no mass factor in a galactica that is generating gravity because all galactica never shows any singularity restriction. Gravity creates mass but mass does not establish gravity. $k=a^3/T^2$; Mass comes about by the reducing of **k** in the case where the Roche limit has been bridged and singularity reels in the dominated ,or claimed space-time. Any object that can slip past the Roche limit value of $\Pi^2/4$ times the diameter of the star, the star would reduce to something no more than mere heat and treat the reduced objections as space-time.

Any object that can slip past the first barrier or safe guard then becomes the dominated and is a $k^0/k = T^2/a^3$. When reading the formula, such reading translates the formula to objects turning the equilibrium around by becoming time-space $k^{-1} = T^2/a^3$ because the time factor at that point has to reduce the space factor since the duplication of the space within the time set does not match. The object lost its independence for one reason only and that is that it could not match the gravity speed set by the dominating singularity because the lesser structure could no produce the independent gravity to secure sufficient heat in its core to drive the lesser object. The distance depends on the position that the orbiting object developed space-time. $a^3 = k\,T^2$;The space depends on the distance the space developed from the centre and the speed the space moves around the centre. $T^2 = a^3/k$; the speed the space orbits around the centre depends on the distance of development and the size into which the space developed. In this view our standing on Earth makes us part of the Earth space-time. The space-time established by the Earth governing singularity has put a claim on us to dominate the independence that makes us as much captured properly as we are part of the electing atoms that elect the governing singularity in place. Singularity running negative or singularity declining motion places the incoming structure under the governing singularity control by diminishing the space value through enforcing the time aspect when the relevant **k** reduces to incorporate all the space-time the Earth will grant. That is described as $k^0 / k = T^2 / a^3$. What this verbally means is that since the speed we have representing all those on Earth and which we all then inherit as being part of the unit the Earth, we secure through motion, we all contribute to a decline of direction compared to distance or space we occupy and that what is also declining. The factor **k** presents in relation to singularity $k^0$. As all independent atoms each have an independent singularity, which stands related to the elected singularity, the independent to elected **k** is in declining growth and this is because our motion $T^2$ lacks space and thereby space $a^3$ declines through the deteriorating of **k**. If by some fluke we gather heat and we increase our velocity the formula which is time-space $k/ k^0 = a^3/ T^2$ we take on then may turn about as we will just start flying

In the light of all this above-mentioned facts Mainstream science still promote the idea is that gravity pulls material closer and even more so in bigger stars. But that gravity pulling can only come from the accumulative effort of every individual atom according to proton mass (number) as a unifying effort of all the atoms in the star in accordance with mass applied. The idea is that mass is the same everywhere and is never changing. Why would there be such huge mass increases in the bigger or should I say smaller stars. What would entice the material inside such stars to grow more massive if mass come about from the pulling of one particle closer to the next particle. If it was about pulling on each other the mass of the particles could not increase through such pulling. Even by combining the mass of two individual atoms the increase is already in the equation. By locking the two into a unit should not change the mathematics when calculating the total mass because 1+1 = 2 whether the two share one unit or two units. There cannot be any mass increase in a star because all material is within the unit the star holds. By fusing atoms the star cannot become more massive using that specific method since it gains no further mass. The two hydrogen atoms was there all the while and then adding one oxygen atom can at best be equal to if not less massive than one neon atom. Though by using this principle the mass of the star cannot grow because the star does not produce mass or work to further the mass it has. The star reduces the mass, which is the thinking of Mainstream science and if we stick to the accepted views that was formulated almost half a millennium ago, we find that those in Mainstream science hold the opinion that the star diminish the mass it has by burning mass as does a coal stove in the old days.

Today we all find we have progressed to the era where we are fortunate and blessed with electric stoves that do not reduce mass to allow a fire to consume the mass the stove gas as was the case with stoves in the Newtonian era. Unfortunately the principles behind the process of driving stars and new stoves using electricity has not yet reached cosmology where the Mainstream scientists find their theories and therefore stars in cosmology still consume material to operate just as coal stoves did in the time so many centuries to the past when modern cosmology theories was brought to book. By fusion protons will only join without further raising the mass they have apart or combined. If the particle has a mass as two units, and the units join in volumetric occupation, they still have the same mass. We can see that some facts about Newtonian science are too astonishing to be real! The above facts are part of Accepted Science and accepted facts, but my theory about gravity being a result of dismissing space to the advancement of compacting matter is not accepted through all my trying to introduce my ideas to Accepted Science. Where space reduces all space reduce. Occupied space might reduce less in relation to unoccupied space, because the atomic individuality has a resistance factor of being forty times more than the gravity applying.

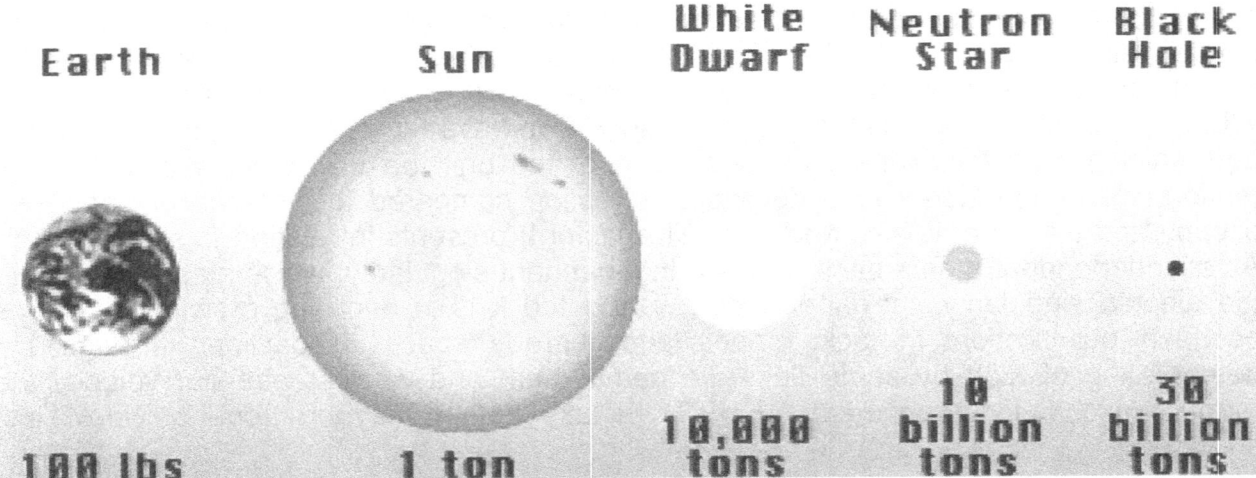

One hundred pounds in mass will be equal to a mass of one ton in the Sun. This does not come about because the mass grew more but it is the result of the overall space being much reduced and as a result it is having a bigger attack on the atomic individuality. The forty times difference there is between the atomic mass fighting for independence and the gravity attacking such independence grew much closer. Therefore the space reduced somewhat and that includes the space granted to material. One cubic meter being one ton on Earth will hold ten thousand tons of material in a star one class more developed than that which the Sun is. In more developed stars the figures rise above human comprehension. But it is so clear that the space diminish as the mass becomes denser. By compacting matter the space reduces in the same process. I have been trying for years to get any professor to admit to that by producing the rest of my theory but with little success.

Our understanding of any star development must lead to our understanding the Big Bang, because a star is the reverse of what the Big Bang was. The heat available during the Big Bang explains the lack of space at the time. Space is heat expanded and heat is space contracted. The one is the reverse of the other. To expand is to bring about excess heat and to contract is to produce gravity by eliminating space. The Coanda effect backs my argument. Where gravity is applying the strongest the heat is the most ancient rated and therefore the available space is the least. As the space develops from the Big Bang this view lends itself to favour space. Seen from the developing one must also see the space that the star reduced in size while space in outer space then remained static. It is not only outer space shifting but by compressing (for the lack of a better word) the star is reducing and the reducing is accumulating heat material from which the star material is growing its singularity.

Coming to this realisation convinced me that mass have nothing to do with gravity. It is all about density in space caused by movement through space in time duration.

the first dot ● space in between ○ the second dot ●

Time became defined when space divided one instant following the next instant by placing space in between. To put partician and seperation in time we find space forming as Π.

In the very beginning heat made singularity move from $\Pi^0$ to $\Pi$, which reduced the overheating as relevancies came in place. But brought on four points as time and as the heat four new points in major singularity heat shifted. With the heat relocating points formed by cooling factor was by the shifting of the came in place that as cold is always contracting. However by the expanding into space and this cold parted from heat and the expanding in singularity serving shifted to form the singularity, the cooled as the displacement of the heat as new singularity, the then brought on heat, the cold that enforced contraction associated with same margin is heat cycle is gravity. As much as heat expands into new territory by hear rising and subsequent motion comes about, there comes the next pulsating action of contracting as the cold becomes eternal within the governing or centre singularity. That is the way the Universe started and that is the way the universe continues.

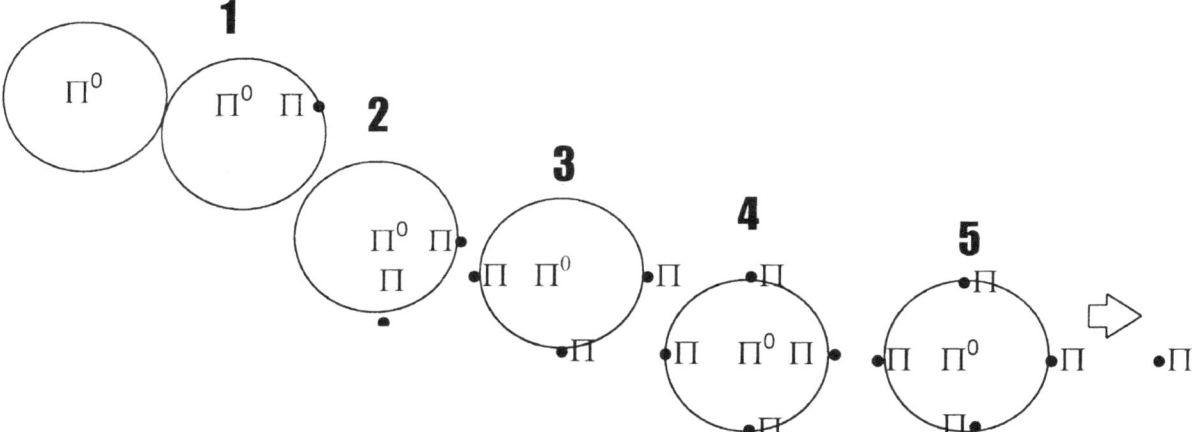

After the four forming the time factor came in place every singularity that spawned as a time factor changed relation to the next that spawned by the heat and the relevancies again changed where each of the four produced a fifth in the four forming a new variation of $\Pi^0$ going on to form four new $\Pi$. The four was a fifth position and the four fives became twenty. Including the centre singularity of one plus the previous singularity of point nine as well as all the others coming before that, the newly grouped again formed the seven position the sphere has in relation to the twenty one point nine, nine, nine and that concludes another sphere.

Therefore the star froze space as much as outer space grew beyond the size the star captured while one is losing what the other is gaining. All three concepts are correct in the final analysis. It is a matter where one whish to place the distinction that brings about any concept. That too is one reason we never were able to understand gravity because gravity never favours any one side and therefore do not produce the type of relevancies we humans need to understand.　　Gravity is the motion producing differences and is dishing out even-handedness.

The line was a continuous flowing of time but on the line singularity expanded from $\Pi^0$ to forming $\Pi$ with only a dimension adding every side of $\Pi^0$.

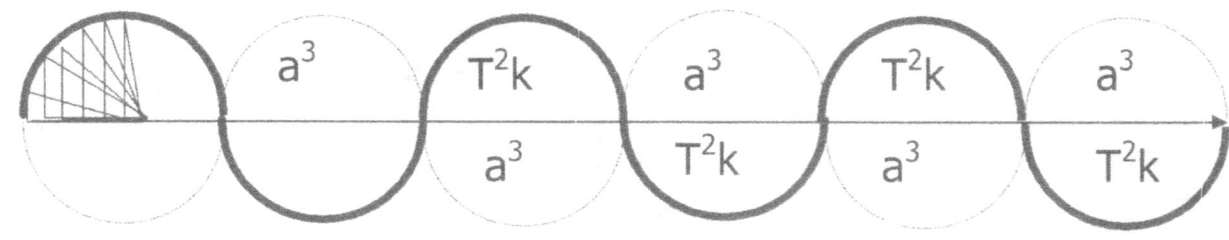

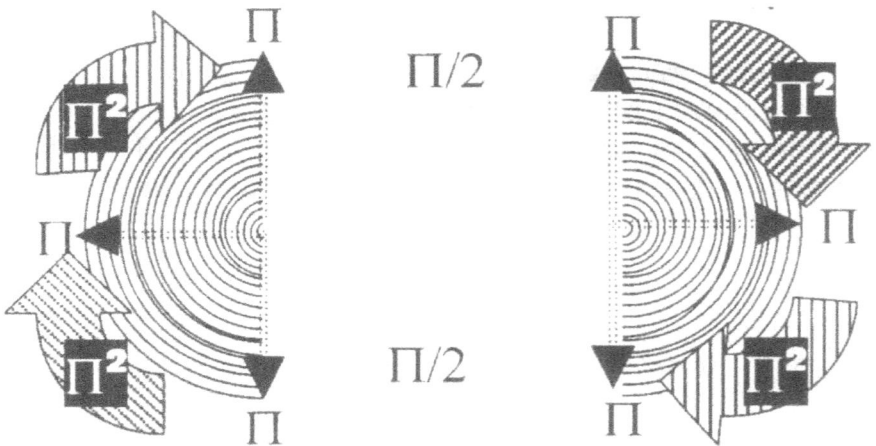

Using the four cosmic laws the cosmos formed. I named it the four cosmic pillars that stand in as the foundation of the Universe. The four Pillars are: The Roche Limit; the Lagrangian points; the Titius Bode law and the Coanda effect. This is all about forming $\Pi$ and applying $\Pi$ to form time $\Pi^0$ that becomes space $\Pi$ by movement $\Pi^2$. By forming 3 + 4 the Universe started because by 3 + 4 the value of $\Pi$ came in place. I have books explaining this process in much detail and therefore I am not about to spend space in this letter in order to explain this process in much more detail than I do.

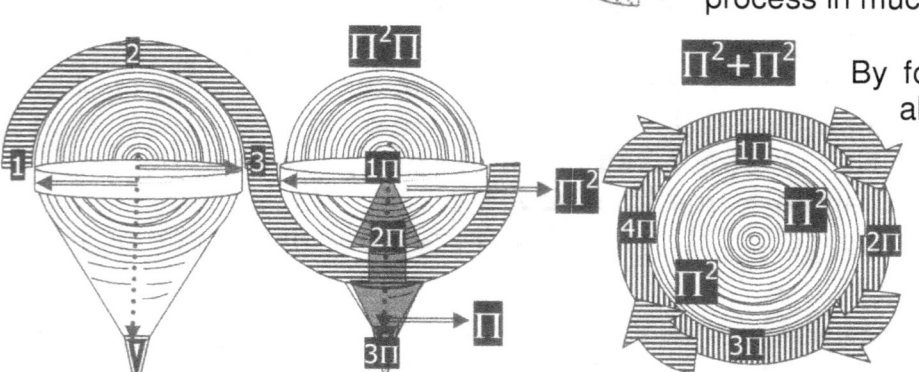

By forming $\Pi$ the atom came about and the atom is the Universe.

The value of the atom is $(\Pi^2+\Pi^2)(\Pi^2\Pi)\ 3 = 1836$ where the description is

The proton $(\Pi^2+\Pi^2)$

The neutron $(\Pi^2\Pi)$

The electron 3 and the displacement is 1836.

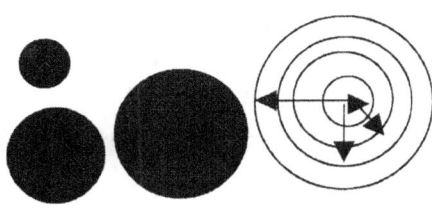

Looking at the affect of gravity it shows the precise quality of no distinctive point, as gravity never seems to end at a point but flows all over, affecting all that holds a position in its sphere of influence. The gravity coming from China meets the gravity coming from America at no particular spot but intermingles without distinction. However, we humans can only understand by applying distinctions.

In the sketch above the circle to the right would come about from a straight line **r** growing influencing the appreciation of $\Pi$, but to influence $\Pi$ would lead to a breakdown in **r** as $\Pi$ and **r** are different entities. The circles to the left shows a continuous growth by extending $\Pi$ every time a cycle of $\pi^2$ completes and since $\Pi$ is the same part as what the previous $\Pi$ was, only extending that immeasurable of an infinite quantity which is many times smaller than what we can understand each time, the circle will truly continue without any signs of a break. In the context of dimensions one find coming from the centre $\Pi^0$ an established eternal flanking of $\Pi$ to six positions since $\Pi^0$ forms the centre to the six sides and not all six sides are having a diameter yet must apply $\Pi$ to indicate specific value. What I try to say in this elaborated effort is that where **k** extended for the very first time when creation started from $\Pi^0$ to the edge of 3D where $\Pi$ begins it had a certain distance.

Such a distance is much beyond our understanding nevertheless, we may never ignore the distance forming every time. We humans are incapable of ever measuring that extending growth but be as it may, it is there. Such expanding is beyond human thought or comprehension but it still had enough value to raise a Universe. Allowing this expanding of **k** that personified as $\Pi$ to continue, and remain uninterrupted as one flow with a continuing growth of $\Pi$ being a line even if that might be ever so small there comes a link to time as well as space. As it is still productive, it will follow one line upon another line producing a cover of the full area of $\mathbf{a}^3$ by forming the time factor $\mathbf{T}^2$ or for that matter gravity $\Pi^2$. Gravity $\pi^2$ will be flowing constantly through out $\mathbf{a}^3$. The factor $\mathbf{a}^3$ will be improvised by the singularity measure of $\Pi^3$ and the time factors $\mathbf{T}^2$ and **k** will then be in singularity terms $\Pi^2\Pi$. Yet, all is controlled by the dot from within the very centre, which I am referring to, where rotation must end or start space depending from what vantage point the relevance is placed.

The very centre forms an eternal divide that will not allow what is on the one side to present an influence on the other side.

$\Pi^0 \Rightarrow \Pi$

$\Pi^3$

$\Pi^2$

It divides spin. It divides direction o spin. It divides all rotation from the outside that one may detect and such divide is there because at one poin spin will run to the left coming from the right and just immediately next to that point must run a direction from left to right.

Singularity cuts by dividing without contributing or participating in segregation. It divides without any form of favour that it shows in forming by favouring sides. That is singularity not having a dimension of space and not having a dimension of time, or a radius connecting the rotating distance to **k**. In the explaining earlier on about the matter of gravity, it was apparent that **k** developed into $T^2$ and $T^2$ was serving a bigger space $a^3$ as **k** interacted with the larger space $a^3$ by intergrading $T^2$. Singularity holds from the beginning the point where $T^2$ breaks down and $a^3$ stops to exist and has never relinquished such a point. It is a point, which we find in all-rotating objects. Every rotating object holds a centre from where the rest of the rotating direction will differ at any and all given points to all other rotating points. That is what provides all in the Universe with independence and with individuality. Not one point is exactly the same, but in the very middle, in the very centre no one can draw, measure or see is a point not in motion. Although the point is beyond detection by our looking at the spinning top, the point is - the point is there all the same.

 $\pi^0$ in the centre runs an axis line that forms the division of rotation. No one human will ever be able to indicate the precise line, but such a line must exist because of our logic telling us about such a line. In the centre, one will always find one more line smaller than the outside but forever also always bigger as it is towards the inside.

The law of Pythagoras is seen by science as a double square forming on two sides of a triangle and the combining of the two squares then becomes calculated as the root of the third side of the triangle. That is correct but again Mathematicians are satisfied with half the story and never seek the rest. The law of Pythagoras is also that what Kepler saw that sides relate to one another by line and by the opposing square. Singularity comes about first of all by Pythagoras where the centre line claims division. By the centre line holding the centre, divide in the single dimension the square comes about from both ends of the singularity circle. Then the square of $\Pi$ forming two points on both sides of the centre line is equal to the centre line becoming equal to the centre line and being in the square. In that the centre line provides the square to which the half circle attaches. In locating singularity, we must return to the spinning top.

The proof we have of this is in the Roche limit. Before space there was form the form was $\Pi$. That was the diameter of gravity from singularity and the motion thereof was $\Pi^2$ circling $\Pi$. Today we find a limit in the Roche still proving the limit that was applying then in being $\Pi^2$ but sine the time position there was where new singularity spawned was at $\Pi^2$ being formed at $\Pi\Pi^0\Pi$ in relation to the then diameter only pure heat could be closer and the pure heat would establish a new location to have singularity form. Anything closer than $\Pi^2 / 4$ becomes heat.

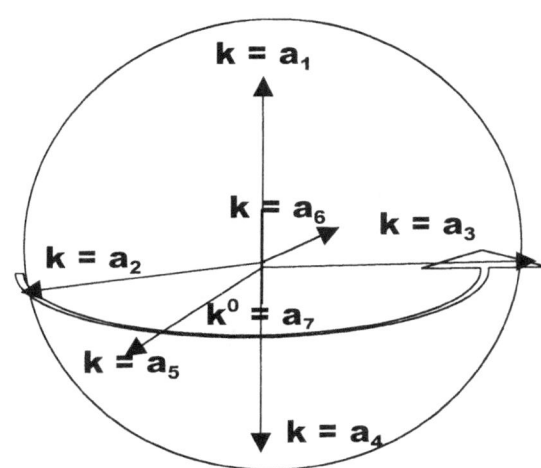

Kepler translated that from singularity $k^0$ came *the three sides by three sides* we receive by motion next to eternity. In outer space outside the sphere the relation is only three sides relating to three opposing sides being $a^3 = T^2k$. In the sphere, the space is secluded by another dimension enforcing a boundary from which the inner material cannot escape freely. The boundary of the sphere has also six sides where three opposes three other sides because singularity maintains very strict form there is one connecting point holding gravity in a part where space disappears. In the centre, a seventh point forms as $k^0$.

Coming from a unit where time as motion divides the circle into equal sectors one unit will form two parts in the square holding singularity as an equal divide. Resulting from the rotation, space forms that positions in a centre, an elected governing singularity charged to secure a dominating and controlling centre... through which comes a point where the sharing of singularity of the entire unit positions the gravity in charge. By that, not one of the two particles can ever be in the space the other claim during the time of the occupying claim. Since it is sharing the mutual singularity, that sharing will bring about an eternal bonding. Although such eternal bonding is in place, the unit tries to escape the limits it shares with the other half by claiming the space of the other half and never reaching the claimed space. In the centre forming a line **k** the line **k** is just a point referring to singularity in proton as well as particles we think of as neutrons. One must appreciate the fact that Pythagoras is mostly about a line **k** if one wishes to use a symbol forming a square, possibly $T^2$ that is symbolising circular motion. In relation to space, which uses another possible symbol, as an indicator in relation to the three sides of the triangle possible symbol used is $a^3$. That is where the Universe started and that is where mathematics started although the rest of mathematics formed some distance down the line. The Universe is mathematics because three dimensions prove equal to two dimensions being equal to one dimension. Although that personifies a sort of mathematics we can only accept but will never full heartedly understand.

Consider the number of turns that brings about new alliances times the number of singularity charge to transport material as the points are activated and see how much singularity can become defined. Every split degree brings innumerable connecting points with innumerable connecting points all carrying singularity $\Pi^0$

Every motion notwithstanding the distance provides a new relevance to any point in motion and spinning of any atom relating to all other atoms brings any possible alignment with no limits. In the entire Universe the number of possible singularity matching is without limitation. Every billionth of a billionth of a degree that the atom turns it aligns with another set of singularity. Every billionth of a billionth of a millimetre of duplication that singularity activates and charge another point holding singularity is infinitely small and yet in that manner duplication is in progress.

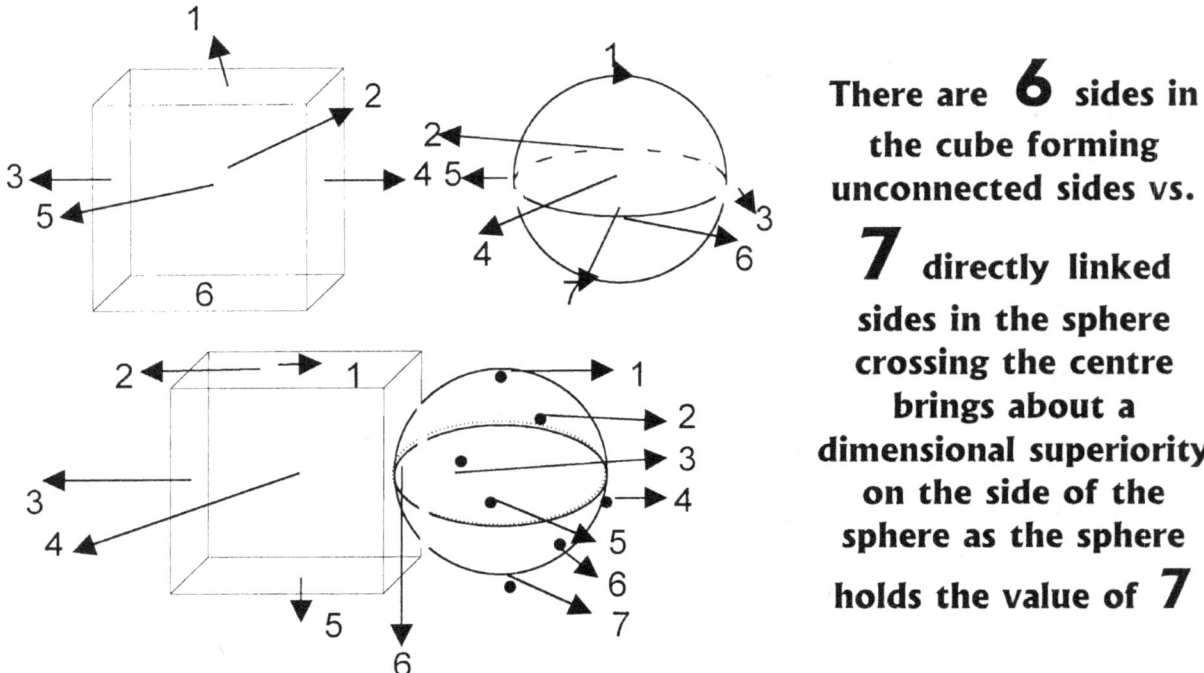

There are **6** sides in the cube forming unconnected sides vs.

**7** directly linked sides in the sphere crossing the centre brings about a dimensional superiority on the side of the sphere as the sphere holds the value of **7**

It is rather obvious that we will not find gravity in the measure of the cube because what will produce and enforce gravity. It is the Newtonians that goes on about a centre being present wherefrom and whereto gravity will pull and push. In the cube the centre shifts as the position shift but in the sphere there is always a permanent centre being the same from any and all points located any where on the surface of the sphere. That is where gravity must be. That is also where space disappears. However Einstein being a Newtonian went lost in mathematics, which incidentally could not be more accurate than it proves to be. It is Einstein that got totally lost in what is the Universe and what is presented by the Universe as space – time $k^0 = a^3 / T^2k$. Listening to Kepler shows that the Universe $a^3 / T^2k$ is singularity $k^0$ and that singularity $k^0$ is pulling flat where singularity holds gravity as the strongest.

The support of one side is literally removed by the centre of the Earth. In the centre of the Earth there is one point where space goes flat and that is the actual position that Einstein saw when Einstein claimed the strongest gravity is drawing space flat. The motion of the object also starts in that point and from such a point gets a direction. There is no pulling on the object but there is removing of space by the centre of that specific point leading the object and the space it is in as well as the space it carries to move to the centre spot. The centre spot can only become activated by the spinning motion of the space occupied by a rotating object that becomes independent from the surrounding space by applying a spinning motion. From that point the spin influences as it produces space-time that stretches the influence of gravity (space in motion) from singularity spinning all the way to where the borders of the sphere form. The spinning motion is activated by a centre that is activated by the accumulating effort of all the atoms within the structure, be it a sphere or whatever, that is in motion and driven by the centre singularity. But in such a very centre the revolving structure it self is motionless in one specific spot in relation to the space spinning around it.

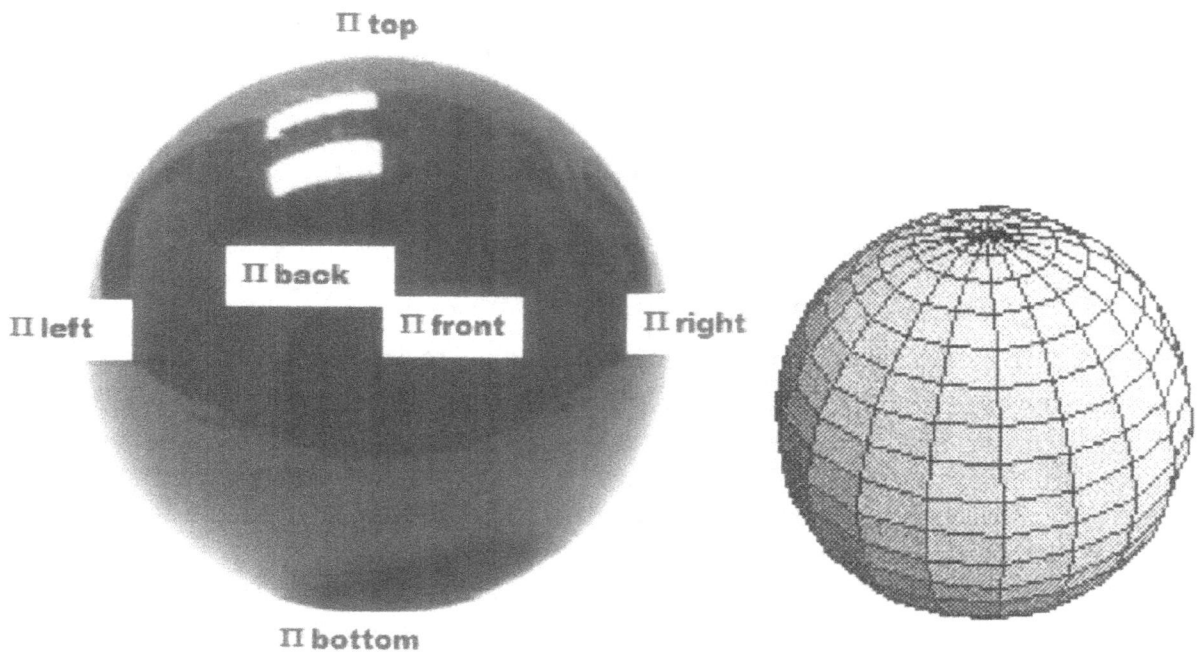

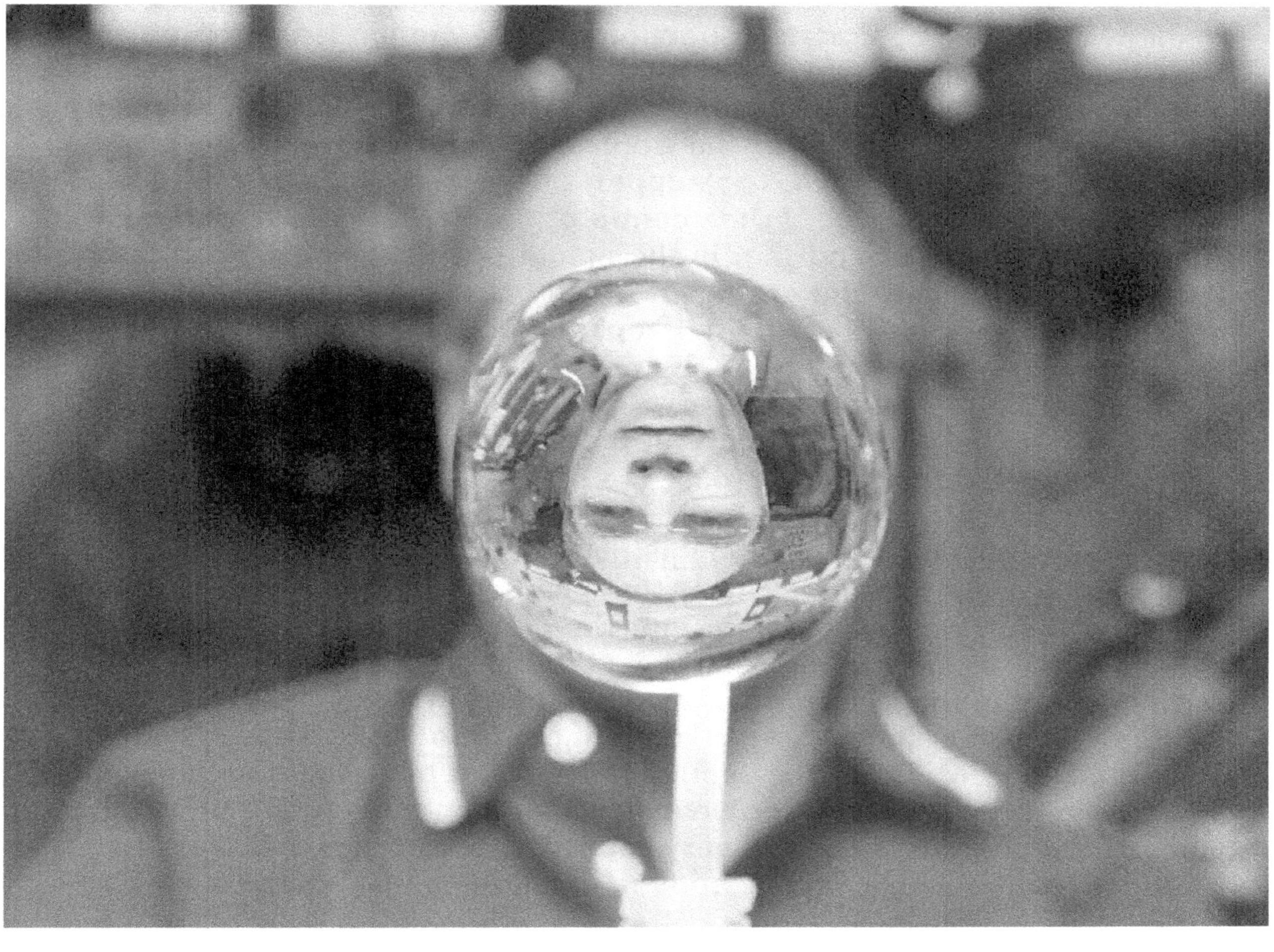

There are the three-sided triangle $a^3$ being $180^0$ and in relation to the square $T^2$ being $180^0$ where they then are equally related to form the line **k** also being $180^0$. In the six sided Universe it does not apply but where dimensions originate (inside the atom core belong the protons) this is true and therefore space $a^3$ can go flat $T^2$ because the space then become equal to the square of time as it moves out the Universe by reducing **k** to the point where **k** becomes $k^0$. In the Universe that we occupy being six sided the method is revealed in order to duplicate space by motion.

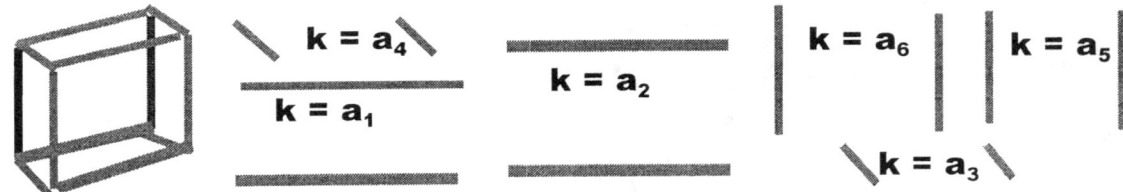

The fact of form proves that the sphere captured all sides that can possibly influence the sphere. The sphere therefore holds $k^0 = a^3 / T^2k$ within the boundaries designated to the sphere. When a body is placed in a location on the outside of such spherical borders, that object seems to float in any direction. There is no control one can establish that will secure movement in any specific direction of preference except by releasing heat to counter act the required motion in a specific direction of choice. We all have seen what happens to any object that comes into the border area of a sphere. The object suddenly is motivated by motion to follow a specific designated direction and the motion leads the object to move towards the centre of the sphere. It is as if the support of the six opposing sides forming the cube has lost one side where the sphere makes contact and took over the control and movement starts in the direction of the Earth centre. The support of one side is literally removed by the centre of the Earth. In the centre of the Earth, there is one point where space goes flat and that is the actual position that Einstein saw when Einstein claimed the strongest gravity is drawing space flat. The motion of the object also starts in that point and from such a point gets a direction.

There is no pulling on the object but there is removing of space by the centre of that specific point leading the object and the space it is in as well as the space it carries to move to the centre spot. The centre spot can only become activated by the spinning motion of the space occupied by a rotating object that becomes independent from the surrounding space by applying a spinning motion. From that point the spin influences as it produces space-time that stretches the influence of gravity (space in motion) from singularity spinning all the way, to where the borders of the sphere form. The spinning motion is activated by a centre that is activated by the accumulating effort of all the atoms within the structure, be it a sphere or whatever, that is in motion and driven by the centre singularity. In such a very centre the revolving structure, it self is motionless in one specific spot in relation to the space spinning around it.

However when the cube comes into contact with the gravity of the sphere the sphere extend that boundary and removes one of the six sides from the cube at any and at all particular points. By removing the space of one of the sides of the cube at a point where the cube and the sphere hold a sharing point and where the cube makes contact with the dimensions of the sphere, the sphere brings contact to that point coming from the sphere centre where we locate singularity, and by superior dimensional contact the object within that point will fall or descend or move towards the centre of the point where gravity is the strongest and space is the least. Gravity is the strongest where space disappears and that is in the centre of all structures holding a sphere as form and has motion and is therefore by motion granting a status of representing the central motionless singularity. That Einstein proved but his lack of studying Kepler as an individual Scientist withheld him from reading his very own mathematic translation accurately.

The proton feeds what the neutron feeds the proton plus the contracted value the proton provides but because the proton has such a high dismissing rate the photon sometimes feed singularity more than the electron can provide the neutron. This is duplication of motion of space but the duplication of motion of space uses a time that provides such motion space-time. This is the Universe that we are discussing. There is no larger Universe but only an accumulation of what resembles a compliment formed by many a Universe. The Universe is the sphere be it in the atom, in the galactica which is a sphere gone oblong or be it a star.

**What Einstein saw was that space disappears. He, (Einstein), then jumped to the conclusion that the space he saw in his mathematical equations was outer space or the space falling outside the parameters of the material that occupied space, which is secluded by dimensional borders. In the sphere, the borders the sphere holds are deliberate and very distinctly placed edges forming a specific distance from the centre. The centre is also proven beyond any debating. The centre of any sphere**

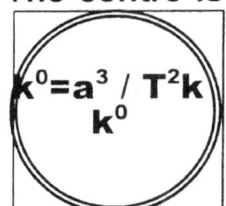

**has to be at the very point where space completely falls away. That will put that space at that point in the single dimension and singularity is the single dimension. Without me trying to teach a Master and fall into the same trap as Newton did, I do have to make a suggestion as I see fit. Newton's observations are very correct except for one condition. It is space that should relate to space. It is the space captured by material that must be in relation to the space influenced by material. It is not space occupied versus space occupied, but space occupied reducing space unoccupied.**

Without me trying to teach a Master and fall into the same trap as Newton did, I do have to make a suggestion as I see fit. Newton's observations is very correct accept for one condition. It is space that should relate to space. It is the space captured by material that must be in relation to the space influenced by material. It is not space occupied versus space occupied, but space occupied reducing space unoccupied. It is space standing relevant to singularity in the formula $k^0 = a^3 / T^2k$ standing in opposing the other space. It is the occupied space going to the place where it will be in the position that is not in direct contact with singularity control. It is space-time $a^3 = T^2k$. The borders extend within the atom as well as beyond the atom. The proof of Pythagoras being part of Kepler is as obvious as creation itself. The factors of Kepler and Pythagoras verifying each other may not be that much part of natural mathematics but the cosmos is not that much part of natural mathematics either.

**The rotation or spin ends on the border where $k^0$ becomes $k$ and steps away from singularity into the very first point proving to form 3D as a factor of $T^2$ where time comes into place**

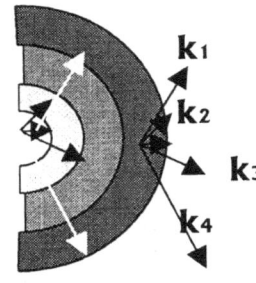

**Every time $k$ reduces (shortens) it changes. It is more than merely the mathematical consequences changing. It is the Universe changing and when entering such space-time, the newly relevant motion will insist on all aspects within such a Universe to change. It is either because the density of heat surrounding the concentration in the space and sharing with whatever is in charge of duplicating promotes duplication more easily with a lot more heat available or contains the time it takes to duplicate because of more heat available.**

**Alternatively, another scenario is that space in need of duplication is less. It effects the time considerably and it does it to the extent that we experience gravity in the manner we do on Earth because of that. That is the reason that the proton is 1836 times more massive while applying the same displacement that the electron does.**

**By using r as a distinction in the circle division is possible but by using, $\Pi$ there is no distinction possible making it a solid flow. Any object being in outer space floats and such floating is seemingly random with no specific detectable interfering favouring a movement in a particular direction. Such a devise is depending on influences not in our scope of detection. Then the object comes closer to the Earth and reaches on specific point where the six dimensions that influence the object in the cube suddenly changes. At one point, one of the six dimensions falls away as it disappears and the object quite literally falls to the Earth**.

Mathematics has a place on Earth and some application in space travel. Take singularity for instance. Singularity was with science from a time before science was science and yet not one mathematician ever noticed singularity while little more could be that obvious. Understanding all I said about space-time being connected intimately is so very important and even more so all conditionally to the fact of accepting that all individual particles in the Universe use motion and therefore spin.

The sphere holds six sides relevant to one centre point in relation to its form that forms a unit and all other forms hold six sides in relevance as does all other shapes and forms. All forms have to have at least six sides indicating different exposures to the Universe. Where gravity has a free choice, gravity always chooses the sphere. As I shall prove elsewhere, gravity is the strongest where the form produces the least evenly distributed space. The first condition for gravity is even-handedness through out the sphere holding the applying gravity and the second is to have most or the strongest gravity located where the space is least. That is where gravity then has a position in the very centre of the sphere and from that centre the gravity produces all the edges or borders that the sphere consist of. In the case of the sphere this factor makes the sphere much more dominating than any other form does. From the centre point controlling all sides is gravity and with gravity applying control the sphere has seven sides to the square in any other possible form having at least six sides. The rotation or spin ends on the border where $k^0$ becomes **k** and step away from singularity into the very first point proving to form 3D as a factor of $T^2$ where time comes into place

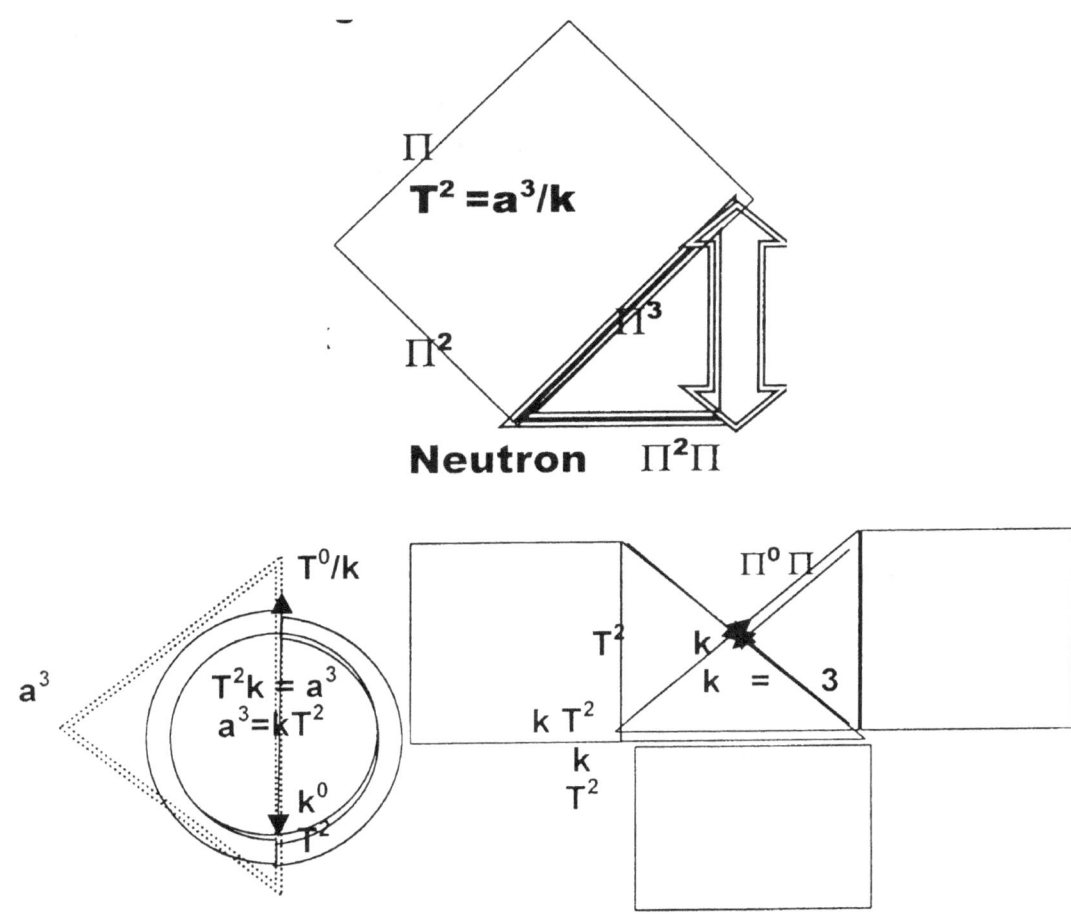

**Neutron** $\Pi^2\Pi$

**Within the circle $k^0 = a^3$ / $T^2k = 7$ points and the cube outside the circle $a^3 = T^2k$. But $T^2k$ is $T^2k$, which means the second space only $a^3$ as three parts are added. This realising is very important in coming to find the laws governing the cosmos being the Titius Bode cosmic principle that is in charge of gravity.**

By placing a connecting circle on the sides of the triangle half a circle forms. By implicating $\Pi$ as a relevancy and not the straight-line r, two values of $\Pi$ applies to each circle, and the straight line is no longer r, but is $\Pi^2$. This will bring about that each circle holds half the square value implicated to the allocated conditions applying to $\Pi$ in that specific instance. By adding the two half squares forming the two half circles and then calculating the square root of the total that then forms the average diameter, an average of $\Pi$ in the connecting line will come about. As both lines are the straight line forming singularity coming from one line being $\Pi$, the connecting line then must be the average of the two lines as $\Pi^2$. That is what **the law of Pythagoras says.**

## Let us again recollect what the differences are between involving r and Π

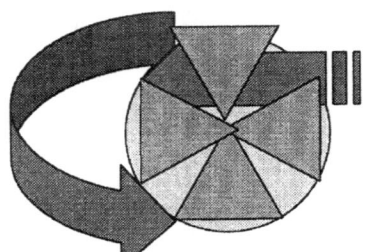

In the circle using $r^2 \Pi$ the r has to have distinctive qualities placing it as a factor apart from Π. Where the growth shows no separate distinction but a continuous flow from the precise centre to the precise edge the flow would become in relation with Π depicting the circle and Π replacing r as reference to any point on the circle.

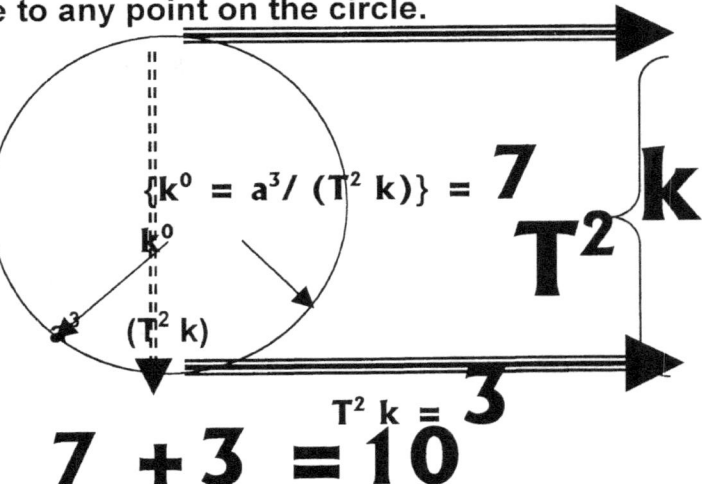

$$\{k^0 = a^3 / (T^2 k)\} = 7 \frac{k}{T^2}$$

$$T^2 k = 3$$

$$7 + 3 = 10$$

In using the formula, Kepler produced and finding gravity in the formula one will get a total of seven factors inside the sphere and three outside the sphere.

Since gravity, also influence the space outside the sphere the space we call outer space has seven plus three points bringing about ten positions of gravity influencing space.

The influence inside the sphere also captures the space outside the sphere.

The reason why the electron receives a value of 3**k** is very incorrect from our stance since the proton shows the three dimensions we find our Universe to be representing our space. Our side receives the values but do not apply

From singularity where there is either **k** being Π or Π being $\Pi^2$ the lines are just **k** X 3. That is why I prefer to use r as a symbol to represent as line because man needs many options and names but nature use one where it may be necessary otherwise it is a repeat of one.

Therefore the sphere holds 7 spots with six being dots and the cube. This point then serves as $k^0 = a^3 / T^2$. The cube holds 6 X $r^2$. I use the symbol r to define the idea of a line weather the line might be used as a radius or to be indicating length breadth or width I do that to avoid the pitfall of using names to hide truths. Where space comes into contact with the sphere the cube loses one of the six dimensions it has to the more dominating seven dimension of the sphere whereby the seven dimension in equilibrium will dominate the six dimension loosely connected by r bringing about that the cube then has 5 sides to the seven of the cube. The space surrounding the sphere takes on the shape of the sphere and not the other way round. The sphere revolves and in accepting the form, the cube is the lesser form. The rotation of the sphere allows one to presume that the form of the sphere is the most dominant of the two choices that allows rotation

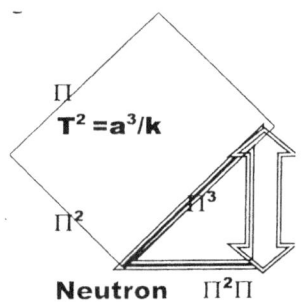

$$\Pi$$
$$T^2 = a^3/k$$
$$\Pi^2 \qquad \Pi^3$$
**Neutron** $\quad \Pi^2\Pi$

The gravity influence we are in search of is something like a woven cloth covering the area and covering all in the area. It is like a silk blanket covering all the aspects. That means the value **k** is running into and past the entire surface as if the surface is one consistency without different particles. That is what motivated me to look for gravity as being heat concentrated in space reduced because everything is space that, consists of heat and if gravity is condensing heat gravity then will be all including. Gravity is all-inclusive therefore it is not possible to draw a precise line that would form a precise ring when using r as an indication of gravity and not cut some atoms in parts. Because where r is used there will always be an atom disallowing the precise positioning of the circle. The circle continues on a solid basis holding a dot in all spots as a positional reference and not r. In every sphere there then are the seven dots relating in precise dimensional and positional equality forming equilibrium to the centre spot as well as to one another by $90^0$ and $180^0$ implicating the dimensional positioning.

.

By using r as a distinction in the circle division is possible but by using $\Pi$ there is no distinction possible making it a solid flow. Any object being in outer space floats and such floating is seemingly random with no specific detectable interfering favouring a movement in a particular direction. Such a devise is depending on influences not in our scope of detection. But then the object comes closer to the Earth and reaches on specific point where the six dimensions that influences the object in the cube suddenly changes. At one point one of the six dimensions fall away as it disappear and the object quite literally falls to the Earth. It changes a stance from floating to falling. It is the point where $k = a^3/T^2$ become $k^{-1} = T^2/a^3$. It is where the Universe swaps ends. While the object remains in outer space the object is floating but that floating stops at one given point and then the falling of the floating objects starts. . The support of one side disappears and the centre point of the sphere takes over the control. At that point the object is under the influence of one centre point in the sphere where the sphere in this case is the Earth and we also know that in such a centre point one will always find the strongest or the controlling gravity.

In the centre of a sphere there is a definitive position where the strongest influence of gravity is located. It is the centre of the sphere where the space is the least. From that centre point gravity extends in keeping the edges of the sphere perfectly true to the form singularity has being $\Pi$ in every aspect. Also such extending continues beyond the specific edges of the sphere where it influences the space surrounding the sphere. We know this by another name given to divert every one from realising that no one has a clue why that phenomenon is applying. We call it the Coanda effect where the process shows that even liquids submit to change of form.

The condition for such a submission is that singularity that is carrying $k^0$ is in place and committing the influenced space. Only by applying singularity through using **k** does the Coanda effect apply. With the sphere being defined by singularity to confirm singularity with the use of **k** as form the sphere is influential enough to remove one side of the square cube, which is, loosely connecting sides to form the cube. The cube is very loosely formed and the sphere is very differently controlled by singularity therefore the sphere is able to dominate the space in the cube by removing all the sides it is in contact with. With the removing of the side the cube in form loose one supporting dimension and therefore will not be able to secure what is in the sphere to the form of the sphere. That is gravity. It is reforming space to the requirements of singularity by reforming form of space through motion. The Universe is a compiling of dimensions forming motion where the motion is coming from the past and runs into the future via the present. Time can and may never stand still in our Universe because then from our point of view we are doomed.

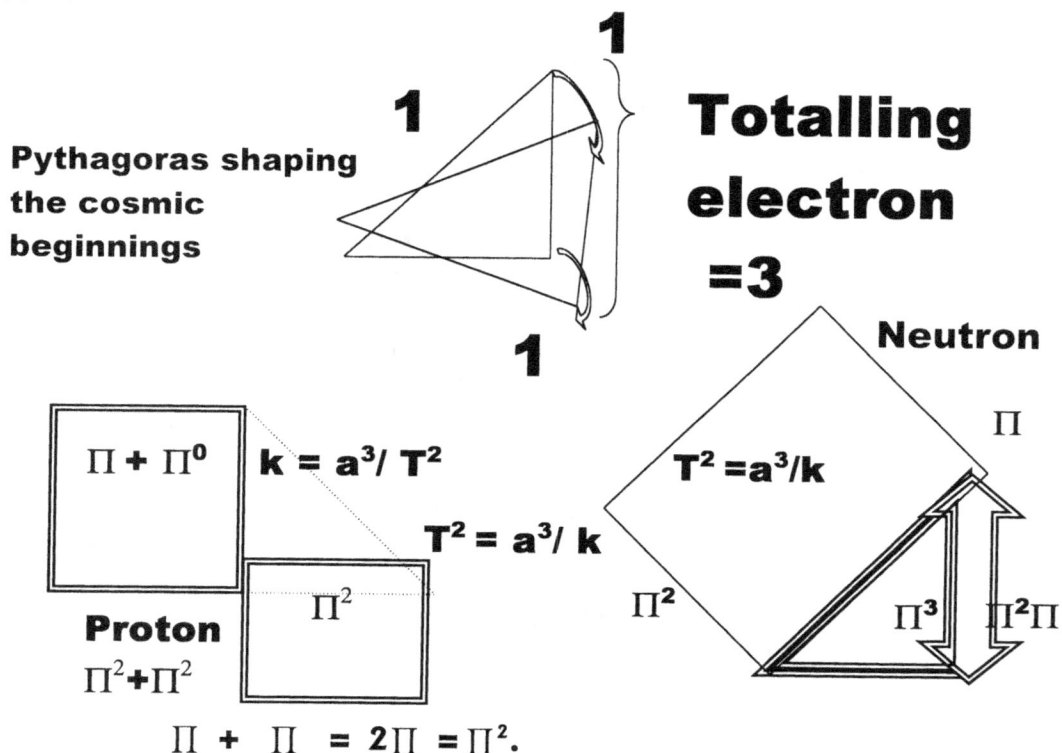

$$\Pi + \Pi = 2\Pi = \Pi^2.$$

The duplication is the result of having no dimensions but commanding other singularity to form many dimensions in support of one another and in maintaining the singularity taking charge. The one only forms support of the other by a time delay activating singularity and exiting by reluctance. Keeping this in mind we lot live in a Universe that is neither solid not permanent but is activated by the split of time progressing from the past through the present towards the future. In this the time delay gets bigger as the time period reduces ever more. It is as Kepler announced that **k** extends by dimensions, as space **a**$^3$ stands related by time **T**$^2$.The form that is in support of gravity is the form of the sphere and without one leading the other point that follows the future point space will collapse as time goes back to being the eternal line. As the rotation brings a sequence that follows the graph very closely innumerable points follow one another as the spin produce space. The spinning space is the result of time forming a backlog of heat where the heat is supporting the one following as much as giving lead to the one, which follows it. Singularity can form anywhere and singularity has no sides. Singularity was matched by singularity and space-time is singularity being activated as the motion presenting the group values it. Gravity is about dismissing or removing heat but removing means vacating and since that which is vacated cannot be represented in the group forming space-time, the vacated and dismissed form a non-existing value in a centre position. It is removing and by vacating something of value goes missing. The value missed is immediately supported by that which is still present so that it means that the missing will shift to a centre where form gives the missing a natural place to be as well as a value to have. As the duplicating is putting redundancy on that which is dismissed the dismissed being vacated transfers to a location held by singularity in a natural position. If that which is vacated must materialise as an absent quantity, the entire Universe will collapse down that absent quantity. That is what a Black Hole presents and that is the direction in which all pulsating stars go not withstanding what name humans use to give classification to size or quantity. Material is supporting non-material by supporting material through the cross reference material has with all sides forming the material. This is better known as being solid or liquid. There is forever a singularity leading another singularity down the path of relevant development.

In the very beginning there was a constant building of points holding singularity where every point found a relevance in time to the one from where the activation was coming going towards where the exiting would follow. This process formed a time delay that later formed material within space. Space is heat that by delay of activating and cooling surges in time delay and that process compacted as time brought about spin and with the principal of the Coanda principle the delay produced units, which today is atoms. The process of time delay just made time delay longer and brought space more but is still an ongoing process forming space-time, which represents the Universe we think of as the Universe.

Long after the establishing of the ground principles did the Universe introduce these measure in space. It was only when the distortion of time became tio much to control by singularity in direct control that space burst onto the scene as part of sin gularity in influence of space

Since the star is the total configuration of the atom's characteristics, the atoms will tell us what we should know about every layer from what is applying in such a layer to what characteristics such a layer would show when it provides the function of what it has to for fill within the star.

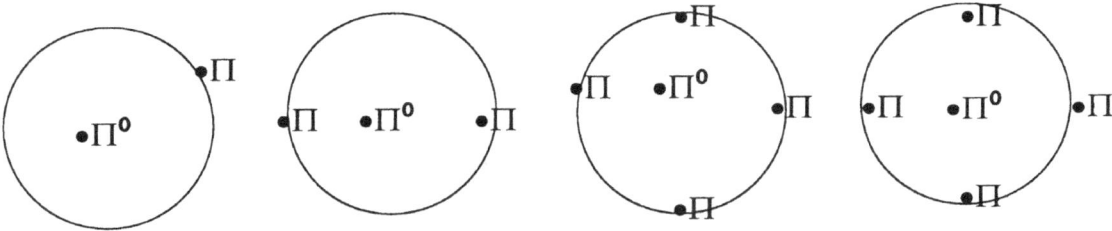

The concept still is about singularity linking to time and that is a distortion of time. At point, five of extended singularity is one outside the rim of time and then is in the distortion of time, which is space by relevance of singularity in specific position according to time.

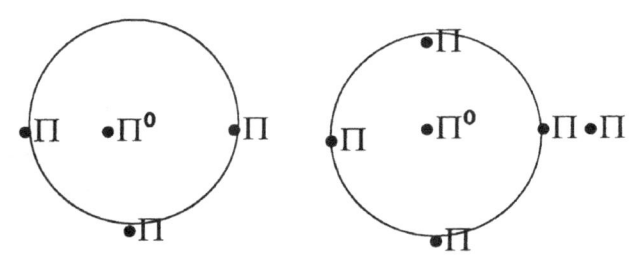

Let us investigate and try to find a way by using logic how a star applies gravity. Therefore it is not the number of dots that is important. It is not the size of the number of dots occupying the position or the size of the space the dots occupy that is prominent. It is the relation in the dismissing of space and the duplicating of space that becomes important. The less space there is the more the favour will be to reduce the space because of the advantage the dots have in securing space-time that will prevent overheating. On the other hand the more space secured will also prevent overheating and therefore those will opt to duplicate space in order to find space to secure and prevent overheating.

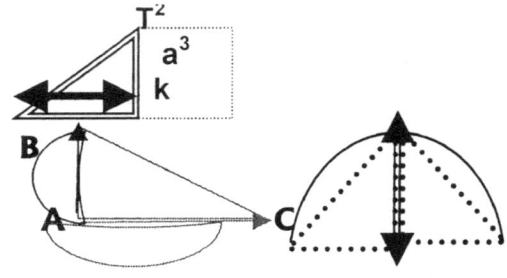

By having **k** in relation to any other side in the square **T²** will produce that area **a³** in three lines meeting committing three lines to space included. Mathematicians only see the two squares adding and when rooted it forms a line again. Yes that is the case but that is the proton part Pythagoras gives to the cosmos. There are so much more than that.

Placing singularity in the position of filling a centre it is two positions added to three positions will conclude the five points coming about. With the normal extending of singularity it will always form the triangle in a half circle whereby $\Pi$ relates to the cube by 5 points to either side of the line singularity forms.

**BC ᴇɪᴛʜᴇʀ ʀᴇʟᴀᴛᴇ ᴛᴏ AB ᴏʀ AC ᴀᴛ ᴀɴʏ ɢɪᴠᴇɴ ᴛɪᴍᴇ ᴏᴄᴄᴜᴘʏɪɴɢ ꜱᴘᴀᴄᴇ ᴀꜱ ᴍᴏᴠᴇᴍᴇɴᴛ ᴅɪᴄᴛᴀᴛᴇꜱ ᴅʀᴇᴄᴛɪᴏɴᴀʟ ᴄʜᴀɴɢᴇ ᴛʜʀᴏᴜɢʜ ᴅɪʀᴇᴄᴛɪᴏɴᴀʟ ꜰʟᴏᴡ**

Thus there are 10 standing related to seven and visa versa. By calculating the 4 squares in the circle with the dimensional changing of space (5) becomes the twenty. It is a triangle forming an area through two squares matching and it is a triangle receiving a three dimensional position in the flat universe by connecting three adjoining lines, It is space coming from time coming from the centre of singularity. The normal flow will allow singularity extending to 10$\Pi$ but when singularity blocks another sphere in singularity the two will form a joint value and by this joining the larger will dominate the space as well as the time of the lesser taking control of the surface and the atmosphere. Through this the Roche lobe comes about with all its other dynamics I describe farther on in the theses. The principle is the same, which we know as the conducting of lightning and Jupiter uses it extensively to implement this action.

### In the manner the Universe was formed with singularity all valued the same, the entire Universe connects like a woven blanket

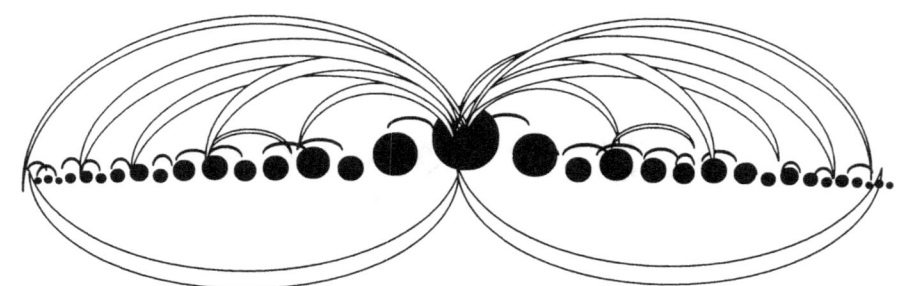

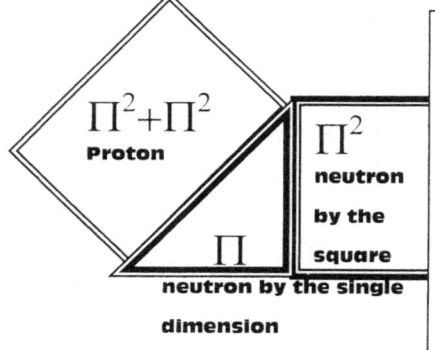

We all know that Pythagoras said the square of the one side would be equal to the total sum of the two squares on the other sides of a rectangular triangle and while that lot is true it is also true that given the value of the one side and the angle of the other side the one square would lead to the value the next square has. If the line was $\Pi^0 = 1$ and the other was **7²** then what would follow is the square of space-time being $50 + 50 = 100^{1/2} = 10$ and there is the square of space.

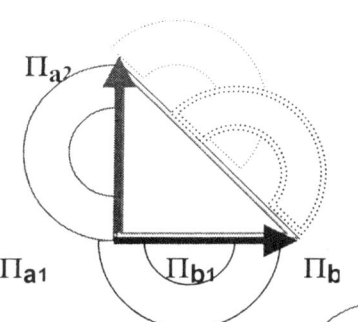

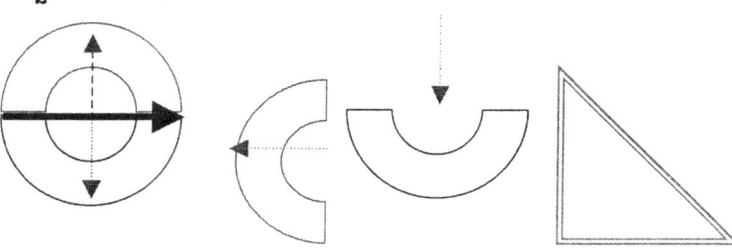

$$(\Pi_{a2} \times \Pi_{a1}) + (\Pi_{b1} \times \Pi_{b2}) = (\Pi^2_a + \Pi^2_b) / 2 = \Pi^2 = gravity$$ and that is proven by Pythagoras. Gravity is the average movement of matter through space in time determent from the position where matter in the sphere meets space in the cube from a point of $\Pi$ to a point of $\Pi^2$ In this the figures of $2(5) = 10$ (space) stands related to 7 from singularity as (matter)

The whole motion concept tying space-time in with gravity connects to Pythagoras like glove fits a hand.

Not long after the law of Pythagoras was introduced mathematics $\boxed{7^0}$ understood where Eratosthenes of

Syene made as big a discovery as the one instance the world took see and Pythagoras did. But in notice because the world could understand and the other instance the world disregarded the findings because the world did not see what the implications was. The same apply to aircraft flying and when the aircraft wishes to escape the earth's singularity hold it has to comply with the laws laid down by the earth. The seven becomes as big a part of the concept as does $\Pi$ as it all interacts

When we dissect the sphere to the bone of singularity we find Pythagoras at that bone. Because the sphere centre is the infinite point such a point will cover the infinite spot of the outside of such a spot where singularity produce matter in the centre extending. From the family of lines there will be three lines being close relatives crossing three lines at the centre and pointing to six edges at the birth point of material. That is then six points forming $a^3$. The factor of k is extending results in 3D lines with six points being material. From the centre of the sphere six but that six is standing in relation to the next six point which are they in a different location in time. From Kepler we know that space in 3D or $a^3$ is produced by time or motion in $T^2$. $k = a^3 / T^2$. Matter then holds six singularity positions where motion puts the six 6 effectively in 3D by the square of motion $T^2$. $a^3 = 6$ and $T^2 = 6^2$

$6^2 = 36$. Then space comes into play. Where six holds the material position the very next position is where space start and that then must be seven

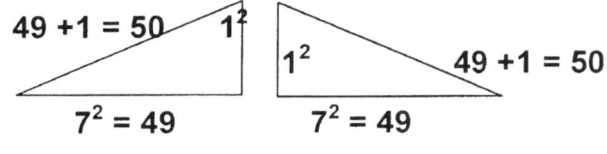

50 + 50 = 100 but that is in both sides of the Universe.

$100^{\frac{1}{2}} = 10$. That places space unoccupied in the factor of 10 being also 7 + 3 = 10

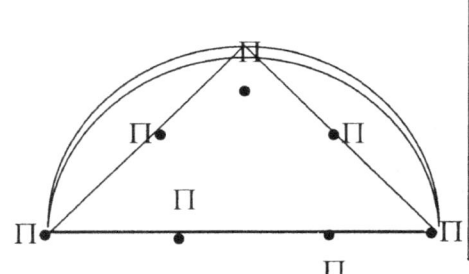

It is at this point where mathematics began. It was where the line was equal to a half circle and the half circle was equal a triangle. There was on value that everything in the Universe could have and that was $180^0$. By being half way between two parts of the Universe were being $90^0$ apart. Everything related to singularity be it in the square of space or in the square of time.

If we take the points the Pythagoras triangle have we fin there are three points in the half circle, which is related to the square having four points in time. That put the three of space in relation to the four of time. Connect that to the two the line follows and add the one in singularity and we find six points being even,

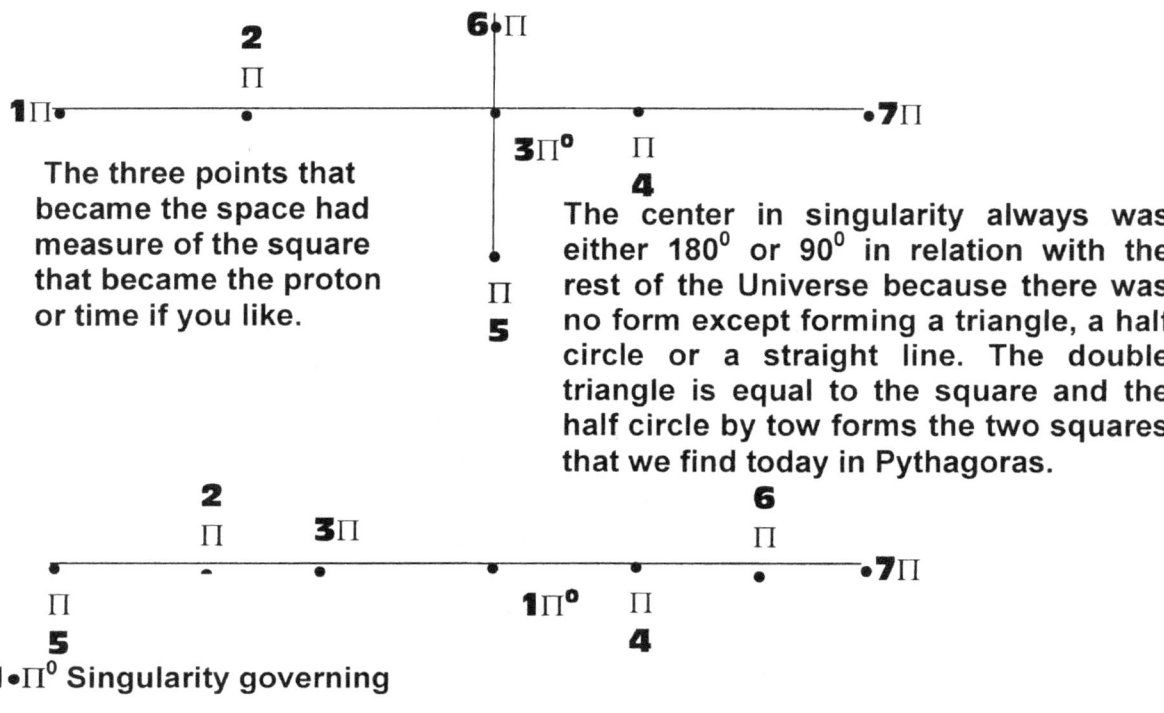

The three points that became the space had measure of the square that became the proton or time if you like.

The center in singularity always was either $180^0$ or $90^0$ in relation with the rest of the Universe because there was no form except forming a triangle, a half circle or a straight line. The double triangle is equal to the square and the half circle by tow forms the two squares that we find today in Pythagoras.

$1 \bullet \Pi^0$ Singularity governing

$2\Pi \bullet$ Singularity in relevance

$3\Pi \bullet$ Singularity in relevance

$\bullet 4\Pi$ Singularity forming motion to become time

$\bullet 5\Pi$  Singularity forming time distortion becoming space.

$\bullet 6\Pi$ Singularity forming time distortion becoming material in space

$\bullet 7\Pi$ Singularity forming time distortion becoming material ending space

It eventually became the shape of things we now see with the full compliment of mathematics that we find but back then there was only form without shape.

The space we now see was only in forming dots on a line forming time and the dots that formed were a continuing of the only available commodity at the time, which was the line of time in eternity. Because the line was flowing and all development was on the line, and even turning around the Universe was following a line of $180^0$ the connection Pythagoras has was not the square at first but the space the line later allowed the square to have with another square on another line. In that everything was on the same line we still find a top can erect by motion, all motion can connect singularity throughout the Universe and in that all space in matter connects to space in time by time in space.

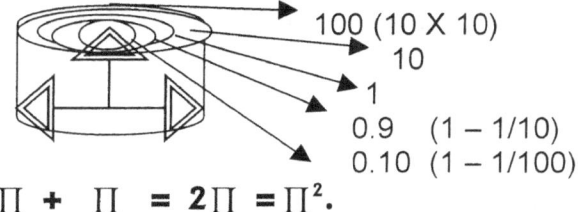

100 (10 X 10)
10
1
0.9   (1 – 1/10)
0.10   (1 – 1/100)

$$\Pi + \Pi = 2\Pi = \Pi^2.$$

**The cube can have square corners; the corners are of less importance because it may come in whatever form there may be. In contrast the sphere adheres to precise measure and behind this principle is all that forms the Universe. The cube has six sides connected loosely and can change form just by changing the relevancy between one side (or more) in relation to the distance brought about by the other sides. The sphere being a complex circle stands related where the sides has to apply precise measure in equality. This becomes a law because in the precise middle one will find the strongest gravity as that gravity holds the object in form and true to form. If there is even gravity spread in all directions the form must be a sphere and the sphere insist on seven points relating to sides or borders.**

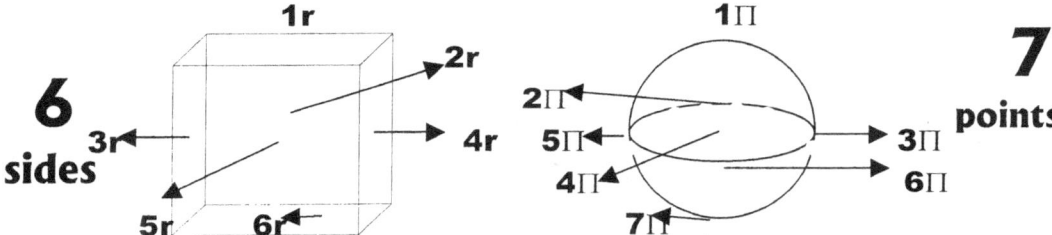

In the sphere, which I am referring to and that one find in cosmology there can be no radius but only the extending of **k** from the centre point $k^0$ that then with motion becomes $T^2$. From the centre $k^0$ in six opposing directions there are precise located points that is crossing a centre and is relating to one another by the square but the factor is remaining **k** because of the unity the matter holds in relating to space. The factor has to remain **k** since the motion takes on the square as $T^2$ and because of this implication there can be no radius in the square or to the third power. In the event of using r the individuality of r as a concept will bring about a specific ring at any specific point where every such a ring will define every time as one ongoing circle. Such a circle has no validity in the Universe since the circle indicates an absence of motion. In the cosmos all space has to have motion. Since gravity is motion of space the use of a radius that will indicate a square is principally flawed. One cannot use such a definitive line as r would be because such a line will have to cut through atoms at some points while running from the centre to the edge. We have to take such a line as **k** and that gravity is **k** that becomes $T^2$ and $T^2$ is securing another space $a^3$ boundary that a new $T^2$ becomes a new k in relation to another object influencing the space that will again become $T^2$ It is a never ending always continuing cycle of interactions. That is the effect one get from gravity. Gravity includes and joins all aspects within the field but such definitive line will have to exclude some and include some because of the definite points and rings developing.

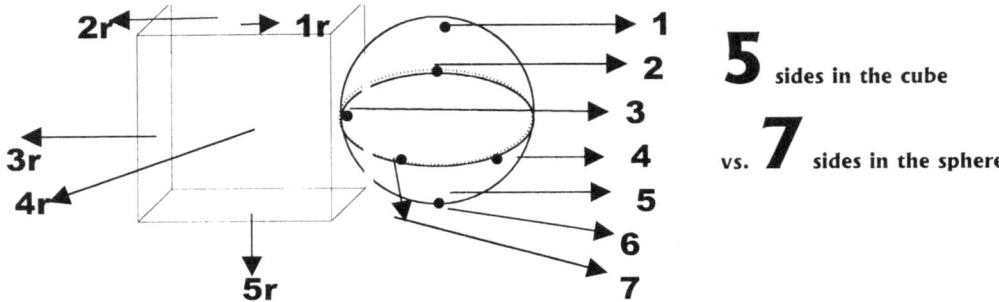

This means that in the cube at the point of contact between the cube and the sphere the cube experience such a contact point as if the "bottom falls out" of the cube and without a "bottom" to support objects they fall to the sphere as objects does fall to the earth. Remember that a body "floats" in space, but at one specific point it starts to "fall" to the earth. That is gravity and it is much more a dimension change than it being any force. Then at any given point the space outside the sphere holds five points where as within the sphere, there are six points relating to the centre spot of the sphere. I shall explain this last remark later on. That too is the Lagrangian system with five cosmic structures holding relevancy to the centre structure where the centre structure stands in for seven positions diverting from singularity and the orbiting structures standing in for five positions in space.

Gravity is all to do with dimensional changing and reforming of forms to re-affirm alliances supporting singularity. It is the reforming of space converting space to more concentrated heat. The Universe is in the three dimensions using twelve dimensions that is visible to us and indefinite number of stages in size differences ranging from the immeasurable small to the immeasurable large where mathematics become a short fall to the next and the previous dimension. There are always 10 in positions running smaller and running larger. Even to us thinking we are at the edge, or since may think we are in a centre that is because we use light as an information source. By our use of mental thought instead of light to obtain information we will find that the Universe are infinitely bigger than what we see and infinitely smaller than what we see.

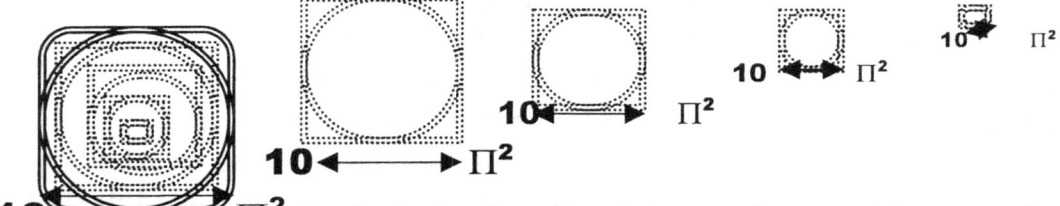

**10** $\Pi^2$ Gravity is the dimensional change of space taking space from 10 to $\Pi^2$. This happens by means of applying the Titius Bode configuration of space adapting form through the seven dimensions interlinking ten dimensions to reform the concentration of the space to heat. In this manner gravity is "building" space by motion of space. From the gravity "building" space through motion the motion of the building leaves an imprint which is detectable in the sequence the Titius Bode law saw in the number arrangement of 3; 6; 12; 24; 48; 96 etc. The incorrect application of the Titus Bode law lies in subtracting the figure of 3 from 10 leaving 7. The other way of reasoning is to add four each time to the firs value of three starting with 3 and so on. One has to see the Titus Bode as two relevancies in the unit bringing across the building of space-time. The true significance of the Titus-Bode law is that it points directly to a circular growth of 7 in the sphere leaving the marks as it grew in stages. The 7 relating to 10 is a precise derogative of the Roche limit or the Roche limit is a precise derogative of the Titius Bode principle because he two systems interlink.

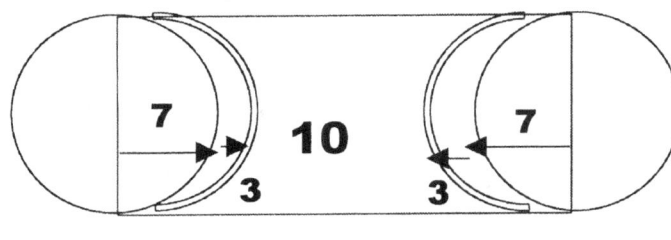

The principle is an indicator of what gravity is and of what produces gravity when material (7) moves in space (3) and through space (10) to leave the imprint of gravity applying on the space growing between the planets.

The sequence the Mathematicians found was undeniably correct. There is a defining relation between all the planets and even the fragmented one shows this tendency too. It is the way space relates with material but because it did not fit Newtonian thought it was dismissed. One should see the Titus Bode law as a vibrating ripple effect leaving marks about those times when collecting heat manifested in material by growth every time it forms a cyclic pattern in the growing space.

If it applied only to one or maybe two planets I can imagine there might be some inclination to doubt the obvious but when all the planets show the same cyclic pattern in all of them such evidence just turns things very much against a Newtonian principal dismissing. Also if this was the only doubt that the cosmos cast against Newtonian principles then some denial was excusable. But the further studies go the more Newton seems to be wrong. At what point will science eventually come to their senses and look beyond the hundreds of years of student brainwashing.

There is a Universe parting Physics that apply on Earth to Cosmology. In cosmology one work with space-time $a^3=T^2k$, whereas on Earth one can work in a fixed and general space with a fixed and general time. However, in cosmology there is nothing fixed and there is nothing general except the four pillars Creation was built on. Using Newton, however one cannot even begin to explain any one of or the combined effort of the above cosmic phenomenon or the four pillars that are all over the cosmos and forms all the laws in the cosmos. Newtonian definition cannot even recognise any of the principles but only Newtonian science are thought to students. No student can have the fortune to disagree with Newton and remain a student. If the student will dare to disagree with Newton it is the end of such a students academic career. By setting this firm condition Newtonian science becomes an institutionalised mind conditioning of the concepts of thought forming in physics. With my saying this I have not made one academic friend but to the same degree did any one in the past proved me wrong. Students are taught to accept Newton and to ignore Kepler and any student doing it the other way around will fail all examinations and any other form of testing at Universities. Students accept Newton or they accept a ticket taking them home. Newton is an institution force fed to each following generation but saying that reserves only resentment towards me amongst Academics

Investigating Kepler is also the understanding of Kepler and that no one did since Newton named gravity and defined what he saw gravity was but gravity is out and out Kepler's discovery. There is a Universe inside the sphere and there is another Universe outside the sphere. One has to see it from the point singularity holds $k^0 = a^3 / T^2 k$ and $a^3 = T^2 k$. Both represent as view from singularity but the two are not the same and neither do they share a principle. That is the relevance bringing about the gravity Kepler discovered. What would Kepler discover that Newton did not discover. Newton named gravity but Newton also missed the chance to discover gravity because Kepler discovered gravity and with all the laws on motion that Newton introduced Newton failed to recognise the work of Kepler as that Kepler introduced gravity. Newton did not see what Kepler introduced as relevance between participating object in motion forming sequence. The gravity Newton saw and the gravity Kepler saw is not the same gravity but since Newton meddled in Kepler's work by changing Kepler's work Newton then being the master on motion should have seen that while we on Earth uphold motion in serving the time factor $T^2$, which produces the space such a space factor will compensate for this inconsequence in the formula $a^3 = T^2 k$. Kepler saw gravity forming a relation that becomes a square when the two bodies interact between the different positions the bodies hold in space that is relevant to the motion each structure performs. That too was what Titius and Bode both observed. The Titius Bode and the Roche limit is only a part of the complete gravity process where gravity forms with motion placing relevancies. Mass does not bring about gravity but gravity produces mass by performing as a resistance to motion. Saying that we better find a means to distinguish the concept we have about more or less particles per unit forming more or less mass, and mass forming the gravity contracting resistance that the unit upholds. In our case we are descending or falling which gives us a negative distance growth. Normally the Kepler's formula read that $a^3 = T^2 k$ therefore $k = a^3 / T^2$. This is normal where the planet is orbiting the Sun in a regular patter and does not bring about a decline to the factor $k$.

From the gravity "building" space through motion the motion of the building leaves an imprint which is detectable in the sequence the Titius Bode law saw in the number arrangement of 3; 6; 12; 24; 48; 96 etc. The incorrect application of the Titus Bode law lies in subtracting the figure of 3 from 10 leaving 7. The other way of reasoning is to add four each time to the firs value of three starting with 3 and so on. One has to see the Titus Bode as two relevancies in the unit bringing across the building of space-time. The true significance of the Titus-Bode law is that it points directly to a circular growth of 7 in the sphere leaving the marks as it grew in stages. The 7 relating to 10 is a precise derogative of the Roche limit or the Roche limit is a precise derogative of the Titius Bode principle because he two systems interlink.

According to Newtonian science space is simply nothing with no qualities but gravity separate space and space does not mingle, as one would expect if space was nothing because space does form borders. Disasters of unprecedented magnitude arise from such borders. The Challenger disaster of February 2003 is much testimony to those borders that was powerful enough to break the aircraft into pieces while the explaining contributed by Mainstream science is evidence of a shocking lack of understanding about what took place as cosmic laws were breached. Let us now start to timely investigate cosmic laws as presented by the cosmos.

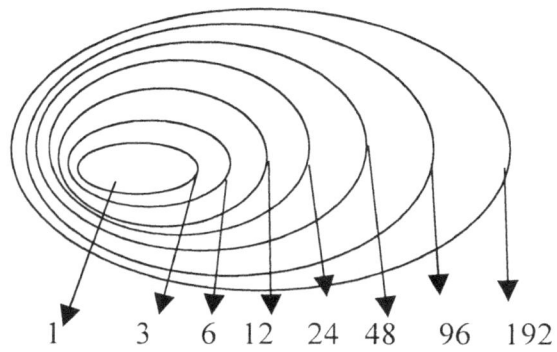

| Planet | Mercury | Venus | Earth | Mars | Ceres | Jupiter | Saturn | Uranus |
|---|---|---|---|---|---|---|---|---|
| Bode's Law dist. | 4 | 7 | 10 | 16 | 28 | 52 | 100 | 196 |
| Actual dist. | 3.9 | 7.2 | 10 | 15.2 | 28 | 52 | 95 | 192 |

**The TITIUS BODE Principle Outside the sphere**

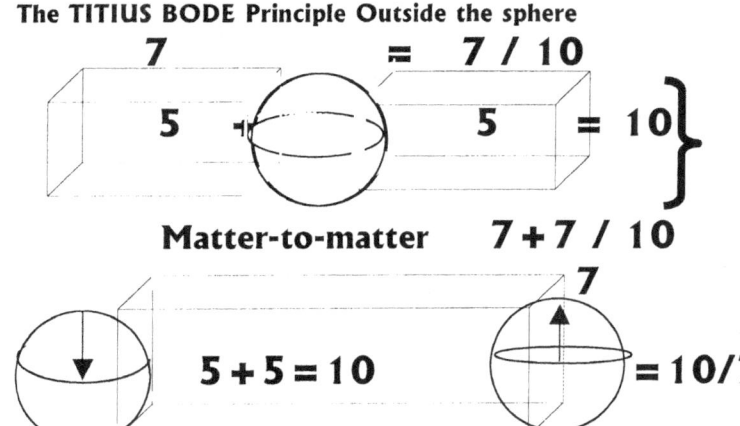

The **TITIUS BODE** Principle is gravity in space. It does no make sense that Mainstream physics denounced the **TITIUS BODE** law. Gravity is motion and motion in space comes about from The **TITIUS BODE** that is in principle gravity.

In the situation we are in and being captured by space-time control we have **1/ k** because we are within a reducing or growing smaller than the **k** the Earth introduce so we submit to the **k** of the Earth. Reduce **k** as it happens in the gravity Newton presented then $k^{-1}$ would produce a smaller space $a^3$ if the time $T^2$ component of such motion relevancy is forced to be equal to both notwithstanding what differences there are in space $a^3$. Newton had available at the time the findings of Galileo and it was Galileo that showed by using a pendulum that swinging pendulum arm will reduce the space to compensate for the establishing of the time. By compromising space $a^3$ the pendulum can manage to uphold the time $T^2$ component because the pendulum is visual proof of $a^3 = T^2k$. In our case the **k** attached to our position in space is in decline or negative which is the formula taken in another relevancy of $k^{-1} = T^2/ a^3$. That is precisely what is happening. We are reducing the space between the Earth centre and us (our singularity) because we are travelling too slowly in relation to the Earth. Again Kepler showed himself being correct and not only that but he proves his brilliance.

We are all aware that the Titius Bode law is about positioning and aligning planets and yes that is true. I can explain that part in detail. However, the Titius Bode law holds much, much more in cosmology that planets finding relevance...

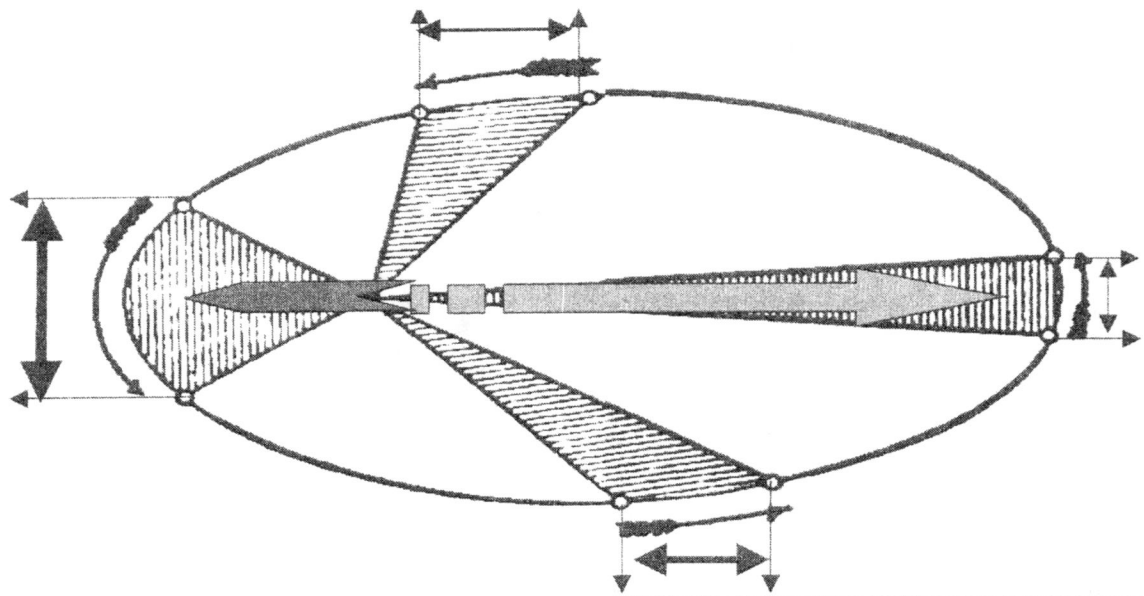

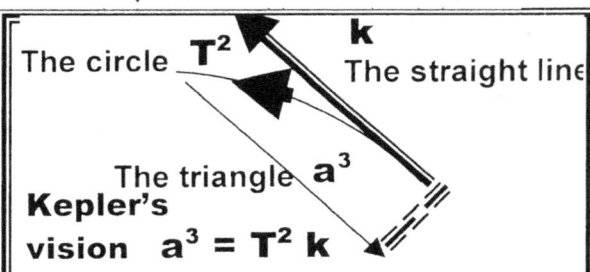

Newton saw in Kepler's work what all other people ever since Newton see, that gave Kepler's work the smallest attention and Newton was the only one that gave Kepler's work the smallest degree of attention. What all other people see in Kepler's work is the incorrectness I see in the work of Newton's idea about the incorrectness in what Newton thought Kepler said. For that I have been dismissed for years. Kepler calculated the movement of the orbiting structure around a centre Sun. Newton did for starters not interpret this rotation as gravity. Secondly he saw what clearly is space, as motion. Therefore Newton was unable to see that Kepler said space in motion is gravity. I challenge who ever is of such an opinion that Newton made an in depth study of the integrate meaning of the work of Kepler, to prove that. Newton made a serious observation about the work of Kepler. Newton saw a circle forming by a planet that is in rotation, running around the Sun. To correct this Newton added $4\Pi^2$ on one side and on the other side he introduced his version of what he saw gravity to be. Newton as a person failed to see in Kepler's study any gravity applying. If Newton did recognise any gravity aspects then he, Newton would have concluded that forming another interpretation by adding is very unnecessary. Newton did not act incorrectly when he added $G (m + m_p) = 4 \pi^2$ but he merely duplicated what he then neutralised as a fact. He brought in his vision he had about gravity and that the two structures in motion is combining a time related value of $4 \pi^2$. He duplicated what Kepler said and by duplicating as well as neutralising the duplicating he brought misconception and he covered information, which I then uncovered. Through my uncovering, I came to the conclusion that what Newton said what Kepler saw is not what the cosmos told Kepler. This action Newton missed completely when he did not see that into the motion of the orbiting object the value of $\pi^2$ is built in and the circle will have to complete in $4\pi^2$ as a rotating action enforced by a centre. The centre is the measure holding $\pi$ to enforce by motion $\pi^2$

From the orbiting structure (the planet) aligning with singularity only one structure, which is the very inside singularity applies as a position of reference and that is reference to the distance applied between points holding the governing singularity.

**This was the prologue letter announcing**
**MATTER'S TIME IN SPACE: THE THESIS**
ISBN   0-9584410-8-1
FROM THE ORIGINAL AFRIKAANS: "MATERIE SE TYD IN RUIMTE" I. S. B. N.   0–620–27041–1
WRITTEN BY PEET SCHUTTE
© KOSMOLOGIESE EN ASTRONOMIESE TEGNIKA

# THIS WAS

# Defining Singularity:
# AN OPEN ACADEMICS  LETTER
# # 1

## NOW TO FOLLW IS

**An open letter**

# Defining Singularity:
# AN OPEN ACADEMICS  LETTER
# # 2

**If it is that simple then why is it complicated.**

BEST WISHES,

PETRUS.  (PEET) S. J. SCHUTTE